materials for
metal cutting

Proceedings of the conference on 'Materials for
metal cutting' jointly sponsored by BISRA–
The Corporate Laboratories of BSC, and The
Iron and Steel Institute, and held at the Royal
Hotel, Scarborough, on 14–16 April, 1970

The Iron and Steel Institute

ISI Publication 126
© The Iron and Steel Institute 1970
All rights reserved

Price £12
ISBN 0 900497 20 3

Text set in 9/10 Monotype Imprint
Printed in Great Britain by Lund Humphries

contents

Conference Organizing Committee

P. F. Tomalin (Chairman)	GEC Power Engineering Ltd
Dr R. L. Craik	BISRA–The Corporate Laboratories of BSC
G. J. Hill	GEC Power Engineering Ltd
T. L. Hughes	The Iron and Steel Institute
F. A. Kirk	Osborn Steels Ltd
E. Lardner	Wickman Wimet Ltd
Dr K. A. Ridal	BSC, Special Steels Division
J. R. Powell (Secretary)	BISRA–The Corporate Laboratories of BSC

preface

The weight of material used for cutting tools in the UK each year is relatively small. The main items in 1969 were 10 000 tons of high-speed steel and 1000 tons of hardmetal. The value of these materials is more significant, £35 million for high-speed steel and £14·5 million for hardmetal. In addition £3 million worth of diamond was consumed.

However, the importance of these materials is very much greater than the above figures suggest. The performance of the cutting tool is one of the major factors controlling the rate of metal removal, and it thus influences strongly the return on the capital invested in the machine tool. Consistency in tool performance is also a significant factor, especially if the optimum productivity is to be obtained from numerically controlled machine tools. Therefore the development of cutting-tools materials has far reaching effects in the engineering industry.

In recent years a number of advances have been made in the composition and in the production processes for cutting-tool materials and these have improved both the properties and the consistency. Furthermore, there has been an increase in the understanding of the factors which influence the tool performance. However, discussions between users, manufacturers, and research workers in this field do not appear to have been very extensive in the past, and this conference was therefore organized to promote an exchange of views on the present technical position and on likely future developments. It was also anticipated that the contacts established at the conference would provide a basis for better communications in the future.

P. F. Tomalin
Conference Chairman

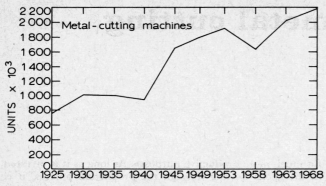

1 Metal-cutting machines in the USA

of finished goods, when added together, are about three times the investment in raw materials. Keeping in mind that the investment in machinery and equipment is about equal to the sum of the investments in raw materials, work in progress, and stocks of finished goods, this means that if you could halve your work in progress and stocks of finished goods by introducing some sort of new production technique, you would have a sum available to invest in new machinery, if that were necessary to make the change, that would be somewhere between 30 and 40% of the current value of your installed machinery. Some people will not agree on this point, because they feel that there is a difference between the money you put into machinery and the money you put into work in progress and stocks of finished goods. They mean that it is easier to convert work in progress and stocks of finished goods into cash than it is to convert the investment in machinery into cash.

They also think that the price they get for work in progress and stocks of finished goods is closer to their own purchase price than the sales price for the machinery would be to the current purchase price for the machinery. I do not think that they are right, because the investment you have in work in progress and stocks of finished goods depends on your throughput time, which in its turn depends on the way you have organized your production. If you do not make any organizational or technical changes, it is impossible to reduce work in progress and stocks of finished goods without at the same time reducing your turnover by about the same per-

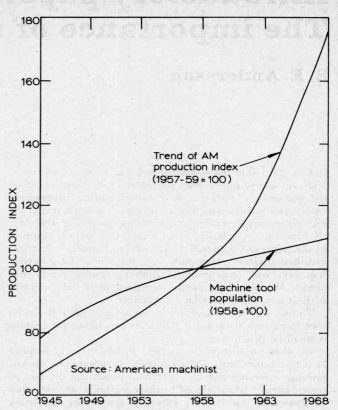

Accelerating productivity of machine tools is indicated by the way metalworking output (as measured by AM Production Index) is outstripping machine tool population

3 Production in the metalworking industry of the USA

centage. If you want to realize half of the capital that you have tied down in work in progress and stocks, you would have to reduce production to about a half. If you wanted to realize all the tied-down capital, you would have to stop production altogether. In a situation where you considerably reduce production or completely stop it I think that it is usually easier to sell your used machinery than to sell your work in progress and stocks.

People who are not familiar with job shop production find

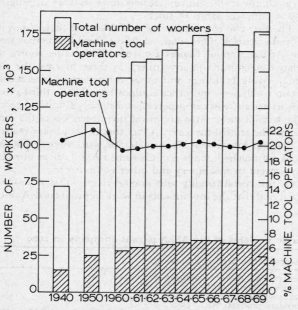

2 Workers and metal-cutting machine operators in companies belonging to the Swedish Metal Trades Employers Association

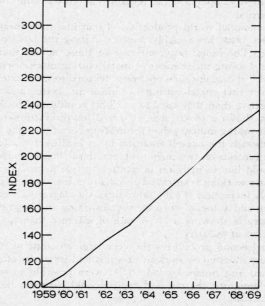

4 Total wage and salary costs per ASEA employee

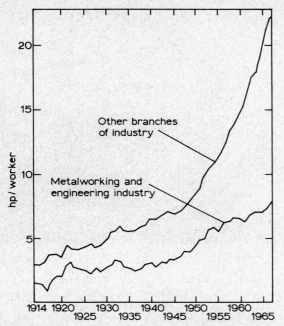

5 Installed capacity of motive power in Swedish industries

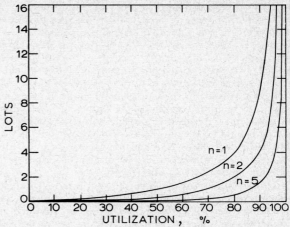

6 Queue v. utilization for different numbers of parallel machines

it amazing that you sometimes have a ratio between through-put time and actual machining time of 10 or 20 and at the same time have your workers being idle for 5–10% of their working hours because of lack of goods to be machined. Figure 6 gives an idea of the reason for this. Should you have somewhere in a factory one specific machine where different lots arrive at random intervals and where machining times for each lot show the kind of statistical variation that is normal in conventional job shops, the graph shows that if you want the machine to be working on the average 80% of the available time, you will have to keep a queue in front of the machine that has an average size of four lots. This means that the throughput time for a lot through that machining operation is roughly five times the actual machining time. If you want to increase the utilization to 90%, leaving the machine idle only 10% of the available time, you would automatically get a throughput time in the neighbourhood of ten times the actual machining time for each lot. But if you install two machines in parallel, you do not have to accept a throughput time of more than five times the machining time at a utilization of 90%. With five machines arranged in parallel, the throughput time at a utilization of 90% would decrease to the comparatively low figure of $2\frac{1}{2}$ times the actual machining time.

If you were to start up a new job shop production and knew in every detail what sort of programme your customers would give you for the first period of production, you could naturally balance your machine capacity very closely with this need. In some cases, in spite of your detailed knowledge of the load you would still have an over-capacity, because if you need 1·2 machines of a certain kind you have to buy two. If you were lucky enough to be able to start up under these well-defined conditions, you would still be sure that the next

production period will be very different from the first concerning the loading of the different machines. This is a natural result of the fact that your customers do not buy products to load your machines, but to satisfy their current needs. This very simple discussion explains why you will very seldom find a job shop where all the machines at a certain moment have the same load waiting to be machined in them.

The fact that you always have a certain over-capacity of machine tools in a job shop owing to the changing product mix means that the problem of keeping down the throughput time by installing many parallel machines of the same type is reduced to having capable men available for those machines that are needed immediately. The simplest way would be to assign one operator to each machine and let him be idle when the machine is idle. This is a very costly way of solving the problem. Experiments have shown that on top of this you will run into psychological problems. Very few people like to be employed in a factory where they have to sit waiting for some work to show up during a good deal of their working hours, even if they get about the same pay when they are idle as when they are busy.

One way of solving the problem is naturally to see to it that your workers can operate more than one type of machine. Then you know that when there is a temporary overflow of work in one type of machine, you can always transfer operators from those machines where you have at the same time a temporary low load and where some workers, if they were not moved, would have become idle. A high worker flexibility is obviously necessary, if you want to cope with the work flow/capital problem in a job shop.

In Sweden most machine-tool operators have a straight piece rate. You cannot expect a milling machine operator to be able to earn the same hourly wage on straight piece rate if he is temporarily moved to, for example, a lathe. To get operators interested in moving we have set extra allowances for the time that a worker operates a machine that he usually does not run. The allowance starts out at a fairly high level the first day and drops off in accordance with the normal learning curve for a specific worker on his secondary job.

To further increase the worker's interest in flexibility we have now set up large tables in each foreman's office in our factories showing each single individual worker's primary job, his secondary job, and a job he would do if neither the primary nor the secondary job were available. This counteracts the normal reaction of a worker who is asked to move to another machine. In such a situation a worker would normally take this as an indication that he is the one in his group that the foreman would lay off first if he had to lay off somebody.

Another way of increasing the workers' interest in flexibility is to create teams. Each team has a line of machines or a group of machines. If the number of different kinds of machines within this line or group is then reasonably well

Table 2 Relative magnitude of investment

	Buildings and other constructions	Machinery and other equipment	Raw materials, work in progress, finished goods
Metal working and engineering industry 1960	28	38	34
Metal working and engineering industry 1968	29	42	29
ASEA Annual report 1969	30	39	31

adjusted to the normal work content of those jobs that the team is going to perform, it will always be possible for the team, by moving around team members between the different machines, to keep the goods flowing rapidly through the group or line.

Another way of reducing the capital you have tied up in work in progress and stores and stocks of different kinds is to reduce your set-up times. As you know, the economic order size depends on your set-up cost. It happens too frequently that an insufficiently good set-up method is accepted. Very often the capital you would have to spend on equipment to enable you to cut down your set-up cost is much less than the capital you will otherwise tie up in work in progress and in different kinds of stocks and stores.

It is interesting to study how the machine-tool industry has been trying to tackle the two main problems that I have been talking about. They have been trying to counteract the rapid increase in wages by automating more and more of the work that was normally done by a machine-tool operator. They have even attempted to reduce the throughput times by reducing the number of set-ups and the cost of each set-up. The worker flexibility has been improved by the introduction of machines requiring shorter learning time than the old ones.

At the end of the 1950's a machine called Flexomatic was designed and built in the USA. As you can see from Fig.7. this was a transfer machine, where different machine tools along the transfer line were individually numerically controlled. The work-holding fixture carried some sort of identification so that each single machine tool could pick out the right programme from a bank of programmes when a specific part was going to be machined. The philosophy was of course that you could send a very mixed production of the kind we see on the trolley in Fig.7 down the transfer line, and that each part would be complete and ready for assembly after having passed the transfer line once. Unfortunately for the designers of the Flexomatic the quality of the control equipment was too poor and the prices too high at that time, which meant that the machine could never become an economic success.

A few years later the same people who had designed the Flexomatic introduced the Milwaukee Matic (*see* Fig.8) on the market. Knowing how it all started it is easy to see that there is the same sort of thinking behind the design of the Milwaukee Matic as behind the design of the Flexomatic. Instead of many spindles they have equipped the machine with one spindle and a tool changer. The machine has an indexing table, which means that the different sides of the workpiece can be indexed to such a position that the machining can be performed on that side. The transfer line is replaced by a shuttle mechanism, which makes it possible for the operator to move in a new pallet with a new workpiece

8 Milwaukee Matic II

as soon as the old pallet is ready to be moved out when the workpiece on this is ready for inspection after machining. Instead of the numerous control units needed by the Flexomatic, you require here only one control unit. Instead of all the tape readers needed by the Flexomatic, Milwaukee Matic requires only two: one reading the tape for the machining of the part that is right now in front of the spindle, the other being loaded with a tape that describes the machining of the next part waiting on the shuttle mechanism.

Figure 9 shows you two views of a frame for one of our low-voltage circuit-breakers. If you study the machining of a complicated part of this kind, you will often find that by using conventional methods you will need somewhere between 10 and 20 set-ups to be able to perform all the machining necessary on the part. If you use a machining centre of the Milwaukee Matic type, only two set-ups are required. The result of the change is very often that the throughput time is reduced to 10% of the original time.

Another interesting effect of the introduction of numerically controlled machines is illustrated by Fig.10, which shows a cylinder for a pressure-regulating device in one of our products. This cylinder used to be made in two parts, which were then fitted together. The reason was that it was very difficult to machine the inside of the part if it was made in one piece. The introduction of numerically controlled machine tools of the kind we have been talking about made it possible to redesign the part. The reduction in cost was quite considerable, the reduction in throughput time was very high, and the problem of leakage was avoided.

It is natural that the designers tend to build complicated pieces of equipment from a comparatively large number of very simple parts instead of trying to make them from a smaller number of slightly more complicated parts. They have been told for many years by production engineers that this design philosophy was the only one that would give an acceptable end result. Now they have to be retrained and told that as long as they stick to the kind of complexity that the numerically controlled machine can master, they should instead build their designs from a few more complicated parts than from a lot of simple ones.

'System 24' is well known in Britain. I have heard that development work on the system side has now unfortunately been cut down in favour of production of the machining units. We know, however, that a good deal of the major machine-tool manufacturers are working along the same basic lines as System 24. It is interesting to note that we have not merely completed the circle back to the old Flexomatic, but have further improved the concept. Instead of the rigid transfer line, more and more machine-tool people are now thinking of a very flexible conveyor arrangement moving parts and their fixtures through all the machines and operations

7 Flexomatic

9 Frame for a circuit breaker

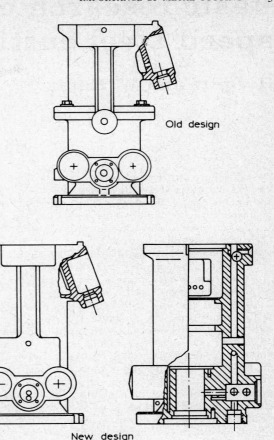

Old design

New design

10 Cylinder redesigned for NC machining

required by each part. Everything in such a set-up is controlled by a process computer of some kind.

Installation of numerically controlled machine tools is one way of trying to solve the two main problems that we have been discussing. They reduce the need for skilled workers in job shop production and they reduce the throughput time, which automatically cuts down the capital invested in the work in progress and reduces the need for capital invested in stocks of semifinished and finished goods. At ASEA more than 10% of all our machine-tool operators use numerically controlled machine tools.

We have not discussed another important problem facing the metalworking and engineering industry: the rapidly increasing costs for salaries. At a first glance one might get the impression that the introduction of numerically controlled equipment would tend to increase the costs for salaries, since a good deal of the work planning is transferred from the shop floor to the offices. A closer study shows that this is not necessarily true. If you want to run a conventional machining shop with reasonable productivity, you have to spend a considerable amount of money on doing the work planning and routing. The cost for doing this is about equal to the programming cost for the numerically controlled machines replacing the conventional ones. The programming cost depends on whether or not you use computers to prepare the tapes for your numerically controlled machines. We have found that the average programming cost is 40–50% lower when we use APT than when we carry out the programming manually for normal turret drill work. For a machining centre with straight-line milling the figure is 30%, and for a contouring lathe the figure is somewhere around 40%. On an average the necessary programming time used by the pro-

grammer is reduced to half of that required for manual programming.

We realize that the introduction of numerically controlled machines is only a first step on the way to automation of the information handling from the conception of a product to its packing and shipping. The introduction of computer-aided programming using, for example, APT is a further step. Several organizations in the world are now working on the problem of getting the computer to do the work planning. They are aiming at making it possible just to feed in information about the shape and size of the workpiece before it is machined and a description of the appearance of the workpiece after the machining. Such programmes are already in operation with, for example, simple shaft production.

The results of this kind of development are that you can reduce the time it takes to perform all the data shuffling and data conversion now required within the company from the day you receive the order from the customer until all the necessary preparations are made before the start-up of the real production. Because the numerically controlled machine is a necessary first step towards the full use of this kind of rationalization it is right to say that in the long run it will enable one to reduce the size of the salaried staff now needed for processing an order in the production department. The fact that automatic processing cuts down the time required between order and delivery means that the amount of capital tied down in stocks of semifinished and finished goods can be reduced.

The two problems I have discussed have some of their main roots in the metal-cutting departments of our companies. I know that the delegates attending this conference by improving the metal-cutting materials do their best to alleviate these problems. I have presented some ideas which, coupled with the progress in the quality of metal-cutting materials, could help our industries to considerably improve their efficiency.

Basic research on the wear of high-speed steel cutting tools

H. Opitz and W. König

Though other cutting tool materials like sintered carbides, ceramics, etc. are used more in cutting operations, high-speed steel has not lost its importance. Table 1 gives an estimate of the quantitative distribution of the applications of high-speed steel and sintered carbides according to machining processes.[1] There are several reasons for this:

(i) no other tool material has both such a great hardness and a high toughness too, which is necessary for all interrupted-cut cutting operations in order to avoid chipping of the tool

(ii) normal high-speed steel is cheaper than carbide

(iii) besides being hot formable, high-speed steel in annealed conditions can be machined by cutting operations such as turning, milling, etc.

Sintered carbides, however, can only be machined by grinding, therefore the fabrication of tools from high-speed steel is easier and cheaper. This is one of the reasons why especially complicated tools like broaches, drills, slab milling cutters, taps, etc. are made from high-speed steel. Furthermore high-speed steel tools are also applied for turning operations under difficult conditions. The range of high-speed steels includes a great number of different grades. Conventional high-speed steels can be divided into four groups:[2]

(i) 18%W steels

(ii) 12%W steels with high V content

(iii) W–Mo steels with 6%W and 5%Mo

(iv) Mo–W steels with 9%Mo and 2%W.

Within these four groups there are different qualities, which differ mainly in the content of vanadium, cobalt, and carbon. Because of their great hot hardness the high-cobalt steels are employed for rough cuts, whereas the vanadium alloyed steels containing wear-resistant vanadium carbides are mostly employed for precision work.

WEAR OF HIGH-SPEED STEEL CUTTING TOOLS

During cutting, tool wear arises because of the mechanical stresses and high temperatures. The most important forms of wear are crater and flank wear (Fig.1). Tool life is normally defined by specified values of these forms of wear on the flank and rake face. The wear zone VB arises on the flank and displaces the cutting-tool edge and, hence, has an effect on the workpiece diameter. At the same time a crater is formed on the rake face by the chip. As cutting time increases, the crater becomes deeper so that the resulting weakening of the cutting edge can lead to breakage. It follows therefore that reliable prediction of tool life as demanded for automatic mass-production machines can only be fulfilled if the wear behaviour of the tools is well known.

Influence of cutting speed

The cutting speed has the greatest influence on tool wear. The correlation between tool life and cutting speed during machining is shown in a general form in Fig.2. From the shape of the curve it can be seen that tool life does not decrease in direct proportion with increasing cutting speed, but for both crater wear as well as for flank wear, tool life maximums and minimums exist. Corresponding to this the wear–cutting speed curves have an inversely related shape, which is mainly influenced by the formation of built-up edges.

An examination of the chip formation at different cutting speeds has shown that at low cutting speeds built-up edges (BUEs) form on the rake face and take over the function of the cutting edge. Between the formation of the BUE and tool wear a direct correlation exists. The flank wear–cutting speed diagram shown in Fig.3 demonstrates the predominant influence of BUEs on flank wear in a low cutting speed range, in this case from 1–40 m/min, when machining normalized carbon steel AISI 1055 with high-speed steels. The micrographs point out the different forms of chip roots and formations of BUEs at different cutting speeds, and prove that the shape of the BUE changes with increasing cutting speed. The increasing flank wear in the range of heavy BUE formation is caused by a continuous sliding-off of BUE particles between flank and cut surface. It has been stated therefore, that these particles coming off with the machined surface are a cause of abrasion on the flank of the tool, whereas the stable part of the BUE protects the rake face against wear. Hence only little wear on the rake face arises in the region of intense BUE formation (Fig.4). With

Table 1 Quantitative distribution of the application of cutting materials in relation to machining processes

Machining process	High-speed steel used, %	Sintered carbide used, %
Turning	20	60
Milling	50	25
Planing	15	10
Spiral milling	10	3
Reaming, broaching, and threading	5	2

By J. Hinnüber and E. Hengler

The importance of high-speed steel tool material for cutting operations has not diminished though other tool materials have been developed. In the first part of the paper the interrelations between the wear of high-speed steel tools and cutting conditions are described. Moreover the mechanism of built-up-edge formation and its influence on tool wear is explained. Because of the great importance of the application of coolants in cutting with high-speed steel tools their wear behaviour under these special conditions is also described. The second part of the paper deals with the possibilities of reducing wear. After brief remarks on the improvements in high-speed steels for better wear resistance, the most important techniques for producing wear-reducing surface layers on tools are presented. These techniques are steam bluing, nitriding, and spark alloying. Some examples of the improvements possible using these techniques are specified. Finally a brief review of different short-time tests is given showing the possibilities for determining machinability without resorting to expensive long-time tests.

621.91.025:669.14.018.252.3:620.178.16

The authors are with the Laboratorium für Werkzeugmaschinen und Betriebslehre, Rheinisch-Westfälische Technische Hochschule, Aachen, West Germany

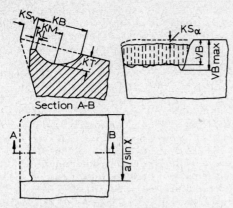

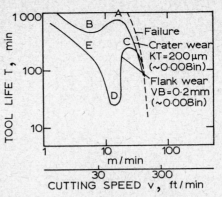

VB wear land; KT depth of crater; KM distance crater centre-cutting edge; KB width of contact area; KL width of crater lip; KS shift of cutting edge (clearance face); KS shift of cutting edge (rake face); b depth of cut; χ side cutting-edge angle

1 Specific data for defining the wear of cutting tools

High-speed steel: S 18–1–2–5; chip cross-section $a \times s = 2 \times 0.25$ mm²; work material Ck55N (AISI C 1055)

2 General characteristics of tool life curves for high-speed steel

decreasing BUE formation, the flank wear also decreases, and encounters a minimum at the cutting speed where the BUE is no longer formed. At higher cutting speeds, flank and crater wear increase steeply till the point where the tool fails. Although up to a cutting speed of 40 m/min mechanical abrasion, and shearing and breaking-off of welded junctions between work material and tool, must be regarded as main causes of wear, the steep increase of crater and flank wear with higher cutting speeds comes from the great loss of strength of high-speed steel at high temperature.

The changed structure in the contact zones of a high-speed cutting tool as a result of high temperature and load can be seen from Fig.5. From the different etching effects the tempering of the high-speed steel, owing to the high cutting temperature, can be recognized. Furthermore, at higher magnification a plastic deformation of the tool material in the crater zone and also at the flank as caused by sliding contact is visible.

Also the tool failure where the high-speed steel suddenly loses its hot hardness in the region of the nose radius of the

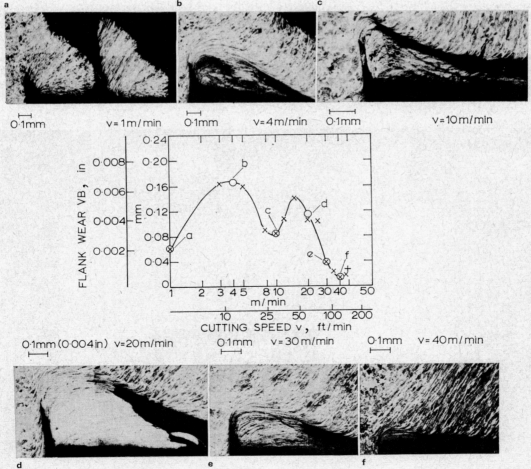

Work material AISI C 1050 normalized; tool material high-speed steel, S 12–1–4–5; chip cross-section $a \times s = 2 \times 0.25$ mm²; cutting time $t = 30$ min; tool geometry $a = 8°$, $\gamma = 10°$, $\lambda = 4°$; $\epsilon = 90°$, $\chi = 60°$, $r = 1$ mm (0.04in)

3 Chip formation and tool wear

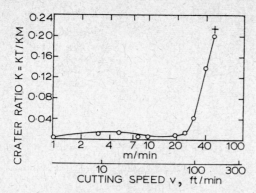

Work material Ck55N (AISI C 1055); cutting time $t = 30$ min; tool material S 12–1–4–5; chip cross-section $a \times s = 2 \times 0.25$ mm²; tool geometry $a = 8°$, $\gamma = 10°$, $\lambda = 4°$, $\chi = 90°$, $\epsilon = 60°$, $r = 1$ mm

4   Dependence of crater wear on cutting speed

tool must be attributed to a thermal overload of the tool material. At this point the tool nose abruptly loses its ability to cut. At the same time great sliding forces arise between tool and workpiece leading to a rapid heating of the tool material and an instantaneous loss of hot hardness. If the tool is not taken out of the cut at this instant the greatest part of the cutting edge is lost as described above. Vilenski and Shaw[3] believe that the sudden increase in wear rate leading to the tool failure is due to the fact that the mean pressure on the wear land exceeds the effective flow stress of the work. According to the results of sliding tests, the total area contact between tool flank and work will cause a steep increase of tool wear and total destruction soon after.

The mechanism of the formation of BUE according to a theory of Shaw[4] can be explained in the following manner. At low speeds and feeds the first layers of work material seized on the tool surface are greatly strengthened by severe cold working and the temperature rise is insufficient to cause any significant reduction in the strength. Furthermore the increasing strength of carbon steels, which appears in the range of blue brittleness over rising temperature, leads to the

formation of BUE. If, therefore, work material is seized on the tool edge, separation during further movement does not occur immediately in the zone of contact which is at the highest temperature, but rather in the flow layer of the chip which, due to the temperature range, has the lowest strength (Fig.6). Therefore a part of the flow layer remains stuck to the tool, and forms the BUE. The stability of the BUE requires that the only load is a compressive stress. The process of building-up on the rake face of the tool cannot go on indefinitely. If the BUE is overloaded, sliding takes place in the direction of main shear stress as shown in the cross section of the BUE in Fig.6. Parts of the BUE are carried away on the work surface and on the underside of the chip. The greatest flank wear develops in that cutting speed range, in which periodically the whole BUE is removed by the workpiece and the chip. The formation of BUE is no longer possible when with increasing cutting speed the temperature of the bottom of the chip lies above the blue brittleness range.

Influence of feed

The results of investigations over the influence of feed on flank and crater wear are shown in Fig.7, in which the wear–cutting speed curves are plotted for three different feeds. Principally the shape of the wear land–cutting speed curve does not change. The locations of the maximum and minimum wear land, however, are displaced with increasing feed to lower cutting speeds. Also the wear land becomes greater with increasing feed. Interpreting the relation between formation of BUE and tool wear it can be concluded that as result of changing the feed the range of cutting speed at which the formation of BUE occurs also is displaced. This can be demonstrated by a view of the bottom side of the chip, because the parts of BUE being removed by the chip can be easily recognized as irregularities in this view. In Fig.8 the bottom side of chips produced at different cutting speeds and feeds are compared. From the smooth surface it can be seen that at a feed of $s = 0.4$ mm/U no BUE is formed at a cutting speed of 20 m/min, whereas at a lower feed such as 0.1 mm/U BUEs are still formed at a cutting speed of 60 m/min.

The cause for the displacement of the range of BUE formation to lower cutting speeds for increasing feed is the cutting

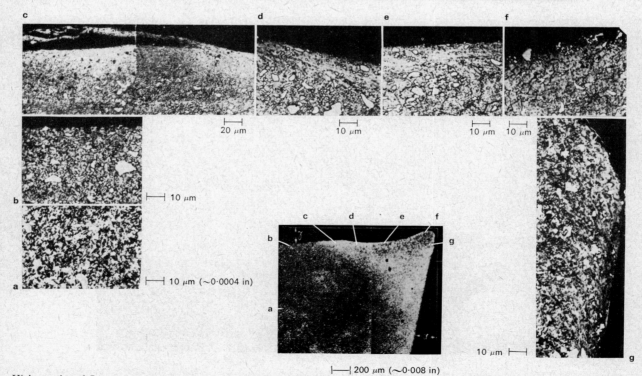

High-speed steel S 12–1–4–5; cutting speed $v = 60$ m/min (~200 ft/min); chip cross-section $a \times s = 2 \times 0.25$ mm² (~0.08 × 0.01 m²)

5   Changes of high-speed steel structure resulting from thermal and mechanical load

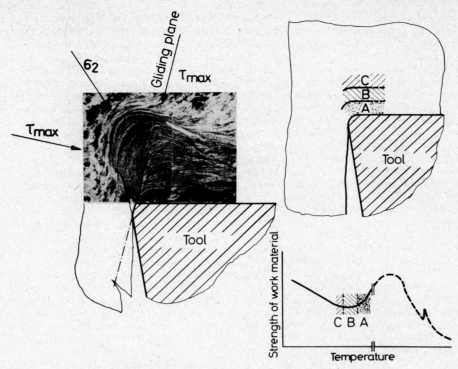

a $\sigma_y=\sigma_2$; $\sigma_x=0$; $\sigma_1=0$; $\tau_{xy}=0$, by E. Schaller; *b* strength in range A> B < C, by M. C. Shaw

6 Conditions for the formation of built-up edges

temperature. With increasing feed a higher temperature is generated. This relation can also be derived from the curves of crater wear (Fig.7) because crater wear is also dependent on the temperature of the contact zone between chip and rake face. The point where the steep increase of crater wear begins moves with increasing feed toward lower cutting speeds. Accordingly the temperature-dependent tool failure also occurs at lower cutting speeds.

Influence of the application of cutting fluids

In cutting operations with high-speed steel tools, cutting fluids are often applied. By cooling, the heat generated in the contact zones of the tool as a result of shearing and sliding will be taken up producing a low cutting-tool temperature. In addition the cooling fluid can also act as a lubricant by which the friction between the chip, the cut surface, and the tool respectively is reduced. In other words, lubrication also leads to a lower cutting temperature. Lubrication, however, is only possible if the fluid is able to penetrate into the contact zones and can form a lubrication film. From investigations on the chip formation, it is known that the penetration of cutting fluids into the contact zone is only possible in the range of discontinuous chip formation and continuous chip formation with BUE. At higher cutting speed in the range of

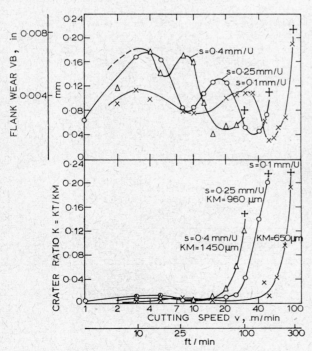

Work material Ck55N (AISI C 1055); tool material S 12-1-4-5; depth of cut *a* = 2 mm (=0·08in); tool geometry *a* = 8°, *γ* = 10°, *λ* = 4°, *χ* = 90°, *ε* = 60°, r = 1 mm; cutting time *t* = 30 min

7 Influence of feed on flank and crater wear at different cutting speeds

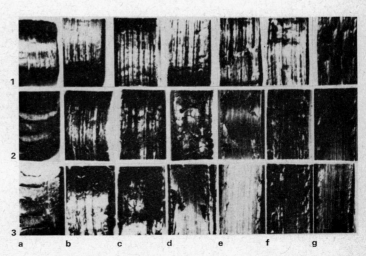

Work material Ck55N (AISI C 1055); tool material S 12-1-4-5; depth of cut *a* = 2 mm; cutting time *t* = 30 min; tool geometry *a* = 8°, *γ* = 10°, *λ* = 4°, *χ* = 90°, *ε* = 60°, r = 1 mm
Cutting speed *v*, m/min: *a* = 4, *b* = 8; *c* = 12·5; *d* = 20; *e* = 31·5; *f* = 40; *g* = 60 Feed *s*, mm/U: line 1 = 0·1; line 2 = 0·25; line 3 = 0·4

8 Chip bottom sides at different cutting speeds and feeds

flow-chip formation the cutting fluid is not able to build a lubricant film because of the total area contact between the sliding surfaces. In this cutting speed range, therefore, cutting fluid acts only as coolant.

In Fig.9 the wear–cutting speed curves for cutting with and without fluid are compared. As it can be assumed from the position of the wear maximums and the range of BUE formation, when applying a cutting fluid, the same temperatures as in dry cutting are generated only at higher cutting speeds. This can also be demonstrated by the view of the bottom side of the chip produced at different cutting speeds. The smaller wear land in the first maximum of the curve is caused by the action of the cutting fluid as lubricant at low cutting speeds, where the fluid is able to penetrate into the contact zones and does improve the friction conditions. With the beginning of flow-chip formation, the lubricant action diminishes and the cooling effect is of greater importance. Then, up to cutting speeds at which the BUEs have their greatest dimensions, the wear land with cooling increases relative to dry cutting up to double the dry-cutting value. Also, greater crater wear occurs when cutting with fluid. The cause of the greater wear may be a result of the changed temperature conditions in the region of chip formation. Schilling[5] has hypothesized that the increased tool wear comes from the increased strength of the work material at lower temperature. Although tool wear in cutting with fluid is partly higher than in dry cutting, the tool failure occurs at slightly higher cutting speeds. Therefore in cutting with coolants greater wear rates will be reached before the tool fails. This is a result of the heat transfer through the cutting fluid out of the tool wedge. Therefore, with coolants, the hot hardness of the high-speed steel is not lost as fast as in dry cutting.

IMPROVEMENT OF THE WEAR RESISTANCE OF HIGH-SPEED STEEL CUTTING TOOLS

From the beginning of the development of tool materials great efforts have been undertaken to reduce tool wear in order to get a greater tool life. This aim can be realized in two different ways; firstly, by appropriately changing the composition of the high-speed steel and/or the production method, tool wear resistance and hot hardness can be improved. On the other hand it is possible to produce wear-resistant layers on high-speed steel tools by means of surface treatment.

Improvement of high-speed steel

On the developments for metallurgical improvement of the high-speed steels only a short review will be given. By changing manufacturing conditions in the melting process (addition of nucleation materials and selection of suitable bar sizes) as well as forming and heating techniques, the quality of high-speed steels has been greatly improved during the last few years. Other than the method of raising the content of vanadium and other expensive alloying elements, an increase of wear resistance has also been attained by the development of high-speed steels with higher carbon content. These high carbon content high-speed steels have been developed from Mo steels with high cobalt content and have achieved a hardness of 70 HRC. As the comparison between some conventional (in parentheses) high-speed steels and their corresponding high carbon content grades in Fig.10 shows, the hot hardness is also increased by the higher carbon content.[6] Therefore these steels have the properties which are necessary for low tool wear.

Wear-reducing surface layers on high-speed steel cutting tools

Ready for use tools from high-speed steel can achieve a better wear resistance by special after-normal-production treatment techniques, which help to reduce friction of the contact zones and lead to a greater surface hardness. A summary of the possible kinds of layers and their influence over the total wear is given in Fig.11.

Bonderizing, sulphurizing, MoS_2 treatment, and hard-chrome plating as techniques for the forming of wear-resistant layers on high-speed steel tools have only been laboratory tested and only in very special cases industrially applied. According to information from literature, the results of these uses have shown in part great improvements in

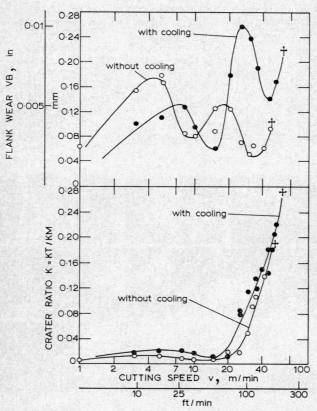

Work material Ck55N (AISI C 1055); tool material S 12–1–4–5; cutting time $t = 30$ min; coolant emulsion (1:50); chip cross-section $a \times s = 2 \times 0.25$ mm²; tool geometry: $a = 8°$, $\gamma = 10°$, $\lambda = -4°$, $\chi = 90°$, $\epsilon = 60°$, $r = 1$ mm

9 Influence of a cutting fluid on flank and crater wear at different cutting speeds

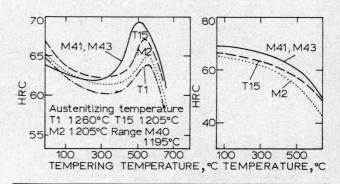

By Fletcher and Wendell

| Nomenclature | | Chemical composition, % | | | | | | |
AISI	DIN	C	W	Mo	V	Co	Cr	HRC
T1		0.75	18.0	..	1	..	4.0	62–63
HiCM2		1.0	6.0	5.0	2.0	..	4.0	66–67
(M2)	S 6–5–2	0.85	6.0	5.0	2.0	..	4.0	63–64
T15	S 12–0–5–5	1.5	12.0	..	5.0	5.0	4.0	66–67
M41		1.1	6.75	3.75	2.0	5.0	4.25	69–70
(M35)	S 6–5–2–5	(0.8)	(6.0)	(5.0)	(2.0)	(5.0)	(4.0)	≈65
M43		1.25	1.75	8.75	2.0	8.25	3.75	69–70
(M34)	S 2–8–2–8	(0.9)	(2.0)	(8.0)	(2.0)	(8.0)	(4.0)	≈65

10 Hardness and hot hardness of different high-speed steels

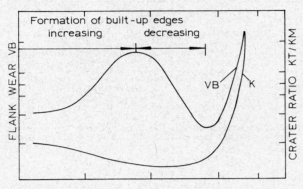

Surface layers	Effect on the total wear			
Phosphate treatment	⊠⊠⊠	⊠⊠⊠	⊠⊠⊠	
Sulfiding	⊠⊠⊠	⊠⊠⊠	⊠⊠	
Mo₂S₂-treatment	⊠⊠⊠	⊠⊠⊠	⊠⊠⊠	⊠
Steam bluing	⊠⊠⊠	⊠⊠⊠	⊠⊠⊠	
Chromizing	⊠⊠			
Nitriding	⊠⊠⊠	⊠⊠⊠	⊠⊠⊠	
Spark alloying	⊠⊠⊠	⊠⊠⊠	⊠⊠⊠	

Improved sliding behaviour
Diminished pressure welding

Increased wear resistance

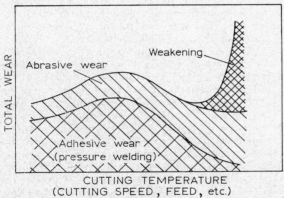

11 Nature of wear, and methods for reducing it by surface layers

tool life.[7-9] On the other hand the steam-bluing technique has found wide application in industry. It is used for saw-blades and other high-speed steel tools such as small twist drills, taps, broaches, milling cutters, reamers, etc. While in the steam-atmosphere furnace during the bluing process a very dense, hard, and seizing blue iron oxide coating of the structure Fe_3O_4 forms on the surface of the tempered and finish-ground tool. This coating leads to better frictional properties and a lower tendency for cold pressure welding between tool and work. The obtainable improvement is dependent on work material.[10] An approximate value can be assumed, that in cutting conventional steels with blued tools an improvement of 2:5:1 is possible. On the other hand, in machining tough materials like chromium–nickel steels greater improvements have been realized.

Just as steam bluing is a good method for improving high-speed steel tools, so also is nitriding. Two different nitriding techniques must be distinguished: salt-bath nitriding and gas nitriding. For tools used in Germany, mainly salt-bath nitriding is applied. During the nitriding process hard nitrides form as compounds of nitrogen with the alloying elements chromium, vanadium, molybdenum, and tung-sten in the surface of the tools. Because the nitriding process is precipitation hardening, dimension altering through the austenite–martensite transition of normal hardening does not occur. The depth of the nitrided case depends on the nitriding time and the temperature of the bath. The necessary depth of the nitrided case follows from the kind of tool load. By application of an intermediate anneal in a neutral bath, the undesired embrittlement of the cutting edge can be avoided and a deeper nitrided case with greater toughness can be attained having a gradual transition to the base material. Nitriding of the tools leads to the following property alterations:

  (i) increase of the edge hardness, wear resistance, and hot hardness

 (ii) improvement of the sliding conditions

(iii) reduction of the tendency to cold pressure welding.

Therefore remarkable increases of performance are obtainable by the nitriding of tools requiring a high wear resistance. From our own experiments, for example in face milling of high-temperature austenitic steels using nitrided high-speed steel tools, a reduction in wear was obtained (Fig. 12). It also became evident, that an additional steam bluing of the tool after the nitriding leads to a further improvement. Furthermore in machining of work materials of lower strength increases in tool life are possible, as can be seen from the results of Rühenbeck's investigations[11] with nitrided twist drills and reamers. The applied high-speed steel grades, work materials, and cutting conditions are plotted in Table 2. One can see that against the non-treated tools, whose performance is arbitrarily set at 1, an increase in tool life of up to thirty fold was in some cases possible.

A further method for the forming of wear-resistant layers on high-speed steel cutting tools is spark alloying, in literature also referred to as percussion welding. In this technique the wear zones of the tool are protected against wear by producing a layer of tungsten carbide or other hard materials by means of a spark discharge. According to information in Soviet literature, great improvements in tool life have been obtained. One problem in the application of spark alloying is that the treated surfaces have a minute crater-like roughness, and this causes a jagged cutting edge. Increased wear is the result of this. The disadvantage can only be eliminated if the quality of the spark-alloyed surface is improved by finish grinding.

Our original investigations with spark-alloyed milling cutters have shown that in face milling of high-temperature austenitic steel an increase in tool life is possible. The results of the experiments are shown in Fig. 13. It can be assumed that spark-alloying wear coating of high-speed steel tools especially for the machining of high-temperature difficult-to-machine materials will bring profit in the future.

MACHINABILITY TEST AND ITS VALIDITY FOR HIGH-SPEED STEEL

Investigations over the causes of tool wear have shown that a prediction of individual tool-life characteristics when machining a special melt is not possible. This is because of the numerous influences dependent on special criterions such as the chemical composition or the mechanical properties of the steel. Exact determination of tool-life characteristics require long-time tests. The determination of wear–cutting time, tool life–cutting speed curves, and the $v_{60:B}$ and $v_{60:K}$ index require at least three to five long-time tests. This implies a long time delay. In order to reduce the test time, diverse short-time procedures to determine the wear behaviour have been developed. In Fig. 14 the effort necessary for a long-time test is compared with that for two different short-time tests.

The short-time procedures either apply the same working conditions and extrapolate short-time wear values to longer cutting times or apply more difficult cutting conditions so that higher wear values are found after a shorter cutting time.

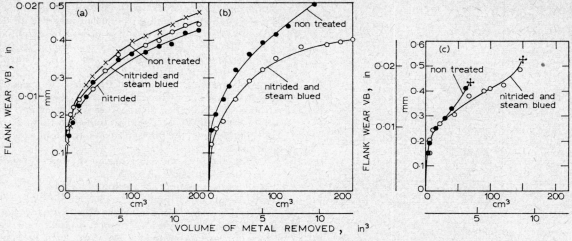

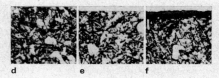

a work material high-temperature austenitic steel; tool material
S 10–4–3–10; cutting speed $v=25$ m/min (~75 ft/min)
b work material high-temperature austenitic steel; tool material
S 18–1–2–5; cutting speed $v=30$ m/min (~90 ft/min) c work
material high-temperature austenitic steel (tempered); tool
material S 18–0–1; cutting speed $v=10$ m/min (~30 ft/min)
d normal structure S 10–4–3–10; e bath-nitrided structure;
f bath-nitrided and steam-blued structure. Chip cross-section:
$a\times s=2\times0.2$ mm²/Z; tool geometry $\alpha=8°$, $\gamma=15°$, $\lambda=-15°$,
$\chi=70°$, $\epsilon=100°$, $r=0.5$ mm

**12 Influence of bath nitriding and steam bluing treatment on
tool wear when face milling difficult-to-machine materials**

The results obtained are related to the desired cutting conditions.

Two short-time procedures whose results correspond with limitations to the results from conventional tests are tests with continuously or stepped cutting speed and the constant feed force method. As the name of the first procedure describes, and as shown in Fig.14, the speed is either continuously or stepped increased until the tool fails. The amount of speed increase runs up to $\Delta v=5$ m/min for every 25 m of cutting length. For comparison of the results of the continuously increased test with long-time tests, the cutting speed at tool failure v_E is used. For the stepped increased test a comparison value is calculated according to the formula:

$$v_{comp.}=v_{z-1}+v\frac{l_z}{lo}\ [\text{m/min}]$$

The speed of tool failure is independent of the initial speed if the initial speed is not too high. Dependent on work material and cutting conditions, the mean speed v_E for tool failure can be determined from three to five tests. A comparison of the results of the stepped and continuously increased speed tests shows a very good conformity, so that for later comparisons with long-time tests only the results of the continuously increased tests are applied. In order to prove the repeatability of the results, investigations over the spread of the results were carried out. The field of trust of the mean values of the tool failure speed for a 95% confidence level is for low-carbon steel $\pm2.45\%$ and $\pm7.65\%$ considering all work materials.[12]

In order to compare the results of the continuously increased speed tests and the results of long-time tests for high-speed steel S 18–1–2–5, the speed for tool failure v_E, the long-time $v_{60:K=0.1}$ index and the v_{60E} index are plotted in Fig.15. From the geometrical similarity of the curves it can be noted that the index v_E can be used to describe machinability.

As another short-time procedure the constant feed force test utilizing a suspended weight to limit the feed should be mentioned. The feed which appears after a short running-in time serves as a machinability figure of merit. This is grounded by the fact that a greater feed appears when cutting a more machinable steel. Recent publications from Sweden show that results obtained by using this method correspond well to the results from conventional tests. Those tests were carried out with high-speed steel drills S 6–5–2 machining various grades of aluminium, brass, bronze, cast-iron, and steel.[13]

Concluding, it can be said that the above-mentioned short-time procedures possess the following positive character-

Table 2 Extending tool life of high-speed steels through nitriding

Tool	Tool diameter, mm	Tool material	Work material	Cutting conditions	Coolant	Efficiency of tools non-treated	nitrided
Twist drill	5–40	High-speed steel ABC 111 (S 3–3–2–0)	16Mn5Cr (SAE 5130) $\sigma_B=75$kp/mm² annealed	$v=18$–22 m/min $s=0.05$–0.2 mm/U	Soap water	1	2–3
Reamer	8–28	High-speed steel D (S 12–1–3–0)	16Mn5Cr (SAE 5130) $\sigma_B=75$–90kp/mm²	$v=5$–9 m/min $s=0.1$–0.5 mm/U	Cutting oil	1	8–15
Twist drill	5–11	High-speed steel ABC 111 (S 3–3–2–0)	Aluminium alloy HB=60–100 kp/mm²	$v=80$–90 m/min $s=0.1$–0.8 mm/U	Without	1	20–30
Reamer	8–28	High-speed steel D (S 12–1–3–0)	Aluminium alloy HB=60–100 kp/mm²	$v=10$–24 m/min $s=0.4$–1.8 mm/U	Without	1	18–28

By A. Rühenbeck

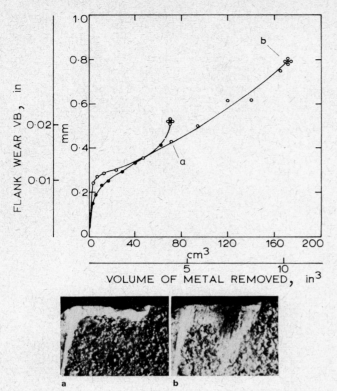

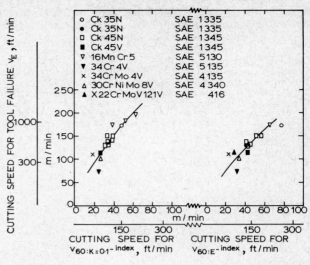

Testing method   Tool material   Chip cross-section
                                            $a \times s$ (mm/U)
Long-time test   S 18–1–2–5 2×0.25
Test with continuously S 18–1–2–5 1×0.1
increased cutting speed

15 Comparison of the results of long-time tests and of tests
with continuously increased cutting speed with high-speed
steels S 18–1–2–5

Work material high-temperature austenitic steel; tool material
S 18–0–1 non-treated, S 18–0–1 spark alloyed; cutting speed
$v = 10$ m/min; chip cross-section: $a \times s = 2 \times 0.2$ mm²/Z; tool
geometry $\alpha = 8°$, $\gamma = 15°$, $\lambda = -15°$, $\chi = 70°$, $\epsilon = 106°$, $r = 0.5$ mm

13 Extending tool life of high-speed steel used for face milling
through spark alloying

Testing method	Method of operation	for	Required time		Total time (including test preparation),	
			Cutting time, min	Measuring time, min	min	h
Long-time test		Determination of the tool-life cutting speed curves and the $v_{60;K}$- and $v_{60;B}$-index	200	450	900	15
Test with (a) continuously (b) stepped increased cutting speed		1 test	<5	⟶0	60	1
		4 test for repeatability	<20	⟶0	220	2
Constant feed force test		Determination of the feed v. feed weight curve for 5 different feed weights	20	⟶0	60	1

x Interruption of the cutting process for wear measurement

14 Testing methods

istics: they have good repeatability, they take little time to perform and correspondingly little material, and most important, they are a good machinability index.

REFERENCES

1 J. HINNÜBER AND E. HENGLER: *Stahl Eisen*, 1964, **84,** 26, 1787
2 H. H. WEIGAND: *Industrie-Anzeiger*, 1968, **90,** 14, 249–53
3 D. VILENSKI AND M. C. SHAW: 'Importance of workpiece softening on machinability', *Ann. CIRP*, unpublished
4 M. C. SHAW *et al.*: *Trans. ASME (J. Eng. Ind.)*, 1961, **83,** 181
5 W. SCHILLING: 'Der Verschleiss an Drehwerkzeugen aus Schnellarbeitsstahl und seine Ursachen', 1966, Aachen, Diss. TH
6 S. G. FLETCHER AND C. R. WENDELL: *Met. Eng. Q., ASM*, Feb., 1966
7 W. SEIFERT: *Neue Hütte*, 1966, **11,** 3, 171–76
8 H. STAUDINGER: Oberfläche, 1968, no.6, 213–18
9 N. DIEDERICH: *Der Ingenieur*, 1967, **79,** 27, 145–158
10 H. D. WECKENER: *Werkstatt Betrieb*, 1958, **91,** 33–38
11 A. RÜHENBECK: *Industrie-Anzeiger*, 1956, **15–16,** 40–43
12 Autorenkollektiv: *Stahl Eisen*, 1963, **83,** 20, 1209–1226 and 21, 1302–1315
13 J. DAGNALL: *Ann. CIRP*, **XV,** 301–361

Wear processes which control the life of cemented carbide cutting tools

E. M. Trent

To be of practical significance an investigation of tool wear must deal with the types of wear observed on tools used under machine-shop conditions. The changes of shape of the cutting edge must be related to the conditions to which the tools were subjected, to the composition, structure, and properties of the tool material, and to the tool geometry.

It has become customary in studies of tool wear to consider only 'flank wear' on the clearance face and 'crater wear' on the rake face. Examination of a random selection of worn tools from different types of operation in different industries makes it apparent that this is an oversimplification because of the variety of conditions to which tools are subjected. Even changes of shape which are superficially classified together as flank wear or as crater wear can be shown to be the result of entirely different wear processes when cutting conditions are changed. In this paper the processes of wear rather than the superficial appearance will be discussed since this approach offers the better prospect of arriving at constructive recommendations for improving cutting tool performance.

EXTERNAL FACTORS ACTING ON CUTTING TOOLS

Major forces

Measurement of the major forces acting on cutting tools has provided a considerable amount of accurate information but this is of little value in relation to tool wear problems. In most cases even the mean stresses on the tool cannot be calculated from published data because areas of contact on the tool have seldom been recorded. The direction and magnitude of the forces indicate that the resolved stresses are mainly compressive and where contact areas have been measured values greater than 50 tons/in[2] have been reported when cutting steel.

The major forces are of more importance in relation to fracture than to wear. Fractures occur most frequently in interrupted cutting, or when swarf makes contact with an unprotected part of the edge, imposing fatigue or impact stresses. It is not easy to distinguish minute fracture of the edge from a wear process. Numerous localized fractures, similar to attrition wear in appearance, may be initiated at fatigue cracks parallel with the cutting edge caused by mechanical fatigue stresses, particularly in milling cutters but also in other interrupted cutting.

Improvements in technology, tool geometry, machine rigidity, and carbide quality, have reduced the incidence of fracture, but a proportion of tools is still likely to fail in this way either through misuse or because there is always an incentive to use the most wear-resistant carbide grade, which may lack the toughness really required for the application concerned.

Temperature

Most of the energy expended in the 'secondary shear zone' or 'flow zone' adjacent to the tool surface, is converted into heat, part of which flows into the tool from the surface of contact. The interface temperatures may exceed 1000°C when cutting steel, iron, and other metals of high melting point. There are very steep temperature gradients of the order of 20°C/0·001in and fluctuations of temperature are extremely rapid when cutting is interrupted. These high temperatures influence the change of shape of the cutting edge in at least four ways of which two will be dealt with later.

The tool material must be able to support the major stresses at the cutting edge temperatures.[1] Cemented carbide tools have adequate high-temperature strength for cutting most commercial materials at the rates of metal removal at present in operation in the engineering industry, and this is a main reason for their superiority over high-speed steel tools.[2] When cutting steels of high hardness, and particularly creep-resistant alloys, the ability of carbide tools to resist plastic deformation at high temperatures imposes upper limits to practical cutting speed and feed rates. Deformation of the tool may lead to rapid wear, often at the nose, or to fracture. To detect the deformation, close examination of tools, often in early stages of wear, is required, but correct remedial action depends on such observations.

In interrupted cutting, particularly in milling, rapid fluctuation of temperature is superimposed on the temperature gradients. The region near the cutting edge is subjected to alternating compressive and tensile stress resulting from the temperature fluctuations in addition to the direct mechanical stresses. Cracks normal to the cutting edge, often starting on the rake face a few thousandths of an inch from the edge, are caused by this thermal fatigue.[3] Frequently such cracks penetrate no more than 0·020–0·030in and have no effect on tool life, but they may act as stress raisers, initiating fracture. The number and spacing of the cracks depends on the cutting conditions. Very numerous cracks most commonly occur when milling hard materials at high speeds with low feed rates, and such numerous cracks may join together so that the edge crumbles away.

Seizure

Except when cutting at very low speed, e.g. a few in/min, the tool and work materials are seized together near the cutting edge for most of the duration of cutting.[4] In this region

The major features of wear on carbide tools used under machine-shop conditions are described, first in relation to external factors acting on the tool, and then in relation to composition and structure of the tool materials. Temperature, temperature gradients, local stress systems at the tool/work interface, and especially the pattern of flow of the work material around the tool edge are major factors determining the types and rates of wear of carbide tools. Interactions between tool and work material influence wear rates by a factor of several times. Cobalt content, carbon content, grain size, and the presence of carbides of elements other than tungsten are the main traditional variables, and their influence on wear in different situations is discussed. A potentially important line of development is the production of composite throw-away tool tips in which thin layers of highly wear-resistant carbide reduce the wear rate at the cutting edge, while the bodies are of a carbide alloy with high strength and toughness.

669.018.25: 539.538

The author is in the Department of Industrial Metallurgy, University of Birmingham

1 Section through carbide tool and adhering metal after cutting steel, showing the pattern of flow around the cutting edge; etched in Nital and in alkaline ferricyanide   ×1500

3 Section through worn flank surface of carbide tool with adhering metal after cutting steel at low speed, showing attrition wear process; etched in alkaline ferricyanide   ×1500

relative motion takes place by shear in the work material, in which a pattern of flow is established under steady cutting conditions (Fig.1). Localized rates and amounts of shear are very great, velocity gradients very steep, and very high temperatures are generated. This unusual tribological condition has a major influence on tool wear in metal cutting.

In flowing over the tool surface the work material exerts a shearing stress. Cemented carbide tools can withstand the shearing stress when cutting most materials, but creep-resistant alloys impose temperatures and shear stresses which are more severe, and, when cutting some very high creep strength alloys, cemented carbide tools are rapidly destroyed by stresses which tear the grains apart (Fig.2).

The tool shape may be changed more slowly by shear or tensile stresses acting on small irregularities in the tool surface over which the work material is flowing. Such irregularities may be the result of initial roughness of the tool surface or may be the consequences of a wear process.

Since cemented carbides are relatively weak in tension, carbide grains, fragments of grains, or groups of grains may be fractured and carried away in the stream of metal (Fig.3). This can be described as a process of attrition.[5] It is most commonly observed when the flow pattern of the work

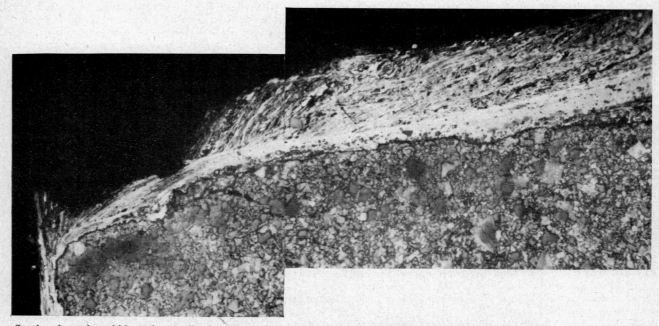

2 Section through carbide tool and adhering metal after cutting nickel-based creep-resistant alloy at 50 ft/min, showing destruction of tool material by shear stresses at high temperatures; etched anodically in oxalic acid   ×1500

material is irregular or turbulent, but it can also contribute to wear at high cutting speeds when combined with other wear processes.

When cutting at low speeds in the presence of a built-up edge, or where lack of rigidity leads to vibrations of large amplitude, the work material may be welded to the tool and intermittently broken away taking with it fragments of the tool. High-speed steel can resist the local tensile and peeling stresses better than cemented carbides and under such conditions carbide tools may be worn more rapidly than high-speed steel.

In relation to wear, the localized tensile or shear stress systems are of more significance than the major compressive stresses normally measured. It would be useful to explore these local stress systems and to determine what relation they bear to the major stresses.

Material being cut

Rates of tool wear are very dependent on structure, composition, and properties of the work material. High-strength materials give rise to high stress on the shear plane. Highly ductile materials give rise to high cutting forces because of the large areas of contact on the tool surface and small shear plane angles which result from this. Tool temperatures are low when cutting metals and alloys of low melting point and most tool-wear problems arise with higher melting point materials. The strength of the work material at high temperatures is of major importance as has already been mentioned. There is evidence that tool-tip temperature is higher when cutting an alloy of high creep strength than when cutting similar alloys of lower creep strength under the same conditions.[6] Serious reductions in productivity result from the low cutting speed and feed rates when machining creep-resistant alloys.

The pattern of flow of the work material around the cutting edge of the tool (Fig.1) determines to a great extent the positions on the tool at which wear occurs, and the wear rates. It has probably the greatest influence on wear rate of any of the factors considered and deserves concentrated study. This flow pattern is a function of the work material, the cutting conditions, the material, and the geometry of the tool.

So far only isolated examples have been studied and the rheology of cutting has not been surveyed as a whole to determine the main controlling parameters. Only a few aspects can be mentioned here to illustrate its importance.

The built-up edge which occurs when cutting most steels, cast iron, α–β brass, and other multiphase alloys is the most obvious example of the importance of flow pattern.[7,8] Under conditions where a stable built-up edge is formed, the material in contact with the tool may be stationary on the rake face for long periods of time while shear in the work material is taking place at a distance of several thousandths of an inch from the rake surface. The type and rate of wear on the flank of the tool are controlled by the flow pattern in this region, where the material is less persistently stationary at the tool–work interface.

When a built-up edge is not present, as when cutting pure metals or when the cutting speed is raised, the flow pattern changes and with it the process and rates of wear. The flow rates very close to the tool surface are increased. The rough worn surfaces with few signs of directional movement, characteristic of the built-up edge regime, give way to smoother wear in which carbide grains are worn through, probably by a process involving diffusion.[5] These worn surfaces usually exhibit markings indicating the direction of flow over the surface. Characteristic of these are ridges, each starting from some structural feature in the tool, around which the work material flows cutting channels on either side but leaving a ridge (Figs.4 and 5). This often gives an apparently scratched appearance to the worn rake and flank surfaces which has led many observers to conclude that an abrasion process is involved in which hard particles plough grooves in the tool material. Close observation of worn carbide tools, however, reveals that the scratched appearance is usually caused by the formation of ridges starting from wear resistant features as described.

It is not pedantic to distinguish between the two types of surface, since the methods of resisting the relevant wear processes are very different, particularly when considering the metallurgical parameters of tool and work material required to reduce wear rates. Attention should be focussed not so much on the presence or absence of abrasive particles

4 Part of worn rake surface of WC–Co alloy tool after cutting steel, showing ridges indicating direction of flow; surface cleaned with HCl to remove adhering metal.   × 100

5 As Fig.4, showing initiation of a ridge at a large carbide grain   × 1500

in the work material as on the metallurgical changes which modify the flow pattern under the intense shear conditions at the tool surface. Two such types of changes are mentioned here.

First, the flow pattern may be altered by the introduction of phases forming surfaces of easy shear under the intense deformation conditions of the flow zone,[5] thus allowing distinct volumes of the work material to slide over one another. Sulphide and some silicate inclusions in steel and graphite flakes or nodules in cast iron influence flow patterns in this way, and the wear rates of tools are sensitive to small changes in percentage and distribution of these phases.

Second, constituents of the work material which segregate to the tool/work interface or react with the tool material may either increase or decrease the rate of wear by a factor of many times. Sulphides in steel, for example, segregated at the tool interface of steel-cutting carbide grades, reduce the built-up edge and greatly improve surface finish, though they may not reduce wear rates in this way. Certain silicates adhere strongly to the rake face of steel-cutting carbide tools when machining steel, decrease the wear rate by many times, and permit much higher cutting speeds.[5,9] On the other hand, increasing the Si content of Al–Si alloys from 13–19% has a quite disastrous effect on the tool life and makes it necessary to reduce cutting speeds by a factor of 4 or 5 times. The coarse primary Si particles appear to segregate at the cutting edge of carbide tools, greatly increasing the wear rate (Fig.6).

There are many other ways in which interactions between tool and work may influence tool life but few have so far been studied in detail.

Atmosphere

The gaseous or liquid environment has a significant effect on the rate, location, and process of wear on carbide tools. When cutting is carried out in vacuo or in an inert gas, the area of contact between tool and work increases greatly and the cutting forces rise correspondingly.[10] Even a trace of oxygen restricts the contact area, and atmospheric oxygen in normal cutting undoubtedly plays this part, thus reducing cutting forces and the incidence of mechanical fracture of tools. It may, however, accelerate wear. The gaseous atmosphere is unable to penetrate completely to the tool edge in most cutting operations but gaseous atoms can intrude some distance from the periphery of the apparent area of contact.[5] Seizure in these regions may be reduced by this contamination of the newly formed surfaces, allowing the partially oxidized surfaces to slide over the tool at two main positions, at the trailing edge of the nose radius, and where the work surface intersects the cutting edge. Accelerated wear is often observed at these positions.

If cutting is carried out in pure oxygen, wear around the periphery of contact is often accelerated. Cutting in oxygen provides a useful research tool to investigate the region to which gases can penetrate. Recent work has suggested that some penetration can take place closer to the cutting edge than had previously been considered possible.[11]

Since the part of the tool close to the edge may be heated to more than 1 000°C when cutting steel, and since the rate of oxidation of carbide is high at such temperatures, serious oxidation of the tool could be predicted. In fact the hottest parts of the tool are normally shielded from oxidation by the action of the freshly cut surfaces which remove the oxygen, leaving protective pockets of nitrogen near the cutting edge. When the trailing-edge angle is very high, however, as with copy lathe tools, a very hot region of the tool is exposed to oxygen on the end clearance face so that the tool shape is altered by oxidation. Thus both oxygen and nitrogen of the atmosphere modify the rate and processes of tool wear.

Cutting lubricants modify the surrounding atmosphere and except at very low speeds probably penetrate only as gases into the contact area. Water-based lubricants can act to reduce the size of the built-up edge at low rates of speed

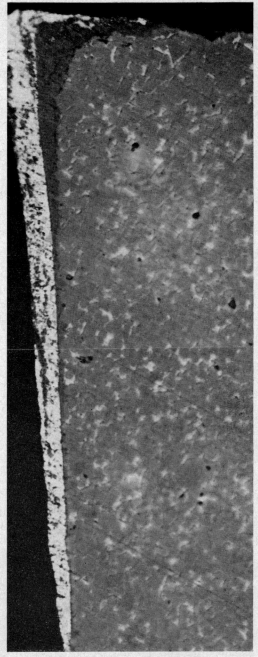

6 Section through cutting edge of carbide tool used for cutting a high Si–Al alloy, with adhearing metal, showing layer of Si built up on worn flank surface   ×1 500

and feed, when they often accelerate wear. As cutting speeds are raised, access to the tool/work interface is restricted and water-based lubricants no longer increase the wear rate. It has been reported that in the higher speed range they may greatly reduce wear rate by promoting the formation of protective layers at the interface.[11]

The character of the wear processes where gases penetrate between tool and work has not been subjected to fundamental research, and work in this area would be rewarding.

INFLUENCE OF COMPOSITION AND STRUCTURE OF CARBIDE TOOLS

The influence of carbide composition and structure on tool performance will now be considered.

Cobalt content

Cobalt is almost universally used as the binder phase in

cemented carbides, and the resistance to mechanical fracture, the compressive strength at high temperatures, and the rates of wear are all related to the Co content. Increasing Co reduces the compressive strength at high temperature[2] and therefore lowers the maximum rate of metal removal when cutting steel and other high melting point alloys.

Wear rate usually rises as Co content is increased. With an attrition type of wear, at low cutting speed, the influence of Co content is small, but the diffusion wear process at higher cutting speeds is more sensitive, the rate of wear increasing rapidly as Co content is raised.

To resist wear, Co is kept to a minimum, but the incidence of fracture is less as the Co percentage is raised. The optimum Co for any application is a compromise and is as low as permissible to avoid high probability of failure by fracture. Commercial cutting tool alloys generally contain between 5 and 11% Co by weight, within which range the optimum tool alloy can be selected. There seems little virtue in going outside this range except for unusual applications.

Carbon content

The carbon content is controlled within narrow limits to avoid the formation of a double carbide of cobalt and tungsten (η-phase) if carbon is too low, and of graphite rosettes if carbon is too high. The former is the more harmful since a sharp drop in toughness accompanies the occurrence of the double carbide. If present in excessive amounts, free carbon increases the rate of wear in an attrition process. Recent work[12,13] on the influence of carbon content suggests that even closer control, within the normally accepted region, may be beneficial, particularly in achieving optimum strength, but there is little evidence that this affects cutting tool performance.

Carbides other than WC

Commercially successful cemented carbides have, with few exceptions, been based on tungsten carbide. For mining tool and other applications where an abrasive wear process is dominant, WC–Co alloys are used exclusively. These alloys have higher tensile strength and more plastic deformation before fracture, so that energy absorption before crack formation is much greater than with any other carbide-metal combination so far tried. It is probably for the same reason that WC–Co alloys have the best resistance to the attrition wear process, described above, which often controls tool life at low cutting speed. Only small amounts of other carbides, generally less than 2%, are added to control grain size.

Tungsten carbide has relatively poor resistance to wear based on diffusion processes when used for cutting ferrous metals. Rapid cratering of the rake face and often rapid flank wear cause short tool life at relatively low cutting speed. Relatively small percentages of other carbides, e.g. 5–20%, greatly increase the resistance to diffusion wear and permit the use of cutting speeds two or three times higher at the same feed rate. The most usual carbide additions are TiC and TaC, though NbC, HfC, or ZrC are also potential additives. All these carbides are of cubic structure and take some WC into solid solution. Their improved resistance to diffusion wear is probably the result of their much lower solubility in ferrous alloys at high temperature compared with that of tungsten carbide. One of the cubic carbides, VC (or V_4C_3), does not enhance the resistance to diffusion wear of WC–Co alloys. This may be associated with the complete mutual solubility of vanadium and iron at high temperatures.

Almost any addition of other carbides to WC–Co alloys lowers their strength and toughness, and it is general practice to use only as much of the cubic carbides as is strictly necessary for wear resistance. The strength and toughness of commercial steel-cutting carbide grades have been greatly enhanced in recent years by changes in composition and by improved processing techniques. Composition changes have included the incorporation of larger percentages of TaC and less TiC, while process changes have aimed at reducing to a minimum the residual oxygen and nitrogen, practically eliminating porosity, and closely controlling grain size.

Carbide alloys containing TiC or TaC have the remarkable property of promoting the formation of stable sulphide and oxide (silicate) layers at the tool/work interface when cutting steels containing these constituents as inclusions. As mentioned above, these can have a very great influence on surface finish and on tool life. The reason why such layers are not formed on WC–Co tools has not been firmly established.

A further advantage of alloys with TiC or TaC is that the compressive strength at high temperatures is greater, thus permitting higher speeds when cutting steel.[2]

The rates of wear of tools based on TiC bonded with Ni or Co are demonstrably lower than those of the best commercial WC-based tool alloys when cutting ferrous metals at high speed. This is probably because of improved resistance to diffusion wear. Their lower strength and toughness has reduced their field of application, but improvements have been made in recent years and they are being used on a wider range of applications than the high-speed finishing operations on which the first such tools were used.

Grain size

Hardness, toughness, strength, and wear rates of tools are all related to grain size of the carbide, which is the most important structural feature of cemented carbides. The attrition wear process at low cutting speed is most sensitive to grain size, and fine-grained alloys, e.g. 1 μm, are more resistant than coarse, e.g. 3–5 μm, the difference in tool life often being several times. Diffusion wear at high cutting speed is less sensitive to grain size but finer grained alloys are usually more rapidly worn than coarse grained. The differences in tool life are not so great though they may be significant.

Fine-grained alloys are harder but less tough than coarsegrained alloys in the range of compositions normally used for cutting tools, so that there are many applications where the finest grain sizes cannot be used unless the Co content is raised to bring the strength up to that required to resist fracture.

Although the hardness at room temperature of a finegrained alloy is higher than that of a coarse grained one of the same composition, the compressive strength at high temperatures is often lower, so that use of fine grain is of no advantage in resisting deformation at high cutting speed and feed.

Surface layers and composite materials

Since cemented tungsten carbide was first introduced, attempts have been made to produce new alloys with the same type of metallurgical structure by modifying the composition and grain size.[13] The other potential interstitial carbides and bonding metals have been thoroughly explored and the limits in grain size of the carbide phase investigated. The introduction of TiC or TaC to replace part of the WC was a major advance in tools for cutting steel, but the numerous other possible changes of this type have offered only minor improvement if any. One reason for this is the still unchallenged strength and toughness combined with hardness possessed by the WC-based alloys. While there are still possibilities for new advances by exploring the now traditional options, a new area of development seems possible in the production of tool tips with highly wear-resistant surface layers and bodies of high strength and toughness.

It has long been recognized that the powder metallurgy process used for production of cemented carbides can be modified to produce composite structures which vary in composition or grain size in different parts of the same body. Such composite bodies have been proposed for use in mining[14] or cutting tools. The introduction of throw-away carbide tool tips has made this line of development potentially

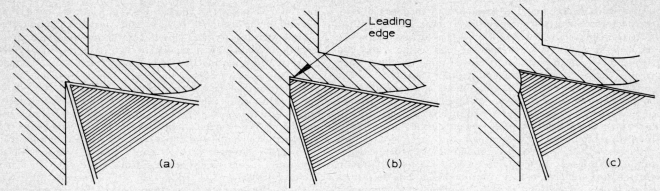

7 Diagram showing relation of layers on tool surface to wear on the tool

much more interesting and there have been two recent commercial developments.

In one of these,[15] a layer relatively high in TiC or TaC is sintered on the rake face of throw-away tool tips, the bodies of which consist mainly of WC and Co. It might be expected that any advantage achieved by these layers, which are usually of the order of 0·010in thick, would be confined to reduction of wear rate on the rake face or only as far down the flank as the layer extends. However, a surface layer containing TiC, only 0·001in thick, on the rake face greatly reduces the rate of wear on the flank even after the wear land has extended to many times the depth of the surface layer. The rate of wear on the flank seems to be largely controlled by the material constituting the leading edge (Fig.7). Under some conditions there was a tendency for the leading edge to be undermined but the possible amount of undermining seems to be small.

A more recent commercial adaptation of the same principle is the coating of carbide tips with a thin layer of TiC by vapour deposition.[16,17] Layers of TiC of the order of 0·0002in (5 μm) thick have been applied to steel tools for some years, but the application to cemented carbide tools is new. The throw-away tool tips are coated all over and not merely on the rake face. In spite of the thinness of the coating, the layers are effective in reducing the rate of wear on the tool flank in continuous cutting operations, even when the underlying metal is exposed on the flank, as long as the layer is continuous along the leading edge of the rake face.

Not only are these developments of practical importance, but the observations on the wear patterns throw light on the nature of the wear process taking place on the flank of tools in general at high cutting speeds. The fact that layers containing TiC reduce the wear rate indicates that the dominant wear process on the flank, as in cratering, is based on diffusion of the tool material into the work material. The fact that a thin layer remains intact on the rake face while it is worn through down the tool flank, suggests that the rate of wear is controlled by the flow pattern, i.e. by the rate at which the atoms diffusing from the tool surface are being carried away. Photomicrographs show that there is often a dead metal zone on the rake face close to the edge, while on the flank the metal moves rapidly at the interface.

The development of layered tools underlines the importance of understanding the rheological problems of metal cutting.

DIAGNOSIS OF INADEQUATE TOOL LIFE

Many factors enter into determining tool life in any situation. The work material, the tool material and geometry, the cutting conditions, type of operation, lubricants, all play their part. To improve tool life or to increase cutting efficiency in any situation it is necessary first to decide which of all these factors control tool life and can be altered to give improvement. A major part in this decision making should be a diagnosis of the causes of wear or failure of the cutting tools.

For adequate diagnosis an experienced observer should examine a representative sample of worn tools and consider the findings in relation to the work they have been doing. The changes of shape are small, and the significant features of carbide tool wear, on which diagnosis is based, are on an even smaller scale. It is one of the aims of this paper to emphasize this. The use of a microscope magnifying at least 30 times is essential and, in many cases, still higher magnification is required. Tools should be examined both as received from the machine shop and after cleaning away the work material from the surface. Without such examination many features of tool wear are never seen.

Sometimes it is possible to obtain all necessary information by examination of tools which have run their full life. In other cases it may be essential to use tools on the operation concerned, withdrawing them for examination before the normal tool life and before evidence of the early stages of deterioration has been destroyed. It should be emphasized that there is no substitute for the trained observer to obtain a sound diagnosis of the factors controlling tool life. There is probably a need for a small illustrated handbook for training and guidance of people engaged in this type of work.

ACKNOWLEDGMENTS
The author is grateful to Wickman-Wimet Limited for permission to include the photomicrographs used to illustrate this paper.

The facilities for continuation of this work in the Industrial Metallurgy Department of the University of Birmingham are appreciated, as is the interest taken by Professor E. C. Rollason.

REFERENCES
1 S. EKEMAR: Paper presented to the 'Machine tool design and research conference', Birmingham, 1966
2 E. M. TRENT: Proc. 8th machine tool design and research conference, Manchester, 1967, 629
3 P. M. BRAIDEN: 'The performance of tungsten carbide tools in intermittent cutting', thesis, Sheffield University, Dept. of Mechanical Engineering, August 1968
4 E. M. TRENT: ISI Spec. Rep. 94, 1967, 11
5 E. M. TRENT: *ibid.*, 179
6 E. M. TRENT AND E. F. SMART: Unpublished rep. 'Machining of Nimonic 115', Birmingham University, Dept. of Industrial Metallurgy, 1969
7 E. C. ROLLASON: ISI Spec. Rep. 94, 1967, 43
8 W. B. HEGINBOTHAM AND S. L. GOGIA: *Proc. Inst. Mech. Eng.*, 1961, **175**, 18, 881
9 H. OPITZ AND W. KÖNIG: ISI Spec. Rep. 94, 1967, 35
10 G. W. ROWE AND E. F. SMART: *Brit. J. App. Phys.*, 1963, **14**, 924
11 W. KÖNIG AND N. DIEDERICH: Paper presented to CIRP General Assembly, Sept. 1968
12 H. SUZUKI AND H. KUBOTA: Planseeberichte 14, 1966, 96
13 H. SUZUKI AND T. YAMAMOTO: *Internat. J. Powder Met.*, 1967, **3**, 3, 17
14 P. SCHWARZKOPF AND R. KIEFFER: 'Cemented carbides'; 1960, New York, Macmillan
15 Brit. Pat. 683, 194
16 Brit. Pat. 1042, 711
17 French Pat. 1,525,512
18 K. BROOKES: *Metalworking Production*, 1969, 113, 31, 28

Some basic mechanisms of wear that may be relevant to tool wear and tool failure

D. Tabor

ADHESIVE WEAR

When surfaces rub together, particularly in the absence of lubricant films, some adhesion occurs at the points of rubbing contact. The friction is primarily the force required to shear the junctions so formed. There is no basic mystery about interfacial adhesion. If the surfaces touch so that they support a normal force, the atoms must be in contact and interatomic forces will become appreciable. The extreme case is when similar metals come together. If clean gold is in contact with clean gold the contacting regions will become a single gold junction, perhaps full of imperfections, but nevertheless possessing strength properties comparable with the bulk strength of gold itself. This process occurs even at room temperature (cold welding) but it will be more marked at higher temperatures where interfacial diffusion will be accelerated. Since high temperatures may be generated at the surface by the sliding process itself, the interfacial interaction may be even more marked. The shearing or breaking of these interfacial junctions is not only the cause of friction. It is also often responsible for the plucking of a piece of metal out of one or both of the surfaces.

In a bearing or in a test rig where surfaces are subjected to repeated traversals, the plucked fragments may initially be attached to one surface but may subsequently be back-transferred on to the other. The net wear rate may therefore depend crucially on the transfer–back-transfer process.* This is very difficult to forecast or quantify, since it may be radically changed by very small changes in running conditions. However, in machining operations this process is probably of very minor importance since fragments plucked either from the tool or work are rapidly carried away from the rubbing region.

For this reason adhesive wear in machining operations is a relatively straightforward concept. The tool is invariably chosen to be harder than the work. If a junction is formed at the metal/work interface it will generally pluck out a fragment from the work. The process of plucking-out will leave the fragment in a very work-hardened condition and it may well be hard enough to score or groove the work. The accumulation of transferred material from the work to the tip of the tool is, of course, the origin of the built-up edge. This nose acts as an extension of the tool, and to some extent protects the tool from wear. However, the built-up edge may occasionally break away with a small portion of the tool itself. This is particularly likely if the tool is heterogeneous in structure so that local regions may be appreciably weaker in tension or shear than the overall strength.

Adhesive wear of the tool is therefore likely to be most marked if the tool is of non-uniform strength. Another factor is the effect of the rate of deformation on the relative strength properties of tool and work. High cutting rates will have two opposing effects. Frictional heating will tend to soften both the work and the tool, making them more ductile. On the other hand high rates of strain will tend to make them more brittle. If the rate at which adhesion junctions are broken is such that the tool material ceases to be ductile, a relatively large fragment may be removed from the tool by fracture. One example of this is the interaction between a soft metal like copper and a hard brittle material like sapphire. Buckley (1967) has shown that under conditions of strong adhesion the copper can pluck pieces of sapphire out of the harder surface. We shall discuss this in greater detail below.

Clearly the best way of minimizing adhesive wear is by reducing the amount of adhesion. The commonest method is by using a lubricant. However, it is still not clear whether the lubricant acts mainly as a coolant or as a means of reducing friction and adhesion. If it acts as a true lubricant it is highly desirable to know how the lubricant gets into the work/tool interface and how quickly it can interact to be effective. Another approach is to allow adhesion to occur but to ensure that the transferred film is very easy to shear. The easiest way of doing this is to incorporate suitable materials in small quantities in the work material itself. It may be that free-machining steels, which contain small quantities of lead, function, to some extent, in this way. Another idea is to make the work relatively brittle so that the removed chip easily fragments and breaks away from the tool face. Silicates in the work probably function in this way, although at higher speeds it is possible that a smeared glass-like film acts, in some measure, as a lubricant between work and tool.

One can of course cover the tool surface with a hard relatively 'non-sticky' type of coating, e.g. hard chromium plate. Provided these films do not become brittle and do not dissolve at high temperatures in the work they can be very effective. However, even here it is probable that any advantages provided by the chromium film are due, in part, to its protective oxide film. We shall discuss this further below.

ABRASIVE WEAR

Abrasive wear occurs if a hard particle cuts or grooves one of the rubbing surfaces. The first criterion for appreciable abrasive wear is that the particle should be harder than the surface being abraded. If the Vickers hardness of the particle is, say, $1\cdot 5$ times that of the surface, abrasion can occur fairly readily. If the particle is smooth, most of the abrasion will be in the form of plastic grooves (with very little material removed) or in the form of chips and flakes if the surface is brittle. If the particle has sharp corners or edges and it is appropriately oriented it will cut the surface. Abrasion then resembles microcutting and the abrasion rates are relatively high. There is a good deal of experimental work particularly

---

* Rabinowicz[1] has attempted to describe transfer and detachment of wear fragments in terms of surface energies

This paper describes briefly a number of basic mechanisms of wear between sliding surfaces. These include adhesive, abrasive, fatigue wear, surface dissolution, surface hardening, surface fragmentation by friction. In addition the effects of running speed, temperature, and surface films are discussed. At various points in the paper the possible connexion between these wear phenomena and the causes of tool wear and failure will be considered.

621.91.025:539.538

The author is with the Cavendish Laboratory, University of Cambridge

by Kruschev in Russia and Richardson in Britain on this aspect of wear. If the particle is not harder than the surface some abrasive wear can still occur but at a greatly diminished rate (Richardson 1966).

Tool wear by abrasion is most likely to occur with work-materials containing hard inclusions. If the inclusions are spherical they are more likely to groove the tool face and flank and the rate of removal of material will be very small. If on the other hand the inclusions are sharp they may produce microcutting and relatively high rates of abrasion. Clearly the best way of avoiding such wear is to use work materials that do not contain hard inclusions or at least to arrange for the inclusions to be smooth. Another approach is to use a tool material that is harder than the inclusions. Alternatively, a softer tool material may be used provided it workhardens under repeated abrasion to give a hard surface layer capable of resisting further deformation or cutting by the abrasive particles. The same effect may be achieved by coating the tool with a very hard skin either by plating or by chemical treatment, e.g., nitriding.

CHEMICAL WEAR

There is another form of wear which may involve both adhesion and abrasion. It occurs if the rubbing surfaces are attacked by the environment to form a removable surface film. For example, in the presence of a sulphurized lubricant a sulphide film may be formed on the metal surface; in the presence of a fatty acid a soap film. More generally, oxide or hydroxide films will be formed in the presence of air. These films may be removed by the sliding process to expose fresh underlying metal which is highly labile and can readily react with the environment to reform the surface film. This type of wear is often slow and generally is to be preferred to the wear that would occur if no surface films were present. However, if extremely reactive lubricants are used, chemical wear or even direct corrosion may become significant.

FATIGUE WEAR

Although fatigue wear can always occur between sliding surfaces it is usually swamped by adhesive or abrasive wear. Consequently fatigue wear becomes important only when adhesive and abrasive wear are relatively small. For example, in well-lubricated systems adhesive wear may be negligible. If hard particles are excluded from the system (this may be difficult because dust can often act in this way) abrasive wear may be small. If then the surfaces are continuously subjected to loading and unloading they may gradually fatigue and pieces of the surface may easily be detached. This occurs in sliding systems where asperities on one surface continuously transmit stresses on to the other even though they are completely separated by a lubricant film. A similar effect may occur in rolling bearings. Fatigue failure is often initated at a surface flaw or crack. An applied stress may open the crack a little: in the presence of a contaminating atmosphere the crack does not 'heal' on removal of the stress. Repeated cycling of the stress will thus gradually produce spalling of a fragment out of the surface. Sometimes fatigue cracks can be initiated at defects which lie below the surface. In this case repeated stressing gradually work hardens the subsurface material. This may be followed by shear or tensile failure.

Fatigue does not usually occur if the applied stress is below a certain limit. To minimize fatigue wear in tools it is desirable to use tools that are considerably harder than the work. The contact pressures which are determined by the yield properties of the work may then be below the limit at which rapid fatigue occurs. It is also desirable to avoid flaws in the surface of the tool and inhomogeneities in its structure.

SURFACE DISSOLUTION

In some cases very high wear rates can occur if one of the sliding surfaces is soluble in the other and if the temperature at the interface is high. Elevated temperatures may occur if the surfaces have to run in a hot environment. On the other hand very high temperatures at the interface may be generated by the sliding process. It is well known that tungsten carbide is soluble above 1100°C in iron. Consequently, if the interface reaches or exceeds this temperature the tungsten carbide will be worn away by a sort of erosion in which the carbide is eaten away or consumed by the hot iron. This is well known in the metal cutting of ferrous materials. At very high cutting speeds the wear of tungsten carbide tools is catastrophic.

This type of wear may be reduced in three ways: (a) by running at lower speeds so that the surface temperatures are lower; (b) by cooling the system so that the interfacial temperatures are diminished; (c) by using tools that are not soluble in the work even at elevated temperatures, e.g. by the use of titanium carbide with ferrous materials.

SURFACE FRAGMENTATION BY FRICTION

This section deals with a concept that is relevant only to brittle materials. It concerns the important role of friction in producing surface fragmentation. Consider, as a simple example, the contact between a hard sphere of, say, steel and a flat glass surface. If the steel ball is loaded normally on to the glass surface the interface is deformed elastically according to the Hertzian laws of elastic contact. Part of the stress pattern includes tensile stresses around the rim of the circle of contact. If these are large enough, cracking will occur and ring cracks will be formed. The important point is that only a small fraction of the normal stress appears as a tensile crack-generating stress. If now a tangential force is applied to the ball, the whole of the surface traction appears as a tensile stress at the rear of the region of contact and a compressive stress at the front of the region of contact. The crack due to the normal stress will tend to close up at the front and open up at the back. Crescent-shaped cracks will thus be formed. Furthermore they will be formed at much smaller loads than those necessary to produce ring cracks in the absence of a tangential stress. Some years ago in a brief paper with Brookes and Billinghurst[2] this was demonstrated by the author (see Fig.1). A normal load of about 10 kg was needed to produce ring cracks. If a normal load of 2 kg was applied no ring cracks were formed, but if a tangential force of 0·5 kg (insufficient to produce sliding) was applied, a crescent-shaped crack was formed. Since then the subject has been further developed in theoretical papers by Goodman and Hamilton[3] and by Frank and Lawn,[4] and also in a more detailed experimental study on glass by Gilroy and Hirst.[5]

In our laboratory we have recently studied the behaviour of a titanium carbide sphere in contact with a titanium carbide flat. Polycrystalline specimens were used, but the contact region was so small that contact occurred within a single grain. The results qualitatively were the same as those for glass. In the presence of a tangential traction the normal load necessary to produce cracking was very much less than when no tangential force was applied. For example, for a specified geometry a normal load of 40 kg was needed: if a tangential force equal to 1/6 of the normal load was applied this was just enough to produce sliding in air (coefficient of friction ≈0·16). Under these conditions the normal load could be reduced by a factor of 10 and crescent-shaped cracks would still just be formed. The effect becomes immensely larger if the surface traction is increased by increasing the coefficient of friction. This may be readily achieved by removing the surface films naturally present on the carbide.

REMOVAL OF SURFACE FILMS

Many hard solids such as diamond, carborundum, metal carbides, and borides have low coefficients of friction in air (μ of order 0·1). Some earlier work in this laboratory by Bowden[6] and Hanwell showed, in the case of diamond, that this was largely due to the presence of extremely thin films of oxygen or water vapour adsorbed from the atmosphere. It is

5 Nomarski interference contrast micrograph of the top face of a production radius tool in the as-received condition ×110

7 SEM of a production tool rejected after use, showing severe erosion of the setting   ×22

which had a top face corresponding neither approximately to a {110} nor to a {111} plane and which was, in fact, closest to a {100} plane. In this tool the position of the nose was such that the main cutting forces were opposed by a 'soft' direction. The tool was rejected after machining less than 100 pistons.

WEAR OBSERVATIONS
Position and extent of wear
Wear takes place in the form of a bevel or land between the top and front clearance faces of the tool. The extent and position of the worn region depends on several factors, notably the depth of cut, the feed rate, and the tool geometry. The depth of cut determines the extent of wear on the leading side of the tool centre and the feed rate, plan approach, and plan trail angles determine the possibility of wear on the trailing side of the tool centre. In the case of the test tools, the wear region extended from the point at which the tool engaged with the workpiece to the foremost part of the edge. For the radius geometry, this was the centre of the nose and for the seven-facet geometry, it was the trailing corner of the centre facet, both tools being used with their axes perpendicular to the workpiece. Apart from the similarity in extent of wear, it differed from tool to tool. Figures 8 and 9 show the worn edges of two radius tools of identical design after each had machined, under identical conditions, 100 equivalent pistons. It can be seen that the shape of the wear region on each tool is noticeably different and also that the wear on

the front clearance face of tool A appears slightly greater than that on tool B. Although the two tools were of the same design, they had different orientations: tool B had its top face nearest to a 110 plane and tool A had its top face nearest to a 111 plane.

Grooving of the edge
A composite micrograph of tool B (Fig.10) shows the wear caused by the swarf as it is removed; in this case a hollow has been created. The deep groove in the edge, seen in the background of this micrograph, is a feature present in almost all worn tools and is formed at the point of engagement with the workpiece. As might be anticipated, this feature is more severe in tools used with constant depths of cut. The groove often becomes deeply worn and is almost invariably the starting point of a shallow groove which is worn in the top face; this can also be seen in the background of the same micrograph. It is believed that grooves of this type on the top face are caused by the edge of the metal chips abrading the diamond as they are removed. Other grooves may also be found at different positions along the worn edge; these are usually less severe than the engagement groove and are arbitrarily situated. In some cases these grooves can be seen to have developed from a chip, either present in the unused

6 Same tool as shown in Fig.3, but when rejected after use
×110

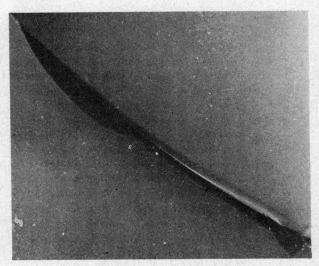

8 (Tool A) SEM of a radius test tool after machining 100 equivalent pistons; approx. plan view of wear region   ×265

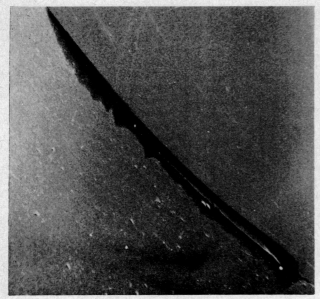

9   (Tool B) SEM of a radius test tool after machining 100
    equivalent pistons; approx. plan view of wear region   ×265

11   SEM of 'ripple' wear on a radius tool   ×1305

edge or removed in the cutting process. A groove arrived at in this way will usually show some evidence of fracture. Many grooves have been observed which show no evidence of fracture but appear very smooth, the ridges between them often being sharp in appearance. The significance of this type of groove has not yet been established but some effort has been made to measure the angles between grooves and to establish the orientations of the groove sides, so far without success.

On a few of the test tools, but not on production tools, another form of grooving has been observed. This is shown in Fig.11 and consists of a series of grooves in the form of a ripple. The grooves are rounded, of equal spacing, and of diminishing depth as the trailing edge is approached, the first groove occurring in most cases around the centre point of the wear region. The spacing of these grooves has been found to be the same as the feed/rev of the lathe. A similar pheno-

menon was observed for carbide tools by Pekelharing and Schuermann.[4]

Chipping of the edge

Chipping of the edge has also been commonly observed and can occur at any point in the wear region, though on facet tools the facet corners often become chipped. The trailing corner appears to be the most susceptible to this type of damage. Most of these fractures are conchoidal but a large cleavage fracture was observed on the flank of a production radius tool, due to, it is thought, an orientation whereby the diamond was in a position to be cleaved when making a large depth of cut. In a few tools, a form of microcleavage of the edge appears to take place. Figure 12 shows a worn edge of this type in which planes or sheets can be seen in the damaged area. This type of wear appears to occur when a 111 plane is in line with the swarf direction and normal to the top face. Small chips in a new edge do not necessarily give rise to further gross chipping, as can be seen from Figs.13 and 14. The sharp edges of the initial fracture have been subjected to a smoothing process.

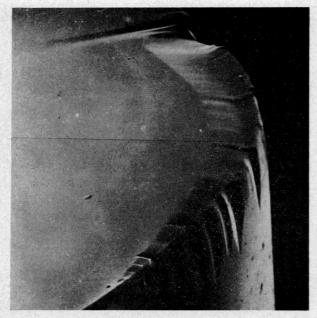

10   Composite SEM of tool B, taken at the same stage of
      machining as in Fig.9, showing the features described in
      the text   ×960

12   SEM of wear damage on the edge of a radius tool which
      appears to be the result of a microcleaving effect   ×5850

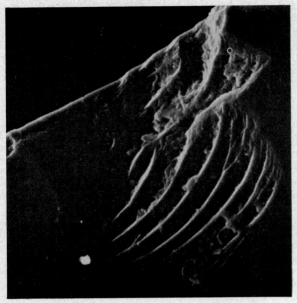

13 SEM of a chip, as seen on the front clearance face of a radius test tool in the as-received condition ×2 680

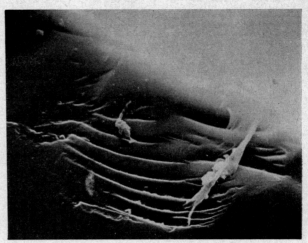

14 SEM of the chip seen in Fig.13, but after the tool had machined 100 equivalent pistons ×2 680

Microstructure

Most of the worn region appears smooth and the SEM has failed to resolve any definite details of a surface structure. It has been possible at times, however, to observe fine abrasion lines in the worn region when using Nomarski interference contrast, such as appear similar to the original polishing lines on the diamond.

CONCLUSIONS
Quality control

Using not too unconventional microscopy, it has been possible to detect cracks, flaws, and other undesirable inconsistencies in the as-received condition of diamond tools. It is likely that these features have been overlooked by the manufacturers due to their being examined only at low magnifications. Very often these features coincide with, and probably contribute to, the eventual failure of the tool edge. Better inspection would give more rejects or entail more polishing time, thus raising the tool costs, but it would save valuable production time.

Orientation and tool geometry

It appears that diamond-tool wear is affected by orientation and it is clear that some orientations might better be avoided; however, as yet, not enough evidence exists to specify an optimum. It is hoped that, with the continued examination of production tools, enough statistical evidence will eventually be collected so that, if a significantly superior orientation exists, it can be specified. If such a specification is made, the degree of accuracy required will determine whether or not it would be economic. A new tool geometry may be desirable for this application in order to avoid some of the inherent faults of the existing geometries, either in tool use or in tool manufacture. Surface finish requirements for pistons have changed considerably since the original geometries were adopted some 20 years ago and it is possible that a geometry

unacceptable at that time might now be acceptable and offer a longer life. Some work is currently being carried out in this direction.

Mechanism of wear

Several proposals exist for the mechanism of diamond wear,[5] in most cases involving wear of diamond on diamond. In this application, apart from the more obvious fracture and cleavage damage observed, it is not possible to state a wear mechanism due to the lack of evidence of an identifiable structure on the bulk of the worn surface; however, it is likely that high temperatures will exist on the surface of the tool edge during cutting and would favour, therefore, a graphitization effect.

Lubrication

Work is being carried out to examine the effects of heavy cutting oil as a lubricant. So far this has proved useful in that build-up of the alloy on the tool is prevented when, under the same conditions in the dry state, considerable build-up had occurred. The work is at an early stage and it is not possible to comment on the effect which lubrication has on the wear rate.

ACKNOWLEDGMENTS

The work described was carried out as part of the research programme of Associated Engineering Developments Ltd and is published by permission of the Directors of the Company.

REFERENCES

1 C. B. SLAWSON AND J. A. KOHN: *Ind. Diamond Rev.*, 1950, **10**, 168–172
2 E. M. WILKS AND J. WILKS: *Ind. Diamond Rev.*, 1966, **26**, 303, 52–58
3 E. H. KRAUS AND C. B. SLAWSON: *American Mineralogist*, 1939, **24**, 11, 661–676
4 A. J. PEKELHARING AND R. A. SCHUERMANN: *Ind. Diamond Rev.*, 1954, **14**, 162, 101–105
5 M. SEAL: in 'Science and technology of industrial diamonds' (Proc. Industrial Diamond Conference, Oxford 1966), 145–159; 1967, London, Industrial Diamond Information Bureau

Failure of carbide tools in intermittent cutting

P. M. Braiden and D. S. Dugdale

When tungsten carbide tools are used for intermittent cutting, cracks often form on the rake face, and tool life is then drastically reduced. This work examines the theory that thermal stresses are responsible for the development of these cracks. This has been proposed by several workers on the basis of experimental observations.

The first reference to thermal cracks in the literature is due to Boston and Gilbert[1] who observed cracks on the rake face of carbide tools used in face milling cast iron. They found that at higher cutting speeds all tools had thermal cracks, and chipping on the cutting edge became more frequent. No justification was given for the use of the term thermal cracks, and no references were given in this paper. Opitz and Fröhlich[2] suggested a thermal mechanism of formation for the cracks which they observed on the rake faces of several carbide tools when milling constructional steels. They performed simulative tests, using a high-frequency generator to heat carbide rods, and were thus able to show that the cracks which formed in their tools could have been due to thermal stresses.

All of the research carried out in intermittent cutting has been of an experimental nature, and the work of Zorew[3,4] is typical of this approach. Zorew compared the lives of carbide tools in a series of lathe experiments, cutting bars of various rectangular cross-sections, hence obtaining intermittent cutting with a series of cutting–cooling ratios. From these and further planing tests, Zorew was able to show that the duration of free-running time had an effect on tool life.

On the basis of this work, Zorew proposed a tensile-stress theory to explain the development of the cracks. This theory suggests that during the non-cutting period the tool cools very rapidly so that the tool surface becomes colder than the body of the tool, and thus tensile stresses are developed.

Okushima and Hoshi[5,6] have performed a number of experiments in face milling and have shown that cracks can develop on both the flank and rake surfaces of tools in intermittent cutting. Cracks on the flank face were only found when a considerable amount of flank wear was present, with its attendant heat generation. These authors have also shown, by means of heating bars of carbide in a spot-welding machine, that localized plastic deformation and cracking can occur in carbide if the temperature gradients are made large by rapid heating and cooling.

Intermittent cutting with ceramic tools

The bulk of the work performed in intermittent cutting with ceramic tools is due to Pekelharing[7] who restricted his investigations to ceramics since he thought that thermal cracking in carbides was a rare occurrence '. . . a phenomenon will be described which has seldom been found with cemented carbides, and never with high-speed steels'. In Pekelharing's experiments the cutting time was usually comparatively long (about 5 s) the experiment being performed by the sudden withdrawal of a tool which was cutting continuously. Pekelharing called the cracks which he found 'comb cracks', since they form a series of fine parallel lines on the rake faces of the tools, and suggested that they were due to the occurrence of plastic deformation in the tool. These cracks were not found to be detrimental to tool life, and the tool would continue to cut satisfactorily with such cracks present. At feeds greater than 0·8 mm/rev (0·0315in/rev) however, a different type of crack appeared, which produced a very rapid tool failure. This crack has formed the subject of further study by the same author.

Simulative experiments

Simulative experiments are attractive in that they permit the mechanical shock, which occurs at the beginning of a cut, and the subsequent cutting loads, to be eliminated. The difficulty lies in the very high rate of heat transfer which is obtained in metal cutting, and that the heat generation takes place over a small area.

Simulative experiments have been performed by Opitz and Fröhlich,[2] Okushima and Hoshi,[6] and Shinozaki and Harada.[8] None of these experiments, however, approach the conditions of metal cutting very closely.

Simulative experiments are commonly used to test the resistance of ceramic materials to thermal shock,[9,10] but the heating rates in these tests are lower than those obtained in metal cutting.

EXPERIMENTAL METHOD

The experimental techniques used to obtain intermittent cutting were similar to those of Zorew, a lathe being used with a workpiece of rectangular section to give intermittent cutting. The workpiece material was a 0·2C steel, containing 0·6–0·8Mn–0·04S–0·5Ni–0·5Cr–0·5Mo, with a hardness number of about 170 HV.

The workpiece assembly was formed by clamping together a number of plates 2in thick, 5in wide, and 48in long. After taking a cut, a screw system enabled the workpiece sections

This paper examines the reasons why cracks are formed in tungsten carbide tools during intermittent cutting. Lathe tests using three grades of carbide showed that cracks were repeatedly formed which were similar to the thermal cracks previously observed in ceramic tools by other investigators. Temperature at the cutting edge was measured by means of the tool–work thermocouple method. Also, thermocouples and thermistors were used to measure temperatures at the surface of the carbide tool tips. These temperatures were used as boundary values for calculating the steady temperature distribution and associated thermal stresses in the interior of the tool tip. When purely elastic behaviour was assumed, the calculations showed that no tensile stress was set up near the cutting edge during the cooling part of the cutting cycle. As the formation of cracks implies the presence of tensile stress, it was necessary to admit the possibility of inelastic behaviour of the tool tip. During the heating part of the cutting cycle, compressive thermal stress may cause local plastic yielding along a strip of the tool face just behind the cutting edge. As this strip subsequently cools under the constraint of surrounding elastic material, tensile stress is set up, which may become high enough to produce the observed cracking.

621.941.1.025.7:669.018.25:539.538

P. M. Braiden (formerly at the University of Sheffield) is with the Department of Mechanical Engineering, Carnegie-Mellon University, Pittsburgh, USA, and D. S. Dugdale is with the Department of Mechanical Engineering, University of Sheffield

Thermal conditions in metal cutting

P. M. Braiden

In general engineering practice, components are normally designed to operate over a small range of working temperature, and however the stress configuration within the component may change, it is rare for the operating temperature to vary appreciably.

In metal cutting, however, the tool must support the cutting forces, and must continue to do so while being heated to temperatures which may be as high as 900°C. In an intermittent process, such as milling, the tool must support the cutting forces at the high temperatures attained during cutting, and it must also endure the cyclic thermal stresses induced by the cutting/non-cutting cycle. It is these conditions of load and temperature which make metal cutting such an interesting study.

In the first part of this paper the research which has been done to determine the thermal conditions in metal cutting is reviewed. The second part deals with the application of this work to the study of tool performance, with particular reference to the study of tools operating under intermittent cutting conditions.

GENERATION OF HEAT IN METAL CUTTING

In metal cutting, heat is generated at two major zones:
  (i) the shear zone, where the major part of the deformation takes place
  (ii) the tool/chip interface, where heavy rubbing under high friction conditions occurs.

A third zone of heat generation develops after the first few seconds of cutting on the flank face of the tool, due to tool wear, heat being generated by the rubbing action of the workpiece finished surface on the tool flank face.

The energy of deformation at the shear plane appears as thermal energy and part of this energy flows into the chip, being retained as residual energy, the remainder being used in heating the chip, tool, and workpiece. At the tool rake face the heated chip is in intimate contact with the tool, and secondary deformation of the chip undersurface takes place, under the action of the high friction forces obtained at the interface.

Thus research has been directed towards two problems in attempts to analyse the thermal conditions obtained in metal cutting, these being:
  (i) a the partition of energy between chip and workpiece
     b the determination of the average temperature, and the temperature distribution at the shear plane
  (ii) a the partition of rake face thermal energy between the chip and the tool
     b the mean temperature along the tool face, and the temperature distribution at the tool/chip interface.
In attempts to answer the above questions, convection and radiation losses are usually neglected, since they will be small during the time in which the chip is formed at the shear zone, and passes the rake face of the tool.

ENERGY STORED IN METAL CHIPS

Latent energy is stored in the chip after deformation, and the work of Taylor and Quinney[1,2] has been used to evaluate this. In their first method, they found that approximately 11% of the energy of deformation was stored in the cold-worked mild steel used. Later work revealed that the amount of energy remaining in pure carbonyl iron was about 15%. When both materials were tested by both methods, the second method generally gave higher results.

These results were used by Trigger and Chao[3] in their analysis of metal-cutting temperatures. These authors take the average of the values for the latent energy found by Taylor and Quinney's two methods and thus assume that $12\frac{1}{2}\%$ of the total energy of deformation at the shear plane is stored in the chip as latent energy. Thus about 87·5% of the total energy of deformation at the shear plane is assumed to be available to heat the chip.

Some doubt exists as to the validity of this assumption, however. Epifanov and Rebinder,[4] drilling annealed aluminium, report a value for the residual energy of 3% and this result is in agreement with later work by Bever et al.[5] These authors, working on actual two-dimensional metal-cutting chips, found values for the residual chip energy of 0·73% for rake angles of 15°, and 2·93% for rake angles of 30°.

These later values are used by Loewen and Shaw[6] in their work after considering the values of strain energy in these experiments, and comparing them with the strain values encountered in metal cutting. The later, improved paper of Trigger and Chao[7] also uses these lower values. In fact both this paper and that of Loewen and Shaw consider that the residual chip energy values are so low as to be negligible. It can therefore be assumed that all of the work supplied during plastic deformation of the chip appears as heat energy.

TEMPERATURE DISTRIBUTION DUE TO THE SHEAR-PLANE HEAT SOURCE

Theoretical solutions for the temperature distribution in the work material have been proposed by several workers, using two general methods of solution. These are:

1 The shear plane is considered to be an oblique band source of heat moving through the workpiece in the direction and with the magnitude of the cutting velocity vector.

The problem of a plane source of heat moving either parallel or perpendicular to the plane of an infinite solid has been investigated by Blok,[8] Jaeger,[9] and Rosenthal,[10] and their results have been applied to the shear-plane heat-distribution problem by several workers. It should be noted that the shear plane is not always a single 'plane', but rather a zone, and heat generation occurs over the whole of this area, as has been shown by Boothroyd[11] and others.

2 Other workers have simplified the relevant heat-transfer equations by making assumptions based on practical experience. This approach was adopted by Weiner[12] who assumed that, at practical cutting speeds, conduction in the direction of motion of the workpiece could be neglected.

Rapier[13] applied relaxation methods to the shear-plane heat and temperature-distribution problems, and was thus able to solve the equations without recourse to Weiner's assumptions. Similar methods were used by Vieregge.[14,15] Some typical results of Rapier's analysis are shown in Fig.1.

The first part of this paper reviews the research which has been done to determine the thermal conditions which exist in the workpiece and chip during metal-cutting operations. In the second part of the paper the performance of cutting tools is discussed, with particular reference to the thermal conditions in the cutting tool during intermittent operations.

621.941.1.011:536.241

The author is Assistant Professor, Carnegie-Mellon University, Pittsburgh, USA

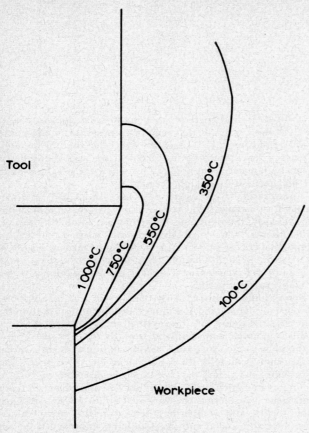

1 **Temperature distribution in the work material (after Rapier[13])**

Hahn[16] employed an extension of the method from Blok and Jaeger, and considered the particular two-dimensional case where the plane of the band source is inclined at an angle to the direction of motion in an infinite medium. This analysis was applied directly to the calculation of the temperature distribution along the shear plane. Hahn employed calorimetry to determine the temperature of the chips produced in an experimental programme, and found good agreement between the experimental data and his theory.

The only analysis which does not require the adoption of a single plane of shear is from Scrutton[17] who assumed the fan-shaped shear zone proposed by Christopherson et al.[18] The temperature field thus obtained was found to be an exponential function of the distance from the tool cutting edge.

Very little experimental work has been performed to check the validity of these theories. Reichenbach[19] performed experiments with buried thermocouples which passed through the shear zone and were carried away in the chip, but the temperature fields obtained show only fair agreement with the theories. However, if Weiner's theory is modified by considering the heat source to be of variable intensity, it can be made to fit Reichenbach's experimental data. Nakayama[20] also used thermocouple techniques to measure the temperature rise of the workpiece, and found that Hahn's theory provided the closest approach to the experimental results, although agreement was poor.

HEAT CONDUCTED BACK TO WORKPIECE
The amount of heat conducted back into the workpiece in any cutting operation will depend on a number of factors, such as size of workpiece, etc. An exact treatment is obviously difficult since the chip about to be removed will have been preheated by the preceding cut. Trigger and Chao quote the results of Schmidt and Roubik[21] in this problem, who report

that, in drilling annealed aluminium at speeds higher than 100 ft/min, and at a feed of 0·0091 in/rev about 10% of the total heat generated is conducted into the workpiece. Work by Leone,[22] who published a method for determining the back-conducted heat simultaneously with, and independently of, Loewen and Shaw's work, is in support of their results. When the temperature distribution along the shear plane predicted by this work is compared with that of Hahn, good agreement is again found.

THERMAL EFFECTS AT THE TOOL/CHIP INTERFACE
As it leaves the shear zone the heated chip rubs along the rake face of the tool at high velocity. Under the intimate conditions of contact which exist at the interface the friction is usually very high, the actual value attained depending on several variables, e.g. tool and work materials.

The thermal effects of this situation have been compared to the theoretical case of an insulated slider moving on an infinite conducting solid. The solution of this problem is due to Kelvin, and is given in texts on heat transfer (e.g. Carslaw[23]).

This work has been applied to the particular case of the tool/chip interface heat-transfer system by Chao and Trigger.[3] The weaknesses inherent in this paper concern the amount of energy stored in the deformed chip, and the percentage of the shear-plane heat which is back-conducted to the workpiece. These points were modified by Loewen and Shaw,[6] whose method is otherwise similar to that of Trigger and Chao.

In their later paper, Chao and Trigger[7] make use of Loewen and Shaw's modifications concerning the shear-plane heat distribution, and continue to expand the ideas from their earlier reference with regard to the interface heat distribution. An anomaly existed in their earlier paper, in that the temperature distributions on the rake face of the tool and the underside of the chip differ, while the heat-flux distribution is assumed to be the same for both chip and tool. Thus, in the paper published in 1955, a non-uniform heat-flux distribution is assumed, the temperature distribution being obtained by a method of successive approximations. This method allows not only the temperature distribution but also the heat-flux distribution at the interface to be found.

THERMAL EFFECTS IN INTERMITTENT CUTTING
When a cutting tool, operating in intermittent cutting conditions, enters the cut for the first time, the tool/chip contact area is heated and expansion occurs, producing a compressive stress in the surface layers of the rake face. During the cutting period the heat generated is conducted into the body of the tool and after several cutting/cooling cycles the bulk temperature of the tool will thus be increased. When the tool emerges from the cut, the tool/chip contact area loses heat very rapidly and the surface layers of the tool contract. Cyclic thermal stresses are thus developed in the tool/chip contact area whose magnitude depends on a number of variables, e.g. the cutting conditions (speed, feed, depth of cut) and the number and duration of the cutting/cooling cycles. Zorew[24,25] suggested that tensile stresses could be developed during the cooling part of the cycle. He found no difference in the lives of high-speed tools when making continuous or interrupted cuts, but this was not true for carbide tools.

Zorew also showed that the mechanical shock which occurs as the tool enters the cut has no effect on tool life. Further tests, in which the tool was given pauses of different time intervals between the cutting periods, demonstrated that the duration of the free-running time had an effect on tool life. In addition, cooling the tool with an emulsion decreased tool life, and heating the tool with a gas flame during the non-cutting period increased tool life. These experiments, together with the observed failures of the tools tested, led Zorew to conclude that the differences in tool life when making intermittent cuts were caused by thermal conditions.

and further, that failure of the tool was due to the tensile thermal stresses developed during the cooling period.

Transient thermal stresses in metal cutting with carbide tools have also been investigated by Okushima and Hoshi.[26,27] In their first report they describe a series of face-milling experiments with carbide tools using various cutting:cooling ratios, and describe the cracks which developed in the tools. These cracks were found on both rake and flank faces, and the authors suggest that thermal stresses are responsible for their formation. Cracks were only found on the flank faces when a considerable amount of flank wear was present, with its attendant heat generation. Experiments at varying speeds and feeds showed that there was a threshold level for both of these parameters below which no cracks formed. When the depth of cut was increased both the threshold speed and the threshold feed were reduced.

Various authors, for example Trigger,[28] and Opitz and Kob,[29] have shown that the tool/chip interface temperature depends on the cutting speed and feed, but is independent of the depth of cut, provided that this is numerically greater than twice the value of the tool nose radius. Changing the depth of cut should therefore make no difference to the tool/chip interface temperature, yet it did affect the cracking experienced in the tool. Two explanations are possible:

   (i) the greater the depth of cut the greater the heated area of the tool surface and thus the greater the quantity of heat flowing into the tool

   (ii) plastic deformation could occur in the tool. For given values of speed and feed, plastic deformation is most likely to occur at large values of the depth of cut at the time when the tool first enters the cut, since a long thin strip of material is being rapidly heated. The presence of the surrounding material means that the centre of this strip is under lateral constraint and plastic compressive stresses could be produced.

The second argument was adopted by Okushima and Hoshi and also, independently, by Pekelharing[30] who confined his work to ceramic tools. Pekelharing observed that the cracks developed some way back from the cutting edge which is the hottest region of the tool/chip contact area. This region is, therefore, the area where the greatest thermal stresses occur, and the area which is most susceptible to plastic flow, due to the reduction of yield point at high temperatures.

The most recent work in this topic is due to Braiden and Dugdale.[31] They measured the cutting temperatures and forces for two grades of tungsten carbide during intermittent cutting and calculated the thermal stresses which were developed in the direction of the cutting edge. Contrary to Zorew's predictions they found that no tensile stresses developed during the cooling periods. Furthermore, the compressive thermal stresses which occurred at the commencement of cutting were sufficient to produce yielding in the tool when combined with the mechanical stresses. The change in the thermal stresses with time is shown in Fig.2.

SUGGESTIONS FOR FUTURE RESEARCH
In this paper attention has been focused on two topics in metal cutting.

   1 *The heat generation in the workpiece.* This is a fruitful area for research as is the more general field of heat generation in plastic working operations.

   2 *The effects of thermal stresses on the performance of a*

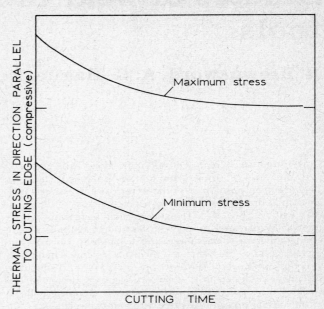

2 **Variation in thermal stress in tool with cutting time**

cutting tool. The study of the stresses seen by a tool material under various cutting conditions is another topic which should yield useful results.

Both of these topics are under investigation by the author.

REFERENCES
1 G. I. TAYLOR AND H. QUINNEY: *Proc. Roy. Soc. Series A*, 1934, **143**, 307
2 G. I. TAYLOR AND H. QUINNEY: *Proc. Roy. Soc. Series A*, 1937, **163**, 157
3 K. J. TRIGGER AND B. T. CHAO: *Trans. ASME*, 1951, **73**, 57
4 G. I. EPIFANOV AND P. A. REBINDER: *Dokladi Academiga Nauk USSR*, 1949, **66**, 653
5 M. B. BEVER *et al.*: *J. App. Phys.*, 1953, **24**, 1176
6 E. G. LOEWEN AND M. C. SHAW: *Trans. ASME*, 1954, **76**, 217
7 B. T. CHAO AND K. J. TRIGGER: *Trans. ASME*, 1955, **77**, 1107
8 H. BLOK: Proc. 'Gen. Discussion on Lubricants', 2222; 1937, London, IME
9 J. C. JAEGER: *Proc. Roy. Soc. New South Wales*, 1942, **76**, 203
10 D. ROSENTHAL: *Trans. ASME*, 1946, **68**, 849
11 G. BOOTHROYD: *Proc. IME*, 1963, **177**, 789
12 J. H. WEINER: *Trans. ASME*, 1955, **77**, 1331
13 A. C. RAPIER: *Brit. J. App. Phys.*, 1954, **5**, 400
14 G. VIEREGGE: *Werkstatt Betrieb*, 1953, **86**, 691
15 G. VIEREGGE: *Werkstatt Betrieb*, 1955, **88**, 227
16 R. S. HAHN: Proc. 'US Nat. Cong. in App. Mech.', 1951, 661
17 R. F. SCRUTTON: *Austral. J. of App. Sci.*, 1962, **13**, 1, 25
18 D. G. CHRISTOPHERSON *et al.*: *Eng.*, 1958, **186**, 113
19 G. S. REICHENBACH: *Trans. ASME*, 1958, **80**, 525
20 K. NAKAYAMA: *Bull. Fac. Eng. Yokohama Uni.*, 1956, **5**, 1
21 A. O. SCHMIDT AND J. R. ROUBIK: *Trans. ASME*, 1949, **71**, 245
22 W. C. LEONE: *Trans. ASME*, 1954, **76**, 121
23 H. S. CARSLAW: 'An introduction to the mathematical theory of the conduction of heat in solids'; 1921, London, Macmillan and Co.
24 N. ZOREW: *Russian Eng. J.*, 1963, **43**, 2, 43
25 N. ZOREW: *CIRP Annalen*, 1963, **11**, 201
26 K. OKUSHIMA AND T. HOSHI: *Bull. JSME*, 1962, **5**, 151
27 K. OKUSHIMA AND T. HOSHI: *Bull. JSME*, 1963, **6**, 317
28 K. TRIGGER: *A. Machinist*, 1966, **110**, 101
29 H. OPITZ AND J. KOB: *Werkstatt Betrieb*, 1952, **85**, 81
30 A. J. PEKELHARING: *CIRP Annalen*, 1963, **11**, 25
31 P. M. BRAIDEN AND D. S. DUGDALE: 'Failure of carbide tools in intermittent cutting', this volume

Studies of wear in high-speed steel tools

R. Brownsword, A. G. Hague, R. F. Panton, and T. Pyle

Cutting-tool steels represent only a fraction of the country's output of steel in terms of the tonnage produced. However, they occupy a position of prime importance in our economy since tools are required for the shaping of most of the goods that we use. The use of higher machining speeds, in the interest of increased production, and the machining of high-strength materials, place increasing demands on tool materials by raising the temperature and the loads generated in cutting. It is this combination of high temperature and load which determines the upper limit of the operating conditions of tool steels. Under the most severe conditions complete collapse of the cutting tip can take place with tool steels. However, the tip continues to change shape even in less severe circumstances by the more conventional modes of wear including abrasion, adhesion, and fracture. The pattern of wear that develops has been described by Trent[1] and Ham:[2] a land forms on the flank and a crater is gouged in the rake face. Flank wear is relatively easily measured for the purpose of tool-life assessment, but is open to error due to smearing of workpiece material over the flank surface and to characteristic lack of repeatability. Furthermore, such measurements offer little information about the metallurgical processes which occur during wear at the tip. If improvements in performance are to be achieved it is essential that metallurgical techniques of examination should be used to elucidate the wear processes that occur at the tool tip. Using such techniques, Trent[1,3] has made a valuable contribution to the understanding of the machining process and in particular the behaviour of carbide tools. For tool steels, little systematic work seems to have been undertaken to examine the tool/workpiece interface in order to relate the structure to the development of wear resistance. High-speed steels, like carbide tools, consist of carbides buried in a supporting matrix; however, unlike carbide tools their final structure is dependent upon heat treatment. The heat-treatment sequence, fully discussed by Hoyle,[4] consists essentially of solution treatment at temperatures in the region 1160–1300°C to dissolve carbides and take alloying elements into solution in the matrix, which is austenitic at these temperatures. Cooling of the steel retains these elements in solution and results in the transformation of the austenite to martensite. This transformation is sluggish in the high-speed tool steels and a certain proportion of austenite, usually 20–30%, is retained in a metastable condition down to room temperature. The steel is subsequently tempered, possibly two or three times, in order (a) to transform the retained austenite to martensite; (b) to develop secondary hardness by the precipitation of alloy carbides within the martensitic matrix; (c) in further tempering treatments, to cause precipitation within the martensite formed from the original retained austenite.

The heat-treatment cycle provides a useful framework within which the properties of the steel may be varied. An increase in solution-treatment temperature leads to an increase in the hardness of the martensite formed. Unfortunately it retards even further the austenite→martensite transformation, thus increasing the proportion of retained austenite present in the structure. Multiple tempering treatments reduce the austenite content but also continue the precipitation of carbides from the martensite formed prior to tempering. The complete removal of austenite from a high-speed tool steel is difficult to achieve. The conflicting effects of heat treatment mean that it is essential for the relationship between microstructure and tool wear to be fully understood if a proper balance between the various structural constituents is to be provided by heat treatment. The work reported below represents a preliminary study of the wear of tool steels which have been heat treated under different conditions, in an attempt to improve the understanding of this relationship.

EXPERIMENTAL ARRANGEMENTS

Machining tests were carried out using tools manufactured from a high-speed steel of the T1 variety, heat treated as indicated in Table 1, with En30B as the workpiece material. The specifications of T1 and En30B are provided in Table 2. Tools were examined after the tests using scanning electron microscopy, optical microscopy, and electron-probe micro-analysis.

Structural examination of the tool steel

Retained austenite determinations were made on samples from each heat treatment using an X-ray diffraction technique in which integrated intensities of austenite and martensite lines were compared.[5] The volume of carbide present in the steels was determined by lineal analysis using a quantitative television microscope. The results of these examinations and the hardness values of the steels are given in Table 1 together with the heat-treatment data.

An examination has been carried out of the behaviour of T1 high-speed steel tools during short-time dry machining of En30B. The tools were given different heat treatments and the wear patterns at the cutting tips were studied using scanning electron microscopy and metallographic examination of sections taken through the worn tips. In addition, the structure of the steel was examined by X-ray diffraction techniques and quantitative television microscopy in an attempt to relate the performance to the heat treatment received. With low speeds and feeds, the familiar built-up edge was noted and wear of the tip was slight. Under more severe conditions, however, crater formation on the rake face was common but adherent workpiece material frequently filled in the worn region. This layer appeared to have resisted, to a large extent, further wear, but at the highest cutting speed, 175 ft/min (875 mm s^{-1}), flow of the tool material was observed below this layer and undoubtedly contributed to the severe wear observed. No simple correlation emerged between the wear behaviour and the heat treatment undergone by the tool tip, in terms of relating tool behaviour to a single metallurgical parameter. Instead it appears that the best performance is achieved from tools heat treated to give an optimum balance between the several microstructural constituents that are present in the steel.

621.941.1.025: 669.14.018.252.3: 539.538

R. Brownsword and T. Pyle are at the Department of Chemistry and Metallurgy, Lanchester Polytechnic, Coventry; A. G. Hague is now at the Department of Production Engineering, Nottingham College of Technology, Nottingham; R. F. Panton is with John Harris Tools Ltd, Warwick

Table 1   Heat treatment of T1 high-speed steel

Type no.	Preheat temp., °C	Solution treatment	Quench	Temper	Carbide, %	Retained austenite, %	Hardness, HV
1	860	2 min at 1260°C	Marquench 560°C air cool	1 × 1 h 565°C	18·4 ± 3·5	2·7 ± 1·0	874
2	860	2 min at 1260°C	Marquench 560°C air cool	3 × 1 h 565°C	15·5 ± 2·6	2·0 ± 1·0	851
3	860	2 min at 1280°C	Marquench 560°C air cool	2 × 1 h 565°C	15·4 ± 2·8	2·1 ± 1·0	864
4	860	2 min at 1300°C	Marquench 560°C air cool	1 × 1 h 565°C	15·0 ± 2·0	3·2 ± 1·0	885
5	860	2 min at 1300°C	Marquench 560°C air cool	3 × 1 h 565°C	13·3 ± 2·3	2·4 ± 1·0	852

Table 2   Specifications of En30B and T1 steels

Steel	Limit	C	S	P	Mn	Si	Cr	Mo	Ni	W	V
En30B	Lower	0·26	0·05	..	0·40	0·10	1·1	0·20	3·9	..	..
	Upper	0·34	0·05	0·05	0·60	0·35	1·4	0·40	4·3	..	..
T1	Lower	0·65	..	..	0·20	0·20	3·75	..	..	17·25	0·90
	Upper	0·75	..	..	0·40	0·40	4·50	1·00	..	18·75	1·30

Machining tests

Machining tests were carried out for 1 min in unlubricated conditions using speeds of 88·5 ft/min (440 mm s^{-1}), 130 ft/min (650 mm s^{-1}), and 175 ft/min (874 mm s^{-1}) with feeds of 0·005in (0·127 mm)/rev and 0·010in (0·254 mm)/rev and the depths of cut 0·050in (1·27 mm) and 0·100in (2·54mm) on a Dean Smith and Grace lathe type 17T.

Tools were machined from T1 high-speed steel having dimensions 0·5 × 0·5 × 0·2in (12·7 × 12·7 × 5 mm) with a nose radius of 0·030in (0·76 mm) providing four cutting edges and held in a standard Wickman tool holder (S 112 M1k3) with a 15° approach angle (Fig.1). The tip was ground on the rake face while held in this tool holder to give a +5° rake.

Scanning electron microscopy

There was little difference in the appearance of the tips after machining at low speeds and feeds. All exhibited signs of pick-up of the workpiece material, although this varied in extent across the rake face, frequently appearing to have a shallow U-shaped profile (Fig.2). As the speed was increased to 130 ft/min (650 mm s^{-1}) a crater was developed in the rake face of tips solution treated at 1260°C; the appearance of tips solution treated at other temperatures was similar to that at lower speeds. Crater wear increased dramatically, in tips from all treatments, on machining at the highest speeds and feeds; Fig.3 shows the appearance of tips, one from each heat treatment, after cutting for 1 min. It is of interest to note that:

(i) under these extreme cutting conditions there is no sign of brittle failure; flow of material across the rake face appears to be the major factor responsible for tool wear

(ii) tools given three tempers appear to wear rather more rapidly than those given one temper (compare Fig.3a with 3b, and 3d with 3e)

(iii) tools solution treated at 1280°C and then given two tempers at 565°C have worn less than those solution treated at 1260°C and 1300°C.

Metallographic examination

Examination of sections of the worn tips revealed several common features in the pattern of wear. Under all conditions adherence of the workpiece material to the tool was evident, but the pattern of adherence varied with cutting speed and feed. At the lowest speeds and feeds there was a prominent built-up edge at the tool tip and towards the rear of this, away from the cutting tip, a thin white-etching layer could be seen between the built-up edge and the tool (Fig.4). A similar layer was also observed at higher speeds, between the adherent layer that was formed in the crater and the tool steel. As the speeds and feeds were increased the adherent layer was extended back from the cutting tip into the crater. At the highest speed and feed, flow occurred of both the adherent material and the tool steel in tips from all except one of the heat treatments (Fig.5). This flow along the rake face was responsible for the severe wear observed in the scanning micrographs (Fig.3).

Microhardness tests were performed, on the section shown in Fig.5, at the points illustrated in Fig.6. Hardness values in the region 600–700 HV were obtained from the adherent layer closest to the tool surface. No obvious structure could be detected metallographically in this layer. However, electron-probe microanalysis showed that it consisted of workpiece material. Further away from the tool where flow lines were evident in the chip, the hardness dropped to 450

Positive rake angle ground on tip when positioned in toolholder

-5°

+5°

f

d

HSS tip

Toolholder

15°

d = depth of cut; f = feed

1   Diagrammatic representation of tool

2   Scanning electron micrograph of a tip after cutting for 1 min at a speed of 130 ft/min (660 mm s^{-1}) and a feed of 0·005in (0·127 mm)/rev   × 40

built-up edge and little evidence of pick-up on the rake face away from the cutting tip. However, the sections of tools shown in Fig.5 suggest that at high speeds, seizure between chip and tool was no longer confined to the locality of the cutting tip but extended over a considerable area of the rake face. In Fig.5, the bottom of the crater follows a smooth line from the cutting point across the remains of the built-up edge down through the crater. This suggests that the underside of the chip followed this path, wearing away the tool by abrasion until the heat generated at the rake face was sufficient to permit welding. Seizure produced a large velocity gradient across the chip and caused shear to take place close to the tool/workpiece interface, approximately along the line AB in Fig.6. The relatively stagnant layer ABCA (Fig.6) separated the tool from the rapidly moving chip and underwent a rapid temperature rise due to the heat generated by the intense shear along the line AB. As movement of this layer was prevented by contact with the tool, a shear force was developed acting on the rake face and parallel to the direction of motion of the chip. The magnitude of both this shear force and the temperature rise in the adherent layer were dependent upon the velocity of the chip and hence the cutting speed. As the flow stress of the tool steel is temperature dependent, flow of the surface would occur when conditions of temperature and structure were developed across the interface such that the shear strength of the adherent layer approached that of the tool steel. In four of the heat treatments this condition occurred and flow of both the adherent layer and the tool steel took place. The strength of the adherent layer could be high enough to transmit a significant shear stress to the tool surface even at the high temperatures existing in this region if work hardening took place faster than recovery. This does not imply that this layer was temper resistant in the manner referred to by Trent[3] for thin white-etching layers. Indeed in this case tempering up to 600°C, after the machining tests, resulted in a progressive fall in the hardness of the adherent workpiece layer.

An important factor which determines the transmission of the cutting forces from the adherent layer to the tool is the bond formed at the tool/workpiece interface. The evidence from electron-probe microanalysis indicates that diffusion occurs more rapidly at the interface than it would under normal conditions. It may be significant that whenever seizure occurred, a white-etching layer existed between the tool material and the workpiece. However, while such layers have also been observed by Trent[3] their precise nature has not been established. Further work is required in order to isolate and identify the structure of these layers before it will be possible to decide whether they do affect diffusion and influence the formation of a strong bond between the tool and the workpiece.

The picture that emerges from this discussion is that the rate of crater wear under severe conditions, where seizure occurs, is likely to be determined by a balance between:

(i) the heat generated in the adherent layer
(ii) the temperature and strain-rate dependence of the flow stress of the workpiece material
(iii) the temperature dependence of the flow stress of the tool steel.

While all these factors may vary in different situations, the upper limit of machining for a particular workpiece will be determined by the hot strength of the tool steel. The complex nature of these steels renders them particularly sensitive to heat treatment and it is evident from the present work that this can have a considerable effect on tool performance. It is not possible to offer a simple explanation for the variation that was observed. The differences that were found in any single structural variable (Table 1) were not sufficiently large to account for the results of the machining tests. It appears that the maximum machining speed and feed will be sustained by a tool having the best hot shear strength but that this property cannot be defined by a single structural parameter. Further work is necessary to clarify the influence of the individual microstructural constituents in order to deduce the optimum combination required for the best performance under severe operating conditions.

CONCLUSIONS

During the machining of En30B with tools made from a T1 variety of high-speed steel, the extent of the seizure between the tool and the workpiece depends upon machining speed and feed. At speeds above 130 ft/min (650 mm s^{-1}) seizure occurs over an extensive area of the rake surface, and the properties of the adherent layer that is formed significantly affect the subsequent performance of the tool. At high machining speeds the conditions of temperature and structure are such that the forces transmitted to the rake surface by the adherent layer are sufficient to cause plastic deformation of the tool steel.

ACKNOWLEDGMENTS

The authors wish to thank the Lanchester Polytechnic, Coventry, for providing the facilities to carry out this work. They would like to express their appreciation also to the Joseph Lucas Group Research Laboratories for assistance with the carbide determinations.

REFERENCES

1 E. M. TRENT: *Powder Met.*, 1969, **12**, 24, 566–581
2 I. HAM: ASTME, Technical paper MR 68–617, 1968, 20501 Ford Rd, Dearborn, Michigan 48128
3 E. M. TRENT: in 'Machinability', ISI Spec. Rep. 94, 1967, 77–87
4 G. HOYLE: *Met. Rev.*, 1964, **9**, 33, 49–91
5 J. DURNIN AND K. A. RIDAL: *JISI*, 1968, **206**, 1, 60–67
6 H. OPITZ AND W. KÖNIG: in 'Machinability', ISI Spec. Rep.94, 1967, 35–42
7 A. G. HAGUE: *Int. J. Prod. Res.*, 1969, **7**, 4, 317–332
8 A. N. RESNIKOFF: *Microtechnic*, 1968, **XXII**, 2, 134–136

Wear processes during machining

A. G. King

The classical approach to the metal-cutting process is through static mechanics where the force vectors are resolved in the tool–workpiece diagram. This approach is applicable only to average conditions of stress, and says nothing about dynamics, or mechanisms of those processes which gradually degrade the tool and render it less serviceable. Progress in metal cutting is inextricably connected to and follows progress in tool materials, improved strength, improved chemical inertness, and improved resistance to wear. Rather than taking the classical approach, I should like to direct your attention to fundamental processes of wear which degrade the tool cutting edge. We will look at these processes from a mechanistic point of view and raise some questions which are posed by this vantage point.

The limiting serviceability of any cutting tool is determined by the rates of the interrelated wear, fatigue, and fracture processes which occur during metal cutting. Ceramics are useful as tool materials because they have inherent chemical properties which minimize the rate at which these processes proceed.

WEAR PROCESSES

When we speak of 'tool wear', we are referring to the sum total of those processes which are transporting tool matter across the metal/tool interface along the direction of sliding. There are four process by which tool wear can occur: (a) plastic deformation; (b) diffusion; (c) viscous flow; (d) fracture and displacement. We will examine each of these processes independently and query the relative significance of each in the ensuing discussion. We will see that the relative importance between wear mechanisms is at least partially dependent upon cutting conditions, and upon the ambient environment.

Deformational wear

The best evidence for deformational wear is the classic demonstration by Duwell[1] on the wear of sapphire as the direction of sliding is crystallographically controlled. Figure 1 shows the six-fold symmetry as slip directions are alternately coincident with sliding and misoriented with sliding. Figure 2 shows the wear rate on the basal plane as a function of minor misorientation. The wear rate under these conditions of relatively low stress and temperature, as compared to high-speed metal cutting, is rate limited by the process of plastic deformation in the alumina crystal. The experimental demonstration of almost three orders of magnitude difference between the wear rates on the basal plane establishes that the wear process is a structure-sensitive property rather than one governed by the surface chemistry on different crystallographic planes. The coincidence between wear rates and slip systems within the sapphire crystal suggests that deformation is the process by which material is transported across the surface.

It is a little surprising at first glance to consider that a very hard and brittle material like aluminum oxide is wearing by a process of plastic deformation, much in the same manner as a metal tool wears during cutting. For identical stress levels and temperatures, the ceramic wears at a lower rate, of course, but the process of wear can be similarly described for materials which have crystallographic simularities in their systems. The work of Scott and Wilman[2] and the continuing work of Wilman teaches us that abraded surfaces on hexagonal metals such as beryllium and magnesium develop surface fibre orientations of the crystallites where the directions of easiest plastic deformation lie either parallel to the surface or along the resultant of the normal and friction forces. The surface tends to recrystallize in accordance with minimum energy sinks pertaining to the wear process, along directions of easiest slip. By structural analogy, we should look for re-orientation of surface layers in alumina after sliding on metal surfaces.

Figure 3[3] presents results from cutting tests using ceramic tools. The wear ranking number is a quality parameter which takes into account both the slope and intercept of the wear–time curves. The number lists only the order from best to worst in which a variety of ceramic tool types fell as determined by the amount of wear measured. Surprisingly, these data correlate remarkably well with microhardness as measured with a Knoop indenter. The correlation suggests that tool wear and point penetration occur by a similar process. This process has been identified as plastic deformation through the motion of dislocations under stress, using etch pit technique on microhardness indentations.

The correlation between microhardness and wear which occurs under relatively modest machining conditions, disappears or becomes obscured as the cutting speed is increased. This suggests that the rate-limiting process is changing from dislocation nucleation or motion to a process such as the chemical reaction rate on the surface. In the case of an alumina ceramic sliding on steel in air, Coes[4] and later Brown et al.[5] have shown that the spinel, $FeO.Al_2O_3$, forms at the interface. The reaction:

$$Fe + Al_2O_3 + \tfrac{1}{2}O_2 \rightarrow FeO.Al_2O_3 \quad \dots\dots\dots\dots\dots\dots (1)$$

creates a cubic material with three slip systems of easy glide. The higher symmetry of the cubic lattice permits the deformation to occur more readily, the rate limiting step now becomes the reaction rate of equation (1) rather than the deformation process itself. The wear process is still probably occurring by a deformation mechanism, but the kinetics are controlled by the rate at which the plastic species is formed. Under these conditions wear rates are higher than they are where deformation of the a-alumina lattice is rate controlling. Figure 4 shows a replica of a ceramic tool wear surface which is smoothed by the smearing which takes place during machining. The material is being transported from top to bottom

Wear implies material transport which occurs between two sliding surfaces. This can occur by plastic deformation, diffusion, viscous flow, and fracture with displacement. Deformational wear occurs at low pressures and sliding speeds. When cutting conditions are more severe, the rate-limiting process is a chemical reaction at the tool/workpiece interface forming the iron aluminate spinel. The possibility of surface diffusion as a wear mechanism is discussed, with the conclusion that the process is possible. It appears that viscous flow is a wear mechanism when a chemical species is introduced which can react to form low melting compounds such as chlorides. Fracture and displacement determines the texture of the wear surface and is a common mechanism in ceramic tool wear.

621.941.1.025.7: 669.018.95: 539.538

The author is with the Zircoa Chemetals Division of the Diamond Shamrock Chemical Company

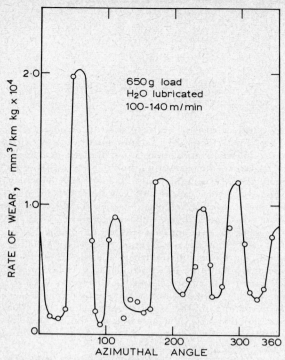

1   **Prismatic wear on sappire (Duwell)**

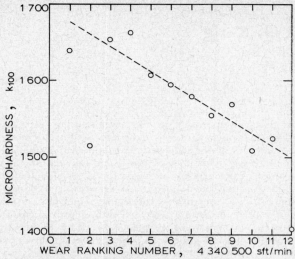

3   **Correlation between microhardness and tool quality for machining 4340 steel at 500 ft/min**

as the wear process takes place. Figure 5 shows the intersection of the original tool surface which shows horizontal grinding marks with the wear surface. The material being transported by deformation is seen to form layers covering the original surface, and pieces of the deformed film are breaking off where they are no longer restrained by the metal/tool contact.

The log of the reaction-rate constant for the formation of the spinel varies as $1/T$, while the yield strength is roughly proportional to $1/T^2$. Therefore, we would expect that wear which is controlled by chemical kinetics to be more important at higher temperatures. In general, this will occur at higher cutting speeds and the correlation of tool quality with microhardness which was shown in Fig.3 does indeed disappear at higher cutting speeds.

Equation (1) also tells us that under conditions where the rate of wear is controlled by chemical reaction, that oxygen from the ambient atmosphere must counterdiffuse along the tool/metal interface in order to form the spinel. The atomically clean metal surface will readily react with this oxygen film and, due to its relative motion, will carry the oxide layer away from the cutting edge. The depth of oxygen penetration should therefore be inversely proportional to cutting speed. This is readily observable by measuring the width of the wear marks on the crater surface. Figure 6 shows this inverse correlation in three atmospheres which were blown on to the tool tip during cutting in the unsuccessful attempt to control the oxygen partial pressure. Figure 7 shows the appearance of the crater wear surface for the four speeds used, with the surprising but understandable result that higher cutting speeds produce less wear. Figure 8 shows the highly polished wear groove at the depth of cut line which was produced after

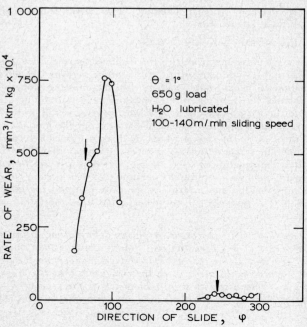

2   **Basal wear on sapphire (Duwell)**

4   **Electron photomicrograph of wear surface**   × 3 000

5 Electron photomicrograph showing removal of detritus at the bottom of the wear land   × 3 000

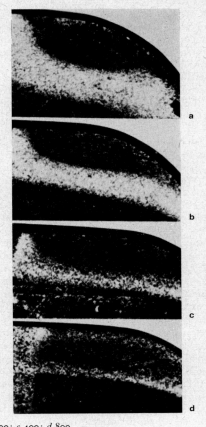

a 100; *b* 200; *c* 400; *d* 800

7 Crater wear surfaces after machining 4340 steel   × 57

machining type A-6 tool steel at 1 200 ft/min. The finish of this surface is so perfect that it is difficult to focus a microscope on the surface because of the complete lack of detail. It is probable that this surface is produced by the rubbing of preoxidized metal surfaces on the tool edge, thereby supplying an abundance of the necessary species to form the spinel $FeO.Al_2O_3$. Smearing of the spinel films forms the highly polished surface.

Diffusion wear

Material can be removed from the tool either by diffusion of surface atoms into the chip or workpiece, or by surface diffusion of tool material along the interface. The relative activation energies of volume diffusion, boundary diffusion, and surface diffusion are in the order:

$$Q_v > Q_b > Q_s$$

It has been demonstrated by Babcock[6] that volume diffusion does occur during metal cutting, but with activation

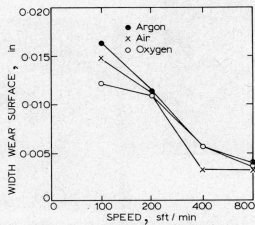

6 Correlation between the width of the crater wear mark and speed after machining 4340 steel

8 Highly polished groove at the depth of cut position   × 57

energies of the order of 50 kcal/mol, the process would be expected to be relatively insignificant for ceramics at metal-cutting temperatures. The flux of tool atoms into the metal would be further reduced by the high compressive load on the interface which would lower the number of lattice vacancies that are available for the process.

Wherever oxygen enters the interface between the tool and workpiece, FeO and Al_2O_3 are in contact and form the spinel as was discussed earlier. Since this is a solid state reaction, the two oxides are interdiffusing at rates high enough to generate detectable amounts of the compound on a sliding surface. This diffusion reaction becomes wear-rate controlling, but is not by itself the wear mechanism, which implies mass transport. As a mass-transport process, volume diffusion is probably not too significant.

The theory of surface diffusion is not as well developed as that of volume diffusion, and an exact evaluation of this as a wear process will have to wait for better theory. Since the activation energy is much lower, and since the surface is under very high shear stress, we might well expect that this process could contribute to wear, particularly in the cutting of high yield stress materials and at high cutting velocities.

The experimental observations in Figs. 4 and 5 can be explained equally well by a surface diffusion mechanism. When experimentally machining 4350 steel at 1500 ft/min, the wear rate of a ceramic tool normal to the surface was $7\frac{1}{2}$ μm/min. If we speculate that a monolayer is being continuously stripped from the tool surface at the sliding speed, the surface would be reduced at a rate of 900 μm/min, or two orders of magnitude greater than the wear rate observed. However, the surface diffusion mechanism does not occur as a continuous shearing of whole atomic planes, but involves jump frequencies for individual atoms. We would, therefore, expect the wear rate to be much smaller than that produced by continuous shear. While the surface diffusion appears to be numerically reasonable and a mechanism of possibly general importance, a better theory with the power to calculate the effects of shear stress will have to be developed before a quantitative estimate is possible.

Viscous flow

The energetics of viscous flow for crystalline solids are at a level where this process does not occur below the melting point. In fact, the shear associated with viscous flow would in itself result in a lattice disordering approaching that of melting. In order to have viscous flow operate as a wear mechanism under reasonable conditions, it is necessary to create a new species at the tool/workpiece interface which has either a high vapour pressure, or a low melting point. When this occurs, the principal shear is occurring along films of fluid materials and there is a corresponding decrease in cutting forces. Shaw's[7] work on machining with tool-steel cutters while using CCl_4 as a cutting fluid showed that while cutting forces were greatly reduced, tool wear was greatly increased.

The reaction of the type:

$$Fe + 2CCl_4 + O_2 \rightarrow FeCl_3 + 2COCl_2 + \tfrac{1}{2}Cl_2 \quad \ldots\ldots\ldots\ldots(2)$$

was probably taking place during machining.

Ferritic chloride has a melting point of 282°C, and a boiling point of 315°C at 1 atm. The temperature pressure conditions on the cutting surface can be extreme, and in general, species which are wearing by viscous flow will follow phase relationships such as shown schematically in Fig.9.

Whenever chemical reaction on the tool surface creates a low melting phase, or a phase with a high vapour pressure, wear can occur by a viscous-flow mechanism depending on the phase relations of that particular species. Shaw's results suggest that wear of this type was occurring during machining and that the primary shear was taking place in a liquid film. Ordinarily we would want to avoid this and machine in ambient atmospheres which do not contain reactants in order

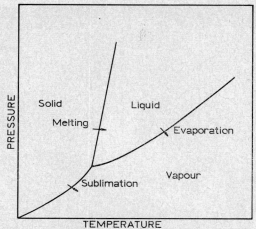

9   Phase diagram, schematic

to minimize wear. Still, this technique could be useful in the machining of delicate workpieces by minimizing cutting forces through use of a reactant which acts as a high-temperature lubricant. The reaction forming WCl_5 or $TiCl_4$ could occur with cemented carbides producing species as shown below:

Chloride	MP,°C	BP,°C
$FeCl_3$	282	315
WCl_5	248	275·6
$TiCl_4$	-30	136·4

The reaction between Al_2O_3 and chlorine would not be expected as the oxide is the more stable compound.

Fracture and displacement

Classical wear theory visualizes chemical bonding taking place at points of contact between two surfaces and fracture of the asperity leaving a small fragment of one surface being carried away by the other. Radioactive tracer techniques have shown that there is indeed a transfer from one surface to the other, and that wear can and does occur by a fracture mechanism. With ceramic cutting tools, the fracture-displacement mechanism can be a major source of wear, depending principally upon the quality of the tool material. This mechanism is on a coarser scale than that visualized by the formulators of the classic theory, as it involved whole grains or aggregates of a small number of grains. I should like to trace the development of this process from the mechanistic point of view.

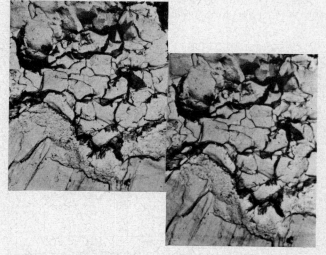

10   Stereo electron photomicrograph showing grain-boundary cracking resulting from material fatigue during machining
×3000

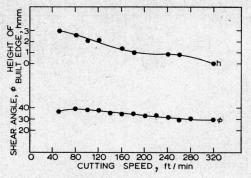

Cutting conditions as in Fig.1

4   Variation of shear angle ϕ and height of built-up edge h with increase in cutting speed

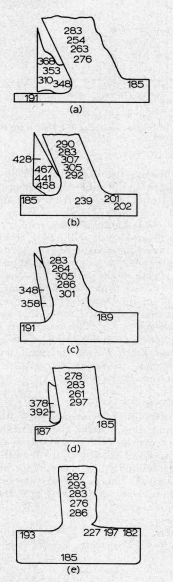

a V = 50 ft/min; b V = 100 ft/min; c V = 180 ft/min;
d V = 220 ft/min; e V = 320 ft/min

5   Micro-Vickers hardness readings of the quick-stop sections at certain selected cutting speeds

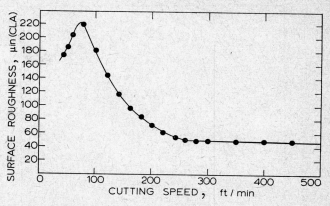

Cutting conditions as in Fig.1

6   Variation of surface roughness with cutting speed

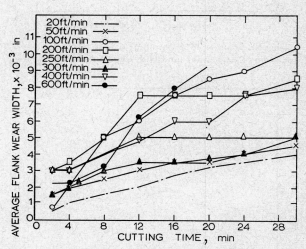

Material En9 (0·55%C); heat treatment normalized; W/TiC P20; width of cut 0·050in; undeformed chip thickness 0·005in; oblique cutting; effective rake angle 15°

7   Variation of average flank wear land with cutting time

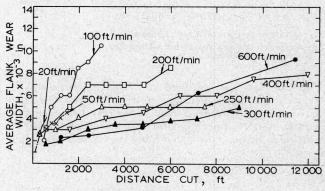

Cutting conditions as in Fig.7

8   Variation of average flank wear land with length of chip cut

Fig.6. The surface roughness reaches a maximum at a cutting speed range of 80–100 ft/min. In this region BUE has not only attained its maximum size, but is also severely work-hardened. This work-hardened BUE breaks up periodically, leaving fragments of BUE on the workpiece resulting in maximum roughness. The surface roughness reaches a con-

stant value at cutting speeds above 300 ft/min which is in fair agreement with the disappearance of BUE in this region.

CHIP FORMATION AND TOOL WEAR

Tool wear tests were conducted on the same material at cutting speeds of 20, 50, 100, 200, 250, 300, 400, and 600 ft/min.

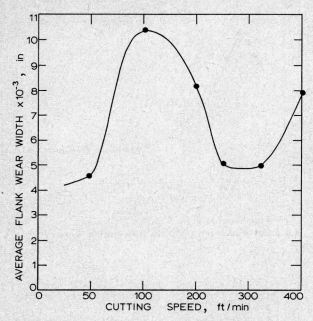

Cutting conditions as in Fig.7; cutting time 30 min

9   Variation of flank wear land with increase in cutting speed

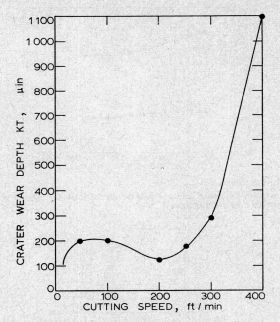

Cutting conditions as in Fig.7; cutting time 30 min

11   Variation of crater wear depth *KT* with increase in cutting speed

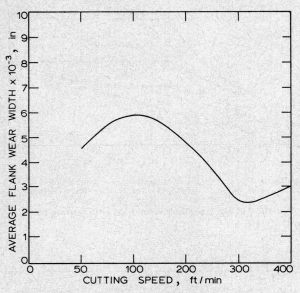

Cutting conditions as in Fig.7; distance cut 1 500 ft

10   Variation of flank wear land with different cutting speeds for the same distance cut

The tool wear was measured at selected intervals and the wear criteria was to machine for 30 min or machine up to 0·01in average flank wear land, whichever happened first. Figures 7 and 8 show the flank wear development with the increase in cutting time and distance cut respectively. Figure

9 shows that the average magnitude of flank wear land after 30 min of machining to be the same for cutting speeds of 50, 250, and 300 ft/min. The high flank wear at 100 ft/min can be attributed to the severe work-hardening of the BUE as shown by the micro-hardness measurements and the subsequent periodical breaking up of the BUE and these fragments squeezing between the workpiece and the flank of the tool. High-speed film taken at a cutting speed of 100 ft/min also showed very effectively the periodic breaking up of the BUE. Figure 10 shows the flank wear development for different cutting speeds for the same helical distance cut. It will be observed that tool wear starts to increase again beyond the speed of about 300 ft/min. The crater wear development at these cutting speeds is shown in Fig.11. It can be seen that the crater wear depth *KT* is not significantly different at cutting speeds up to 300 ft/min, but at 400 ft/min, there is a significant increase in crater wear.

CONCLUSIONS

From the results of the experimental work, it is apparent that when machining En9 material the formation of a built-up edge is detrimental for both surface finish and tool life criteria. At the speed of 300 ft/min at which the built-up edge disappears, optimum machining conditions prevail, *viz*: flank tool wear is at a minimum for a given distance of cut (Fig.10); surface roughness is at a minimum (Fig.6), and cutting forces are at a minimum (Fig.1).

REFERENCES
1 M. C. SHAW: in 'Machinability', ISI Spec. Rep.94, 1967, 1–9

Influence of tool geometry on machining performance in some turning and face-milling operations

W. T. Clark and B. W. Ludwig

The performance of a cutting tool measured in terms of tool life, surface finish, and machining rate depends mainly on tool material and tool geometry. And the two are interdependent in so far as the selection of tool material can limit the use of certain tool shapes and that certain tool shapes permit the use of brittle tool materials.

The geometry used in any application will depend on the tool material and the workpiece material and on the operation whether it is roughing, finishing, or forming. All these factors set bounds to the geometry which can be employed; nevertheless, quite a large degree of latitude exists for varying the geometry in many applications even when the tool material and operation is fixed.

Below is given first a number of examples in turning which illustrates the role played by tool geometry and its interrelation with other factors. Then a detailed account is given of how geometry affects performance in face milling. This operation is selected as representing an instance where not only the individual tooth geometry is of importance but also the overall geometry of the tool.

ROLE OF TOOL GEOMETRY IN METAL CUTTING
General
Two factors which control tool geometry are tool life, which depends on breakage or wear, and surface finish, which depends on wear, metal flow properties, and vibration. Consideration of these factors therefore influences the selection of the tool geometry.

A distinction has to be made between single-point tools and multiple-point tools as far as geometry is concerned. This is because in the latter case two geometries are important: firstly, that of the individual cutting tooth and secondly, that of the complete tool. The overriding requirements of the complete tool geometry therefore partly govern the individual tooth geometry.

Some limitation on geometry and its use may be imposed by throwaway-tip tools. It is sometimes the case that the best tool geometry for a given application cannot be used because it is not suitable for use with throwaway-tip tools; nevertheless these would still be employed because of their inherent economic advantage.

Effects of geometry in turning
Since the introduction of carbide tools and the recent developments in greater wear-resistant carbides, together with ceramic tool materials, geometry has played an important part in their use. By making best use of the high compressive strength of such materials, negative rake geometry enables the relatively brittle carbides to be used on heavy and interrupted cutting. As improvements are made in the strength of carbides so more positive rake cutting is possible. Nevertheless, negative rake geometry is still used extensively and especially for the more brittle carbides and ceramics such as the titanium carbide and aluminium oxide tools, although even for these materials, positive rake is used for light cuts.

In general, negative rake geometry leads to larger tool forces than positive rake and consequently results in higher temperatures and shorter tool life. It can, however, be more economic than positive rake when used in throwaway-tip tools since it allows more cutting edges per tip to be used and also has the advantage of being less susceptible to breakage.

High positive rake angles are often used with high-speed steel tools and this has the advantage of reducing tool forces, improving surface finish, and improving tool life. However, a limitation exists to the amount of positive rake which can be used with carbide tools due to the cutting edge becoming too weak.

Brazed-tip carbides often have an advantage over throwaway tips in so far as their geometry can be modified to suit the application. Rake angles, clearance angles, approach angle, and trail angle can be independently varied to give optimum performance. For instance, it has been found for turning work-hardening metals such as high-nickel–chrome alloys, especially in the cast condition with a rough surface skin and using tough carbide grades (ISO K20–K30), that a large approach angle ($\sim$45°) is desirable in order to reduce notching of the cutting edge at the depth of cut. High positive back and side rake angles are also desirable since the forces on the tool face are reduced; this minimizes the plucking out of carbide grains which is so prevalent when cutting these materials. In addition, a low plain trail is most suitable if a good surface finish is required.

Effects of geometry on ceramic tools in turning
Ceramic and cermet tools are very susceptible to fracture failure, and rigid machining conditions are required to obtain satisfactory results with them. However, careful selection of tool geometry can go a long way towards improving tool life and extending the application of such relatively brittle materials. Negative rake is recommended for resistance to edge breakage since this utilizes the high compressive strength properties. Larger negative rake is possible than with carbide

Ways in which cutting-tool geometry affects machining performance are discussed. Emphasis is placed on the interrelation between geometry and such factors as tool materials, workpiece material, form of tool holding whether inserted blade, brazed, or throwaway tip, and the limitations imposed by economic considerations. Specific examples are given of the role of tool geometry in turning and how it contributes to the best use of tool materials notably carbides and ceramics. By referring to a specific machining operation it is shown that for complex tools with multiple cutting edges, not only is the individual tooth geometry significant but also the overall geometry of the tool. Results of research on the face milling of cast iron are presented which illustrate the dependence on tooth geometry, overall tool geometry, workpiece material, methods of tool tip holding, and cutting conditions for attaining optimum machining performance.

621.941.1.013:621.941.1.025.7:669.018.25

The authors are with the Production Engineering Research Association of Great Britain

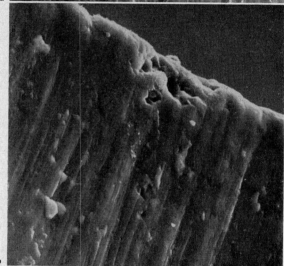

a cutting speed 1 000 ft/min, feed rate 0·005in/rev × 600; *b* cutting speed 1 500 ft/min, feed rate 0·005in/rev × 1 050

1   Cutting edges of aluminium oxide tools after turning plain carbon steel for 10 min

tools since this has less effect on surface finish. This is because there is less tendency for a built-up edge to form on ceramic tools and also because the ceramic material can withstand the higher temperatures generated. Tools should have the largest possible nose angle and maximum corner radius to give the strongest edge geometry (the limit is set by vibration), and large plan approach angle to spread the chip load. Tool life is improved, especially under severe cutting conditions, if an additional negative land is honed on the cutting edge. The size of land depends on the material being cut; the usual recommendation is that it should be equal to half the feed rate. But much wider lands are found to be useful for machining hardened steel.[1] The improvement is attributed to such negative lands causing compressive loading of the cutting edge which retards the development of fracture. Evidence that this is so can be seen in Fig.1*a* and *b* which show the cutting edge of aluminium oxide tool tips after machining plain carbon steel. The edges of these tools were initially sharp after diamond lapping. The cutting edge is undergoing fracture and grains are being removed causing avalanches down the flank face thus increasing the flank wear land. The radius of the worn edge on Fig.1*a* and *b* is less than 0·001in. By honing a land on the tool edge the wear process will be modified.

The more brittle the tool material, the more critical is the geometry for successful cutting. This is especially so for certain new ceramic materials such as those composed of borides and nitrides; these are at an experimental stage but have shown promise for cutting high-strength steel and other difficult-to-machine alloys.[2] One geometry which has proved successful with these tool materials is 7° negative back rake, zero side rake, 45° approach angle, and a 0·045in nose radius.

Effects of geometry in face milling

A face-milling cutter, being a multi-toothed tool, has a complex geometry and its performance is governed by a number of factors. These are the cutting tool material, the workpiece material, the individual tooth geometry, the total number of teeth, and the accuracies to which they are set or ground. Most of these are interrelated and geometry cannot be considered in isolation.

Carbide throwaway-tip cutters are preferable to brazed-tip cutters since they eliminate regrinding time and costs, but their use imposes some limitations on the variation of geometry which can be used. Efforts are made to devise 'universal' geometries which will be suitable for the largest number of applications. The use of universal geometries, however, relies very much on carbide grade selection to cope with widely different workmaterial characteristics. This is not always satisfactory and brazed-tip cutters are used in order to obtain suitable tooth geometry for a number of workpiece materials. A good example is the face milling of cast iron.

Face milling cast iron

For certain cast-iron components such as engine cylinder blocks and heads, stringent requirements are placed on the quality of the face milling. These are high-quality surface finish, flatness, and no edge breakout. Edge breakout or 'frittering' is a well known problem in machining cast iron and it occurs at the exit side of the cut, especially along sharp edges, bores, or cavities in the components machined. It is the result of small pieces of cast iron crumbling and flaking off at the edges. In order to eliminate or reduce breakout it has been found necessary to use positive rake, low approach angles (as low as 15°), and to restrict the feed per tooth to low values (0·005in/tooth/rev). These requirements have necessitated the use of carbide brazed-tip cutters, because a large number of teeth are required to give acceptable machining rates (because of the low feed per tooth). A further reason is that low approach angles can be readily ground on the teeth, and axial and radial setting differences between teeth (run-out) can be controlled to very small values (0·000 1 –0·000 3in overall). This ensures a good quality surface finish.

Throwaway-tip cutters have the disadvantages that fewer teeth can be packed in a given cutter circumference, approach angles are too large, and tooth run-out is, at least, 0·000 5in overall. It is also difficult to maintain small run-outs especially as the teeth wear and are indexed.

Improved throwaway-tip face-milling cutter for cast iron

A programme of work was recently successfully completed at PERA aimed at developing an improved throwaway-tip face-milling cutter for the precision machining of cast iron.[3] The main requirement was to overcome the edge breakout.

Initially, single-tooth milling tests were carried out in order to study the effect of tooth shape and tooth loading, i.e. the feed/tooth/rev on edge breakout. These were followed by tests on a multi-tooth cutter.

SINGLE-TOOTH CUTTING

Tests were carried out using a single hexagonal-shaped tooth in a 10in diameter cutter. Keeping the approach angle constant at 45° and cutting speed at 300 ft/min the effect on breakout was determined for a range of axial and radial rake

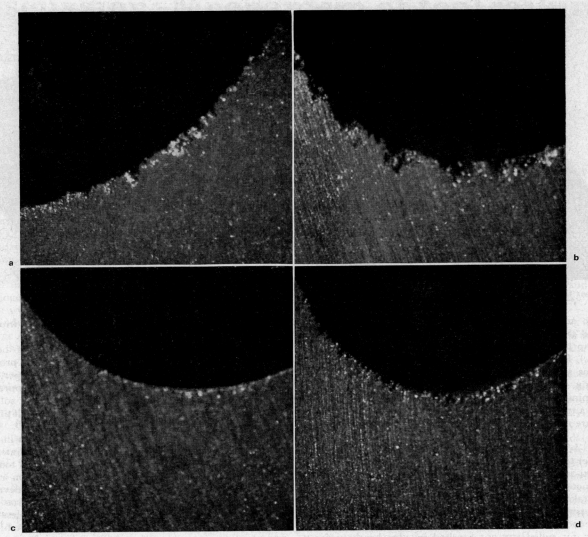

a hexagonal tooth, nose radius 0·060in, feed rate 0·010in/tooth/rev; *b* hexagonal tooth, nose radius 0·060in, feed rate 0·016in/tooth/rev;
c circular tooth, radius 0·75in, feed rate 0·010in/tooth/rev; *d* circular tooth, radius 0·75in, feed rate 0·016in/tooth/rev

2 Typical edges of cast iron produced in face milling   ×5½

angles and for various values of nose radius for a range of feed rates and depths of cut.

It was found that edge breakout was not appreciably affected on changing radial rake angle from 0 to 10° positive, but negative radial rake angles gave rise to increased breakout. Edge breakout was also least for axial rake angles between 5° and 10° positive. For zero or small nose radii, edge breakout occurred for feeds in excess of 0·005in/tooth/rev and was very bad for a feed of 0·016in/tooth/rev (Fig.2a and b). However, by increasing the size of the nose radius the edge breakout was reduced.

This work led to the idea of using a circular-shaped tooth held at 5° positive axial rake and 0° radial rake. This geometry has the effect of gradually reducing the chip thickness

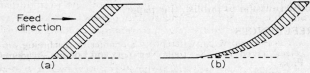

a straight-sided tooth; *b* circular tooth

3 Comparisons of sections of cut for hexagonal and circular roughing teeth

4 Roughing tip and holder showing method of location

Laue diffraction patterns might be economically justifiable, because of the high cost of tools and the importance of reproducible tool life. In particular it seems that a {100} plane orientation of the top face must be avoided and {110} or {111} favoured.

There is a natural emphasis in these papers upon high-speed behaviour of tools, since this imposes more severe conditions, and tool life is usually the factor limiting further increase in cutting speed, though we may question whether speed is always the limiting factor. Cherry and Prabhu show the advantage of choosing the speed at which the built-up edge disappears. The importance of rake-face temperatures in promoting reaction or the formation of surface films has been extensively discussed in the past by Trent and Opitz and others. Trent, in showing that the presence of a seized layer reduces wear appears to disagree with Cherry and Prabhu. Even with diamond, Keen mentions the possibility of wear due to graphitization (reversion to the more stable carbon lattice) at high temperatures. Perhaps this might also happen with borazon (BN). In the present papers a spinel formation by reaction of FeO and Al_2O_3 in the presence of oxygen is described by King. The rate of formation, and consequently the rate of wear, is controlled by chemical kinetics. The reduced access of oxygen at high speeds impedes the reaction and consequently reduces the wear, but unfortunately his attempts to displace atmospheric oxygen by other gases were not successful, though Trent and ourselves have found that displacement by a jet of nitrogen is possible at lower speeds. King concludes that surface or boundary diffusion may contribute to wear, and calculates a removal rate of 900 μm/min on an assumption of monolayer removal. The actual rate is about 7 μm/min. Brownsword, Hague, Panton, and Pyle, however, found from their microanalyser studies that the scale of diffusion was too small to account for gross removal of tool material; hardly surprising since their cutting speed never exceeded 175 ft/min. They do not compute temperatures, though they do give evidence for a 'white-etching band', beloved of metallurgists, with high hardness (600 HV compared with 450 in the rest of the swarf), which indicates diffusion at least in the swarf. They consider that the main forms of wear are abrasion or adhesion at low speeds and plastic deformation at high speeds. Again reference may be made to the optimum speed found by Cherry and Prabhu. Tabor also describes adhesive transfer, which is the basis of much of the frictional work of his laboratory, but he considers that adhesion will readily occur between rubbing metals in the absence of lubricant. Brownsword et al. infer that the heat produced by abrasion is necessary to raise the rake-face temperature sufficiently to cause welding. Tabor does not discuss diffusion explicitly but he does refer to high-temperature rake-face wear which he associated with a Latin concept of erosion, gnawing away. I suggest to him that 'cratering' derived from the Greek bowl used for mixing wine, is more appropriate to this conference.

Little is said about lubricants and their influence on wear. Tabor briefly mentions the use of boundary and extreme-pressure lubricants. King refers to a curious explanation of the action of CCl_4, associated with viscous flow of molten $FeCl_3$, and concludes that wear by viscous flow can occur whenever chemical reaction on the tool surface creates a low-melting phase or a high vapour-pressure phase. No explanation of the significance of vapour pressure is given, and the observation that CCl_4 is most effective at very low speeds is ignored. Perhaps more attention might have been given to coolants and lubricants, which are of very great commercial importance in machining because of their influence on tool life and to some extent on surface finish. They are discussed briefly in the paper by König. There may also be more subtle effects on the rake-face conditions, particularly the extent of adhesion. Tabor refers to a 1966 paper showing that the friction even of diamond is very much influenced by absorbed films. I must agree with this, having published the same information myself from the same laboratory in 1954.

SELECTION PARAMETERS

A major conclusion from the present papers is that the mechanical properties of the tool material are of great importance, with some slight indication of the significance of tool chemistry, but it is less obvious which properties should be controlled. Brittle failure of ceramic and sintered carbide tools is well known and is mentioned several times. Keen finds it to be important in diamond tools; as chipping, cleavage, or conchoidal fracture. Braiden and Dugdale show that thermal and mechanical fatigue stresses are particularly important in intermittent cutting. Clark and Ludwig also emphasize the fracture of ceramic and cermet tools and conclude that tool geometry is very important, recommending negative rake angles with the largest possible nose angle and maximum corner radius. They suggest that an additional negative land honed on the cutting edge gives further improvement. Their explanations are brief and could be questioned in detail, for example about the modifications to chip flow and wear introduced by the land.

Plastic flow is clearly of major importance in tool failure, especially at high temperatures. The present papers refer to this mostly in connexion with high-speed steel tools, but Trent has shown its importance also in setting an upper limit to the use of sintered carbides. Brownsword et al. have shown photographs of the wear of tools given different heat treatments, but they were not able to correlate the wear directly with the known effects of tempering. They conclude that at present it is not possible to say more about selection than that the maximum speed and feed will be sustained by the tool having the best hot shear strength. It is not obvious how this should in fact be measured. Braiden and Dugdale make the practical suggestion that tools should be warmed before starting, to reduce the magnitude of the thermal cycle. Tabor suggests that the improved hot hardness of certain ordered carbides, especially VaC–TaC should make them more effective for machining under extreme conditions. We may refer also to the paper by König, and his quoted conclusion that massive wear occurs when the mean pressure on the wear land exceeds the effective (undefined) flow stress of the work material. He also makes several suggestions for practical improvement.

CONCLUSION

Perhaps we could ask the authors of the present papers what practical conclusions they would draw from their work. One paper, by Clark and Ludwig, concludes very practically, with a design of an actual milling cutter. It is a little unfortunate that, having emphasized the virtues of negative-rake land tools and the importance of geometry, the designers are forced by the characteristics of their workpiece to use positive rake angles of as much as 15°, and to choose their overall design on the basis of many extraneous considerations.

It might also be valuable to reflect in general upon the relationship between theoretical studies of metal cutting, experimental work under selected conditions, and the practical problems encountered by manufacturers and users of cutting tools.

Several of the papers emphasize the importance of improved mechanical properties: fatigue, hot shear, strength, ductility, hardness. Undoubtedly these are desirable, but how should they be combined, and how achieved?

The Chairman
Professor König dealt not only with theoretical aspects of tool failure, but also made practical recommendations for improving tool life, and I would like to thank him on your behalf.

I would described Dr Trent's lecture with the words 'informed enlightened commonsense,' which is a very rare quality today. I thought that Dr Trent did himself less than justice when, after having described a number of phenomena, he talked about them as 'anomalies'; I would think they are the essence of the situation.

Dr Rowe will have ensured a lively discussion by provoking the reactions of some of the contributory authors of the papers for which he acted as rapporteur.

We have had a brief theoretical treatment of the subject, and it is fairly obvious that this has provided enough scope for hypotheses for tool development and cutting stock development in any direction anyone would like to choose. We might later discuss to what extent one can extrapolate directly from theory to tool development as such.

Dr K. A. Ridal (BSC, Special Steels Division)

Dr Rowe is to be congratulated for provoking and stimulating at least the academics, by his excellent review of the papers. I would like to stimulate the tool users and toolmakers a little, because it seems to me that this conference is really a coalition of the various facets of the business, and it would be wrong if the conference followed its programme and dealt with the academics' view, the steelmaker's view, and then the toolmakers' and tool users' views. The discussion should be very much a free-for-all.

One might be forgiven for commenting that the standard progressively deteriorates and becomes less objective as the conference proceeds. However, toolmakers and tool users should not get too depressed, because as you see, the academics are not perfect and all that can be magnified is not necessarily correct. It is gratifying, in the last ten years, to see the increased use of modern techniques. The scanning microscope and the electron probe microanalyser feature very strongly in the academic papers. Forgive me for being cynical and noting that if the instruments are available in the laboratory then people sometimes feel they are obliged to use them. I could not help noticing in Brownsword's paper that he plotted elegantly the diffusion contours for chromium, nickel, and iron and never mentioned carbon which, due to its much higher diffusion coefficient, would be expected to play a much greater part in the hardening of the workpiece. Perhaps next time we have a conference we shall hear more about field emission electron microscopy and low-energy electron diffraction, which are the current fashions.

One of the points I would have picked out from the papers we have heard so far is the importance of the tool/workpiece interface. It was rather disturbing to hear that Dr Trent thought that interfaces were not terribly important in his carbide tools, while Professor König made the point that interfaces were extremely important with high-speed steels. Here is the dilemma; we have progressed from a purely empirical approach to machinability and tool wear and have now got physical explanations of various aspects of the process. What we really want now is for the more important features in any one situation to be described and made practical use of. It is always endearing to hear the academics make a plea that they really are trying to help industry. This again is a very great fashion of our time, but at the end of the day may I be forgiven for thinking that the tool buyer does not always think in this way. He is probably more interested in his discount than he is in the more obtuse aspects of technical efficiency.

I think the point made in the first paper about surface treatment of tool steels is particularly relevant. I hope that tool users and toolmakers will correct me, but it seems to me that nitriding, steam bluing, and sulphurizing are relatively cheap processes, and if they really do cause the improvements suggested by Professor König, I do not see why they should not be used more widely. Why does not some ambitious toolmaker try these techniques? Perhaps he will still have to offer a discount, but he may increase his market share.

Dr N. Gane (Cambridge University)

I feel that I should rise to the defence of Dr Tabor, in his absence, although I do not know that I can say adequately what he would have wished to say.

The first point Dr Rowe mentioned referred to the relevance of the formation of ring cracks when a hard slider slides over a brittle surface. I think that the point Dr Tabor was really trying to make was that a very small increase in friction can give rise to a very large increase in tensile stress round the back of the slider. In fact, a nine-fold increase in friction gives rise to a 600-fold increase in tensile stress. Therefore, I think the point was that it may be a relevant factor in machining operations because of this very large increase.

The second point Dr Rowe made concerned the very high frictions obtained in vacuum where there is no gaseous lubricant. I am not really familiar with Dr Rowe's work in 1954. The work to which Dr Tabor refers, carried out in 1960, was, I venture to suggest, done in much more stringent experimental conditions; a vacuum of 10^{-10} torr, without prior heating of the specimen to degas it. I do not think UHV techniques were in that stage of development back in the early fifties.

Dr Rowe

Very briefly, the same result was obtained, nevertheless.

The Chairman

There is obviously a divergence of views between Dr Trent and some of the other contributors on the importance of adhesion.

Dr King

I was very interested in Dr Trent's presentation. We have done some work on ceramic cutting tools and we have found, or have deduced, a similar phenomenon at the workpiece to cutting tool surface. When you look at the topography of the cutting tool with the electron microscope you see the curvatures where the metal has curved around various features at accelerations which are hard to believe. If the metal were sliding past the surface at anything like a cutting velocity, these are something like 4 billion cm/sec/sec, which I was not prepared to believe. We reached a similar conclusion based on the premise that the material is primarily shearing out in the workpiece. This was somewhat surprising for ceramics but it has rather broad implications as to what we should do to develop better cutting-tool material because, in an analogous way, the cutting-tool, when it is cutting through metal, is somewhat like an aerodynamic sample in a wind tunnel where these flow patterns going round the tool edge are present. The details on the microscopic scale over distances of 1 1 000in or less have tremendous bearing on the local stresses. We have estimated stress gradients as high as 25 million lb/in²/in occurring on a cutting-tool edge. There is therefore a tremendous opportunity for tool designers to achieve microscopic tool design in an aerodynamic sense so as to distribute these stresses in an optimized way. From some of our very crude attempts to do this by putting different lands on at different angles or at small radii, 0·001–0·002in, we found that this profoundly affected total tool life, as much as 3 or 4 fold.

Dr Trent

I was aware of Dr King's work on the subject of ceramic tools; in particular, his very clear and informative stereo photomicrographs should be seen stereoscopically to be appreciated. I agree with his evidence that the metal flows around very small radii at the tool surface where it must be flowing very slowly. With ceramic tools seizure is less likely than with steel or carbide tools, but Mr King's evidence shows that, using ceramic tools under these conditions, seizure does occur. This is confirmed if you look at microsections, through the chips produced. The undersurface of the chips produced by ceramic tools is in fact sheared in the same sort of way as in chips produced by carbide tools. Movement over the tool surface has taken place mainly by shear in the work material. This is where the forces are being generated, and this is where the heat is being generated that we are concerned with in the tool. What would be of great interest, I feel, is for the physical metallurgist to tell us exactly what conditions can exist at the interface between the tool material and the work that is being cut, and in the flow zone on the underside of the chip, where such extraordinarily severe strain rates occur. This sort of information must be of interest to people who make tools which have got to resist these strain rates.

If I can reply also to Dr Rowe in relation to Professor Cherry's paper, in which he found contradiction between my remarks concerning wear under conditions of seizure and Professor Cherry's finding that the optimum machining condition was just after the built-up edge had disappeared. The suggestion that there is a contradiction here appears to be based on the misconception that seizure occurs when a BUE is present, but seizure conditions disappear when the BUE disappears as the cutting speed is raised. In fact conditions of seizure persist when the BUE disappears, and wear characteristics typical of those related to seizure occur, but average tool life often increases because damage caused by breakaway of the BUE is eliminated.

Mr A. I. W. Moore (PERA)

I would like to ask Professor König and Dr Trent if they would comment on some phenomena which we observed some years ago at PERA. Although I cannot recall all the details of the tests, an interesting hypothesis was put forward by one of my colleagues for some wear behaviour that we observed.

The tests concerned conventional single-point tools cutting some of the 'difficult-to-machine' alloys. The type of wear involved is commonly described as 'grooving at the depth of cut' and 'grooving at the nose', but in fact the grooves occur beyond the depth of cut and at the end face just clear of the tool nose cutting edge. My colleague observed that these conditions were occurring at relatively high speeds when there was very little flank wear in the normally accepted sense. His hypothesis was that grooving occurred as a result of a flow of metal parallel to the cutting edge, as distinct from the usual chip flow normal to the cutting edge.

I would like to ask the authors if in fact they have observed this kind of wear, and studied chip flow as a three-dimensional phenomena rather than purely as a two-dimensional problem.

Professor König

I remember that Professor Shaw published a hypothesis similar to that mentioned by Mr Moore some years ago.* He observed a flow in this direction too, especially when machining high nickel alloy materials. Even when using carbide tips, you can still observe these grooves. This can be avoided, in our experience, by increasing the inclination angle of the tool to 15°.† Even in machining ordinary materials with carbides, similar grooves are often observed and can be directly influenced by an oxygen jet.

Due to this temperature rise oxidation takes place resulting in a complex iron–tungsten–cobalt–oxide. The weakening of the carbide leads to a 'grooving wear' at the rake face of the tool as well as at the side cutting edge. For cemented carbides with a higher tungsten carbide content the 'oxidation wear' on the side cutting edge very often leads to a breakdown of the tool nose and thus to total tool failure. By increasing the speed and feed, which is necessary when using

*M. C. Shaw et al.: Plasticity problem, involving plane strain and plane stress simultaneously: Groove formation in the machining of high-temperature alloys, *Trans. ASME*, 1966
†H. Mütze: Beitrag zur Zerspanbarkeit hochwarmfester Werkstoffe, *Diss. TH Aachen*, 1967

capital-intensive machines, the cutting temperature will increase.

Dr King

As we know, metal cutting is a very complicated process, and we have difficulty in reconciling the academic world and the practical work. What needs to be done is to have criteria by which to recognize what processes are going on in an industrial type of operation. This knowledge gives you a rationale by which corrective measures can be taken. The edge grooving is perhaps one example where I think we might know what is happening. We have quite a different hypothesis. We have not seen this lateral flow along the top edge of the tool. The evidence we have seen is that the wear striations are along the top face of the tool in the direction of chip movement. I have not seen the wear striations going in the other direction so I would not suspect that this is happening.

The grooving at the 'depth of cut work' is the locus where the outside cylindrical surface of the workpiece is contacting the tool. At high cutting speeds, the rate controlling mechanism on wear is the chemical reaction forming the iron–alumina spinel.

The wear pattern produced on the tool under these conditions reflects the locus where the three species, Fe, O_2, and Al_2O_3, come into contact. This is along the flank and crater areas. However, at the 'depth of cut mark', the metal surface is already oxidized, and the reaction rate is not dependent upon counter diffusion along the tool/metal interface. The reaction can proceed more quickly, and a deeper wear groove results. Under certain machining conditions, cutting A-6 tool steel, 50 HRC, at 1 200 ft/min, the 'depth of cut' mark shows a very highly polished surface. The appearance of this surface suggests a wear mechanism which is not structure sensitive.

It is suggested that the mechanism is formation of the spinel, and plastic flow of these thin films under the extreme conditions of pressure, and at the modestly high temperatures. The locus of other wear marks on the crater and flank surfaces of the tool also suggest a similar mechanism. The counter diffusion of oxygen along the tool/metal interface places into contact the three species necessary for this reaction to occur. We therefore see a crater wear mark displaced back from the cutting edge, and the width of this wear mark is inversely proportional to cutting speed. At first glance the result is surprising, greater amounts of wear occur at slower cutting speeds! The wear mark reflects the locus of oxygen penetration along the interface. Since the oxygen is quickly reacting with the hot and molecularly clean metal surface, the depth of oxygen penetration along the interface is expected to be inversely proportional to cutting speed, and narrower craters are produced at higher speeds. The cutting edge itself is often relatively free from wear.

Mr Clark

I would like to make some comments on the type of wear known as 'spiking'. In cutting carbon steels at speeds below 500 ft/min a 'spike' is often observed at the edge of the normal flank wear. This spike reduces in size as the cutting speed is raised and is often completely absent at high cutting speeds. Reducing the feed rate also decreases the size of the spike. From experiments at PERA we have concluded that spiking is a phenomenon dependent on the presence of a built-up edge and an oxidizing atmosphere. The presence of a built-up edge protects the tool thereby reducing the flank wear. However, oxygen can penetrate the cutting zone at the edge of the chip causing local oxidation of the tool which cannot then support a built-up edge. Thus in this region the tool flank is no longer protected and increased wear causes the formation of a spike. Increasing the cutting speed or decreasing the feed tends to reduce the built-up edge and hence reduces the spike.

Dr Trent

I agree almost entirely with everything Dr King said. I think the ideas of oxygen penetration and of chip spread are not incompatible. That there is some chip spread is undeniable. It is not so obvious in hard steel but with pure metal it is very obvious indeed, and every practical machinist knows it when he works in pure copper or pure aluminium. Dr Milner and others have written a paper which deals with the formation of a 'collar' at the work surface.

I think that the excess wear on the tool at the point where the tool edge contacts the work surface is largely due to oxygen penetration at that point. If you look at Fig.4 in my paper, this illustrates the point. This was part of the rake surface of a carbide tool which had been used for cutting steel at quite high speeds. At the top is the cutting edge and the outer edge of the chip moved over the tool surface near the right hand side. Towards the left near the edge is a region in which there was complete seizure of the surfaces and there were no signs of any flow along that surface at all in that region. Further from the edge there are ridges and valleys showing the flow direction where the work material was in motion at the tool surface. Near to the right hand side the flow lines tend to diverge outwards, indicating some spread of the material there. Also, the flow lines come right up to the edge of the tool, indicating that the material was flowing over the tool surface at this position, although it was not flowing over the surface further to the left. This could be the result of oxygen penetration, preventing seizure at that point. Accelerated wear could occur where the conditions of seizure are prevented.

Dr D. H. Houseman (University of Technology, Loughborough)

I would first like to comment on a matter that has already been raised several times, namely, the interaction between academic and industrial people, and to draw attention to the fact that there is a vast amount of factual information available in the literature that is never used in industry. The metal-cutting industry is no exception to this. A recent survey that I made shows that some 15–20 papers relevant to this industry are published every week. So failure to make full use of the information is not the fault of the production people; if they read all these papers they would not have time to do anything else! What is needed, I believe, is not more experiments performed in academic laboratories, but for academics to work together with industrialists to bring to the latter's notice what information is available. This requires the establishment of mutual confidence since, all too often at present, the academic tends to see the industrialist as a person more interested in short-term profits than long-term viability: on the other hand, the academic often appears to industry as a person whose interest is primarily in getting money to support his investigations and, of course, neither of these pictures is wholly true. In the USA this collaboration is called 'creating a think-tank', and my belief is that there are far too few 'think-tanks' in this country at present.

Next, I would like to make a few comments based on information already available in the literature, on King's interesting paper on the wear processes that occur when using ceramic (alumina) tools for ferrous machining. One does not need a crystal ball to predict that production engineers will increasingly call for faster machining rates (implying higher temperatures) and that the bulk of this machining will be carried out in the normal, humid, workshop atmosphere. Under these conditions, the oxide-type tool must have a better resistance to the total environment than any possible metal or carbide formulation, so it is worthwhile to look at the potential of this type of tool.

King points out that, in ferrous machining, at high cutting speeds, the spinel iron aluminate ($FeO.Al_2O_3$, hercynite) will be an interaction product. He suggests that the rate-limiting process is for oxygen to diffuse along the interface to form the spinel. Molecular oxygen certainly comes in, but its diffusion is not the limiting process, since this oxygen diffusion will be, at the slowest, a grain-boundary process and, as such, is likely to be a couple of orders of magnitude faster than a solid state diffusional process. Furthermore, iron oxides are certain to be present both from the inherent oxide film and by production in situ as the work heats up ahead of the cutting process; a situation which King himself seems to appreciate, since he refers to 'pre-oxidized metal surfaces on the tool edge, supplying an abundance of the necessary species to form the spinel $FeO.Al_2O_3$'.

In the formation of a spinel from its constituent oxides in the solid state, the compound will be formed at the interface between the two oxides by a diffusion mechanism in which the two oxygen lattices serve as a 'reference framework' through which the smaller metal ions diffuse, from opposite directions, to form the spinel. The rate-controlling factor will be the diffusion rate of the slower of these two cations, in this case, probably Fe^{2+}. While the temperature conditions in cutting are not known, using King's suggested temperature range of 800–850°C, the rate of movement of the spinel interface into the alumina would be of the order of 1 μm in 10 min, which is adequate to form an appreciable layer on the tool face.

However, further consideration shows that such a spinel is likely to be short-lived under the conditions described, and perhaps this is just as well, since the appreciable density difference between the spinel and the alumina would lead to intolerable stresses at the interface. Also, in intermittent machining, the thermal stresses set up between the two phases in cooling from, say 850°C to room temperature, would be those resulting from a contraction difference of about 0·3% linear, about the same as those at the cobalt/tungsten carbide interfaces in a hardmetal tool. These are known to give rise to undesirably high interfacial stress concentrations, and this in a system having inherently higher ductility than the ceramic one.

What is almost certain to happen, in high-speed cutting conditions, is that the spinel will oxidize. Hercynite is known to be amongst the most easily oxidizable of all spinels; oxidation is appreciable at 450°C and by 850°C it is proceeding at a rate of the order of 1 μm/s, say two orders of magnitude faster than the spinel can itself be produced.

The reaction may be represented as:

$$2FeO.Al_2O_3 + \tfrac{1}{2}O_2 = Fe_2O_3 + 2Al_2O_3 \ \dots\dots\dots\dots\dots (1)$$

The products crystallize in two distinct phases, (a) alumina with a little ferric oxide in solid solution, and (b) ferric oxide with some alumina in solid solution. Providing reducing conditions do not afterwards obtain, these will be the stable end products in the conditions described.

From equation (1) (since the densities of the two oxides are similar) it can be seen that the ferric oxide phase occupies about $\tfrac{1}{3}$ of the total volume. Looking now at King's Fig.11, the top half shows dark areas which are said to indicate the penetration of the mounting plastic into cracks. These dark areas occupy almost exactly $\tfrac{1}{3}$ of the total area and, while such close agreement is probably just coincidence, this result is in harmony with a wear mechanism in which ferric oxide (itself a spinel structure, having the same number of easy glide planes and the same 'smearing tendency' that King attributes to hercynite) is torn out in heavy machining conditions, leaving solid-solution strengthened alumina behind.

Furthermore, as King himself points out, we should expect surface reorientations to occur, so that the directions of easiest plastic deformation are either parallel to the surface or along the resultant of the normal and the frictional forces. These reorientations will not occur as a result of temperature rise alone since, as is well known, alumina does not flow plastically below about 1250°C. However, they might well occur, under superimposed stresses of the order of 10^5 lb/in²

(King's figure) during the separation of the alumina and ferric oxide phases following oxidation. Thus, it is reasonable to look for orientations in which the basal plane of the alumina is parallel to the working surface, or at some constant angle to it.

Much more could be said about the details of this deformational wear process. My hope is that I have said just enough to illustrate my original point, namely that by taking advantage of what is already contained in the literature, we should be able to make postulates, capable of being tested, that will enable us to improve the formulation and performance of a given class of materials, in this particular case, to develop alumina into a really outstanding material in cutting applications. The extent of the information now available is such that this will prove a quicker and cheaper way than the trial-and-error system in common use today.

Mr H. Gardner (British Timkin)
My firm are users, and over the past ten years we have done a lot of experimentation in the production form, in the sense that all tool testing is done on production machines, to create an economic tool life, from the point of view of creating a standard tool change time.

With regard to the comments that have been made on tool wear, it is rather interesting to look at this interface welding or built-up edge, which is quite a problem.

The way we overcame this problem was through a series of tests, over about ten years, and we have ended up with the top rake of the tool always negative. We do not use any positive or even zero rake at all. I do not do any experiments to the extent that I take photographs of what happened. We only do the results of so many hours of work on the economics of the cutting edge of the tool. Generally speaking, the chip breaker is a ground-in chip breaker in a throwaway tool, and this phenomenon of wear is the surface that is getting the most load and the biggest amount of pressure, which causes breakdown of the tool.

We have run a system equating wear land with so many hours of the machine's operation. The tendency is if the chip breaker shape, length, and depth, is proportional to the flow of the chip, with a copious supply of coolant, the zone of load is more in one area than another. It seems to be that we are able to get a standard tool change time of 4 h. With presetting tool application we take one out and put another in every 4 h, knowing that the tool will probably run for 7 h, but we change it at 4 h just to make sure that we are not getting any machine-tool breakdowns, which may be more costly than taking the tip out on preset tools.

Dr Trent mentioned this galling up of this particular surface. I can only say that this point seems to take most of the load; this is about 10° negative, 30 thou radius, 19 thou deep, and somewhere in the region of 65 thou wide. Therefore this rather congested ground-in chip breaker is probably the reason for this area successfully taking the load.

Mr E. P. Riley (Midland Rollmakers Ltd)
I represent perhaps a different aspect of the utilization of carbide from most people. I am from the heavy engineering industry, and for years we have been applying carbides to our particular machining processes.

Over the last decade we have not gained any real improvement in machining from carbide, in the heavy machining industry. There have been no dramatic developments that allow one to say, 'I am going to improve my lot, I am going to get a better cutting speed and a better cutting feed than was applied 20 years ago'.

We have an entirely different problem. All our cutting is in the dry state. No lubricants are added, no coolants are added. Dr Trent has been down to our factory but he has probably forgotten this.

We in the heavy industries feel that liaison is lacking between the scientists at universities, the carbide manufacturers, and the users. We have no evidence anywhere of people being able to come to us and give us a better method of utilization of carbides. I have heard people talk about the built-up edge, the crater, the various aspects that virtually control the whole of this function. We have played about with all sorts of cutting angles, from positive to negative, and it might surprise you to know that we have certain materials where we go as high as 25–30° negative to produce the pieces we want. This is rather exceptional, I know. We talk about plastic flow from the tool point. We get this and we run at temperatures on the tip point of perhaps 900°C.

What is our problem? Our problem is tool breakage, tool wear, and continuous breakdown of cutting edge. We have tried ceramics, we tried abrasive methods of machining, and we come back to the old carbide, and I say 'old carbide' because this has been with us for a tremendous number of years without any basic improvements.

Therefore I have a problem, and although I listened very carefully to the scientific discussion, I want to know what it means to me, what I am going to get out of this, what improvements I am going to get? It would seem that we are getting very little. We have problems that are associated only with our particular trade. I want better tools, tools that will stand up longer and wear better and in effect reduce my costs, and so I go to the abrasive field to solve some of the problems, which we are doing reasonably well. Over the last three years, we have made trials on the shop floor, perhaps unscientific compared with the laboratories, but nevertheless trials which have given us information. We have not found a solution. We have invited the carbide manufacturers to come into the plant and work on the shop floor, and I think that this is the essence of the whole business. We have machining problems, we have vibration problems, of which you are well aware, associated with heavy machine tools, so when one runs carbide at £800/month through the factory one would welcome any attempt to help by university professors, scientists, or even the carbide manufacturers, and invite them to come down on the shop floor to try to solve the problems, because they are real problems to us.

I have gone through the whole range of positive, negative, zero rakes and approach angles and, in 20 years, we have made no dramatic improvement at all. It is in your hands as well as ours. The facilities are there. We want your assistance in a more practical form, more practical than the form we have been given today. Any comments from any of you will be very welcome.

Dr King
I could not be in more sympathy with the last speaker. I think the industry is guilty of a great deal of misdirection of priorities. The advances in cutting-tool materials, the advances in machining and stock removal rates, the advances in economics of machining, are very much tied in with cutting-tool materials and any machine tool is limited to what the cutting tool can withstand.

Of the total amount of money spent on machining research, only a small amount is directed towards the cutting tools themselves, and this is almost exclusively done by carbide manufacturers. Very little has been done with ceramics which, thermodynamically, offer the best prospect for the future. A lot of the work on carbides is in making small modifications of existing systems, without branching out broadly and looking for the new types of materials that this speaker is needing on the shop floor.

I think we should, both in the USA and throughout the world, review our assessment of priorities of what we are doing, and I personally would advocate a very large increase in materials research to impart to these tool materials those properties necessary to do the job.

In particular, we need ceramic materials with improved fatigue strength, and improved tensile strength at elevated temperatures. When we develop these materials, then we can help to solve the multiplicity of problems on the shop floor.

The Chairman

The point that everyone in the discussion has made is the problem of integrating materials aspects with the machining theory aspects. Something one would hope for from the theoretical papers are pointers as to the direction in which materials development should go. Would anyone care to comment on the organizational aspects of this?

My impression during the discussion is that one gets isolated contributions from people who are looking at part of the problem. There are very few organizations looking at the situation as a whole from the users' point of view. Would anyone care to pursue this difficulty from the point of view of their own organization and what they think needs to be done nationally, or elsewhere to improve the situation? Would anyone disagree with the point Mr Riley made? He claims he is getting no help except in a marginal sense.

Mr Clark

I would like to suggest a means of helping Mr E. P. Riley who complained about lack of progress in tool materials as applied to his problem of turning rolls. A solution may lie in assisting the tool by using a hot-machining technique.

Various methods of hot machining have been tried in the past, but PERA has recently developed a technique using a plasma arc which enables much higher cutting speeds to be used. For instance, whereas the cutting speed for Nimonic 115 using carbides is conventionally 30 ft/min, with our technique, speeds of 500 ft/min can be achieved with satisfactory tool life. Similarly, with a high-manganese steel, the cutting speed can be increased from 80 to 500 ft/min. Moreover, interrupted cuts can be accommodated.

There is no doubt that companies can benefit from new developments such as the one described, providing that when a specific problem arises it is drawn to the attention of an organization such as PERA.

The Chairman

Essentially, in discussing mechanism of failure of tools, we are talking about analyses of systems as they are. One of the things that could usefully be discussed is whether one should look at these fine points of a highly developed system with only marginal scope for improvements, or whether really major changes in the system, of the kind just described, are needed if the technology is going to improve. Can we predict from the analyses of the mechanisms of failure as they have been presented, whether there is much scope for improvement, or whether we are in a marginal situation already? Is the scientific approach trying to catch up with the state-of-the-art, or can it really improve the art significantly at this point in time? Someone must want to comment on this because, after all, it underlies the whole philosophy of the research which has been reported here today.

Dr R. L. Craik (BISRA, the Corporate Laboratories)

I would like to draw attention to development of new tool materials. I think in this country we adopt the art of super-rogation which means 'the Lord helps those who help themselves'. The only person to whom we can pray is the Ministry of Technology, and in answer to the Chairman, I hope the Ministry of Technology's representative might have a few words to say, because he is supposed to have funding for development for national needs.

To the academic members I would say that they have to face the consequences, because it seems to me that Dr Ridal identified one very important issue, the question of interfaces. But really, is it interfaces? We are starting off with surfaces in sliding contact and we only see the consequence, not necessarily the mechanism of interaction.

I would like to point out one or two experiments on wear of surfaces in sliding contact which have been done in the engineering field. If one puts two dissimilar metals in rubbing contact, taking ordinary mild steel, the consequences are that

you can develop reaction products, debris, and the debris will be finely divided. It is this initiation mechanism we have to look at if we are to make significant improvement in the whole development of tool steels. It may well be that the initiation mechanism is such that you can do very little about it, but I think something is beginning to emerge from Dr King's philosophy. If the reaction is chemical and the thermodynamic products are oxides, you will start off with oxides, and here is an interesting line of development.

I want to make a small point to the academics about substitute tool steels. I would like to ask Professor König with regard to the build up of this built-up edge, in which he gets quite high hardnesses and shows that flank wear reduces because of this built-up edge, etc., and he goes on to spark alloying. Why must it be a tool-steel base? Can it not be any other material? Can he not use a cheap steel base and use spark alloying? Have any experiments been carried out in this direction? This would be a point in substituting tool steels which in themselves are very expensive.

I would like to ask Dr Trent a question with regard to seizure. He talks about metal build up, which he says depends on the rheological character of metal being machined. Is it the rheological character of the debris or the rheological character of the metal he is cutting?

A thing which seems not to have been stressed is the question of the lubricant which is supposed to cut down or improve the area of contact in which the heat arises. Are there any forms of improved heat removal during cutting operations? We have not explored many of the things we should be looking at.

I would like to finish by saying that this question of materials research and development seems in this country to boil down to one issue, how much one is prepared to invest. Various approaches have been made throughout the cutting industry in relation to tool materials. The Ministry of Technology have two types of money because they have two forms of contract, defence and civil, and it is the civil area at which we have to look, where, for every pound one puts in the Ministry are prepared to put in a pound, and it is up to yourselves to become organized if this finance from national resources is to be obtained.

Professor König

With regard to the spark alloying of high-speed steel tools, in the last two years we have obtained a lot of information on spark alloying from Russian literature. They have reported good improvements in tool life. We therefore decided to investigate this and get more details on it. In doing so, we find – and we have shown you some results – that there is a possible improvement in the tool life of high-speed cutting tools, but at the moment we are far from knowing the mechanisms and the possible advantages of such a procedure. I cannot answer your question as to whether it is possible to use this method for less expensive materials, such as mild steels, because we do not know the process yet. But at the moment it seems to be less successful.

The Chairman

What about plasma spraying, for example, with titanium carbide?

Professor König

We used plasma spray but did not succeed with it. However, a new development for carbide tool tips is the coating technique. Layers of about 5 μm, consisting of TiC or TiN, are produced from the vapour phase. Depending on the cutting conditions, the improvement in production rate using these 'coated tips' is remarkable.

Besides this I would like to point out that not only are improvements in tool life possible by improving some properties of the tool material, but it is also possible by altering the steel. I would like to draw attention to the fact that the

The effect of temperature (higher sliding or cutting speeds) influences the wear in two different directions. In the first case we may obtain increased wear due to the fact that at sufficiently high temperatures the dominating wear mechanism will be plastic deformation and diffusion processes between the worked material and the tool (cratering). Oxidation and thermal fatigue processes also increase the tool wear. On the other hand, in certain systems thin oxide layers or high-temperature refractory coatings such as TiC in Secotic grades (*cf. Metalworking Production*, 1970, 15 April, 52) decrease the wear rate as the diffusion processes are hindered and the surface hardness of the tool material is increased. Wear-decreasing effects are also obtained with thin metallic layers or lubricating films in the surface contacts.

The wear-decreasing mechanisms due to the thin TiC coatings on cemented carbides are discussed in the paper presented by Dr P. O. Snell in the symposium on wear mechanisms mentioned above. The most prominent features are the decrease of temperature by 150°C between the contact surfaces due to the decrease of the frictional forces, the blocking of diffusion processes between the tool and the worked material, and plastic flow of the TiC layer at working temperatures (which in the case investigated here was 950°C).

The general description of wear processes given here is naturally a very simplified one. It gives a certain frame to the discussions, at the same time indicating the really fundamental processes involved in wear mechanisms. Much more work is of course needed before we can state any more quantitative laws in this area.

From the users' point of view some criticism was directed to the producers of cemented carbides, i.e. that no real advantages have been achieved during the last 20 years. This is a slight exaggeration. We certainly have made considerable advances. This includes the development of new cutting grades, and a quantitative knowledge of the factors which determine the properties of the cemented carbides. Our paper is one of the examples of that work as it gives one of the last hitherto quantitatively immeasurable quantities needed to allow the prediction of the cemented-carbide properties. Secotic alloys and other TiC coated alloys mentioned by Mr Persson during the discussions on this conference are other examples of the advances made possible by research work during recent years with a 3 to 5 fold increase in tool life. In percussive rock drilling, considerable increase in tool life has been made possible due to extensive research work on carbide microstructure, as has been shown in Fig.1 of our paper.

It seems to us that the problems on the user's side often seem to be a question of proper information. Despite the fact that we have made considerable efforts on this area, it is evident from the discussions during this conference that this work is either insufficient or that it is not utilized properly. Certainly we producers must look at this side of our work more carefully and increase our efforts in this area.

High-speed steel technology: the manufacturers' viewpoint

F. A. Kirk, H. C. Child, E. M. Lowe, and T. J. Wilkins

COMPOSITION

The number of available high-speed steel compositions has increased considerably during recent years largely due to widespread efforts to keep pace with demands for increased cutting performance both in the use of high-output automatic machines and in the machining of high-strength and complex alloys. The current situation is well illustrated by the 26 AISI grades listed in Table 1.

As may be noted, the tungsten and molybdenum groups of steel follow a somewhat similar alloying pattern. Cobalt has been introduced to certain steels to improve cutting efficiency at high speeds, while in other grades the carbon and vanadium levels are increased to provide greater wear resistance. The more recent developments listed under the heading 'super high-speed steels', represents various approaches to the use of complex alloying in basically molybdenum types to effect improvements in all-round durability. The high carbon contents of the M40 series reflect attempts to achieve a near stoichiometric balance between carbon and major carbide-forming alloy content[1] which, aided by the effects of cobalt in delaying precipitation during secondary-hardening treatment, enables hardness values up to 70HRC to be obtained.

Many of the available grades are in effect relatively minor variations about a common base, and the problem of steel selection, has, therefore, become increasingly perplexing for the majority of users. At the same time, the demand for numerous different types, often in small quantities, has presented both technical and economic difficulties to the manufacturers. British high-speed steel producers have, therefore, made serious attempts during recent years to rationalize by concentrating on those grades in regular production which seem likely to satisfy the widest range of applications. A selected list of 12 such grades grouped according to usual area of duty is given in Table 2, where typical heat-treatment and working-hardness data are also presented. An AISI symbol is used where appropriate and the proposed British Standard designation[2] is also shown. Of the 15 high-speed steels listed in the draft standard, 11 have been selected plus a free machining grade which is provided for as an optional requirement in the standard.

PROPERTIES

Even a reduced list of steels presents problems in selection and further difficulties arise when the claims of newer compositions, which will be considered later, need to be taken into account for a specific duty.

It may be considered that cutting tests should provide the best guide as to what may be expected in service, but experience has revealed great difficulty in obtaining meaningful results, especially when greatly accelerated tests are employed. The large number of variables which are met in both the tool and the workpiece, also the difficulties which exist in accurately determining the end point of the tests, generally give rise to a considerable amount of scatter. In addition, the differences found between grades are relatively small, seldom exceeding 10% on the basis of useful cutting speeds. Trials under actual service conditions are normally subject to even wider variations, and unless very great differences in performance are involved quite extensive experience is required to reach any useful conclusion. In these circumstances, therefore, it is appropriate to examine the relative properties of the grades of high-speed steel concerned for guidance as to their potential cutting performance.[3]

The balance of properties required for a particular duty will, of course, vary and while other factors no doubt play a part, it is felt that an assessment of wear resistance, red hardness, and toughness will be useful. It is by no means suggested that laboratory tests can replace service experience in a specific application which must always be the final criterion, but that helpful guide lines to selection can be obtained.

With regard to wear resistance, no laboratory test is considered to be available that would provide data having sufficient relevance to the wide range of conditions met with in service. Our assessment of wear resistance is, therefore, based on room-temperature hardness in the hardened and tempered condition. Since the types of carbide present in a particular grade have an influence on wear resistance, and taking into account practical experience, it is felt that some allowance must be made for the beneficial effect of the large amount of extremely hard vanadium carbide present in many modern compositions. This factor is noted when the data is presented in a comparative manner.

The second major property, that of red hardness, is of particular significance in view of the very considerable heat generated during metal-cutting operations. We have taken the residual room-temperature hardness after tempering at 650°C as a basis for comparison, in consideration of the likely significance of relative rate of overaging of various grades of steel, also the very real effects on the cutting edge of such exposure when repetitive operations are involved. Hot hardness (hardness at temperature) has been considered as a

Heat-treatment and property data for a selected list of British high-speed steels are offered as a guide to suitability for various kinds of duty and the likely role of certain newer compositions is discussed. The effects of manufacturing variables on such major quality aspects as carbide distribution and decarburization are described. The VDE method of carbide assessment is discussed relative to bars up to 3in diameter. Drill test data suggests that finer carbide distribution does not necessarily result in superior performance. Manufacturing considerations impose limits on carbide breakdown in bars over 3in diameter but special techniques are discussed which may be used to provide material suitable for unground form gear hobs. Close control over hob manufacture is essential for optimum dimensional stability. Developments in the manufacture of high-speed steel sheets of superior flatness and with reduced surface decarburization are outlined. The quality of high-speed steel may be improved by the use of electroslag refining (ESR) but economic considerations are the main limiting factor as regards the wider use of the process to provide greater product consistency.

669.14.018.252.3 : 539.4/.5 + 621.785

Mr Kirk is with Osborn Steels Ltd, Mr Child is with Jessop-Saville Ltd, Mr Lowe is with Firth-Brown Ltd, and Mr Wilkins is with BSC, Special Steels Division

1000 1050 1100 1150 1200 1250 1300   500 550   600   650   700
AUSTENITIZING TEMPERATURE , °C   TEMPERING TEMPERATURE , °C

a effect of austenitizing temperature double temper 560°C; *b* effect of tempering temperature, austenitized 1230°C

3 Properties of M2 high-speed steel

Table 1   AISI high-speed steels

Grade	C	Cr	V	W	Mo	Co
Tungsten series						

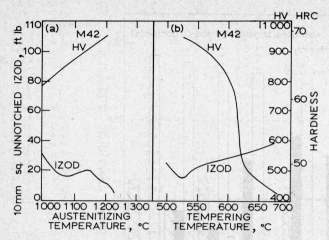

a effect of austenitizing temperature, tempered 530°C × 3; *b* effect of tempering temperature, austenitized 1 185°C

4   Properties of M42 high-speed steel

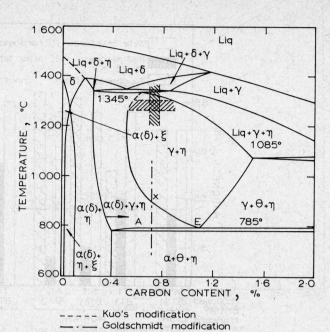

- - - - - Kuo's modification
— · — Goldschmidt modification
▨ Normal heat-treatment range
▧ Normal composition range

5   Constitutional diagram for high-speed steel; binary section through quaternary Fe–W–Cr–C system at 18%W and 4%Cr as published by Murakami and Hatta[8] and as modified by Kuo[9] and Goldschmidt[10] respectively (courtesy of the ISI)

finding some favour at the present time. Of these three steels the chart (Fig.2) shows a preference for T1 where red hardness is important, e.g. toolbits, but since in the majority of applications, wear resistance and toughness such as to allow retention of a keen cutting edge are more important, M2 is favoured by the authors as a standard grade of high-speed steel for normal use. The validity of using M2 rather than T1 as twist drills is illustrated by the cutting test data presented in Table 5. A sulphurized, free-machining grade of 6–5–2, M2S, is produced for the manufacture of unground form tools such as gear hobs where a smooth tool finish surface is important.

Newer materials currently being considered for use in this potentially wide area of application include M2H, a 1%C version of 6–5–2[7] in which improvement of red hardness to the level of T1, also some increase in wear resistance (about 1–2 points HRC) are attractive features. M10, which is a 2%V version of M1 and its 1%C variety, M7, (*see* Table 1) which grade offers similar benefits to the high-carbon M2, are also receiving attention. There would appear to be a case for further work on M2H and M7 grades to determine their cutting potential as compared with M2.

Notably the 'fast cutting' steels are all cobalt-bearing tungsten grades in which high red hardness is the predominant property. T4 appears to justify its place by offering improved performance at a moderate increase in cost, but there is a good case for rationalization in the case of T5 and T6 which are so similar as to warrant the consideration of a single steel to the leaner composition.

Such newer materials as M33 (*see* Table 1) and also those of the hard cutting group discussed below are considered as potential replacements for T4, 5, and 6, and may do so when tougher workpiece materials are involved, but where rapid

continuous cutting is the major criterion the tungsten–cobalt grades are difficult to replace.

The four steels listed under the heading 'hard cutting' were introduced primarily to deal with the increasingly tough high-strength and strongly work-hardening steels and super-alloy materials employed in the aircraft and missile field. In the case of T15, M15, and 9–3–3 a high content of vanadium carbide combined with high working hardness levels have provided an edge-retaining advantage in a number of applications such as intricately shaped broaches and form tools.

The most recent introduction for this kind of work is M42 which develops hardness up to 70 HRC, but which contains considerably less vanadium carbide than the other steels and, therefore, is very much easier to grind. Consequently, tool manufacturing costs are often reduced and there is less risk of damage to cutting edges due to heat developed in grinding, providing normal care is taken. On the other hand, since M42 depends essentially upon hardness for its abrasion resistance and since as a high molybdenum–cobalt grade it is rather susceptible to decarburization, special care is required during heat treatment to ensure that the best results are consistently obtained. M42 can, however, be heat treated

Table 5   Comparative drill tests using T1 and M2 high-speed steels

Test stage	Speed, rev/min	Feed, in/min	Drill nos, T1 high-speed steel						Drill nos, M2 high-speed steel					
			1	2	3	4	5	6	1	2	3	4	5	6
1	650	8·8	18	18	18	18	18	18	18	18	18	18	18	18
2	650	10·6	18	18	18	18	18	18	18	18	18	18	18	18
3	650	12·3	18	18	18	18	18	18	18	18	18	18	18	18
4	785	14·1	18	18	18	18	18	18	18	18	18	18	18	18
5	830	15·8	18	18	18	18	18	18	18	18	18	18	18	18
6	870	17·6	18	18	18	18	18	18	18	18	18	18	18	18
7	960	19·4	18	18	18	15	18	18	18	18	18	18	18	18
8	1 050	21·1	55	40	11	..	13	3	100	3	99	11	9	21
9	1 050	22·9	..	..	..	..	..	..	11	..	..	..	..	..
Total holes:			181	166	137	123	139	129	237	129	225	137	135	147
Average:			146						157					

Drills $\frac{7}{16}$ in diameter, manufactured from $\frac{9}{16}$ in diameter bars; blind holes $1\frac{7}{32}$ in deep drilled in standard test material (BS328); coolant 'Dixol' soluble oil in water (1:10)

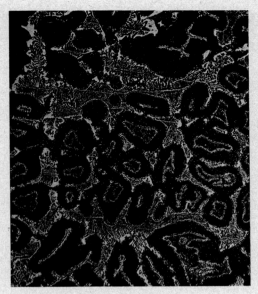

6 As-cast microstructure of T1 high-speed steel ×100

quite readily to a wide variety of conditions as the heat treatment–property chart in Fig.4 shows, and this versatility is an attractive feature.

Newer contenders for this type of duty are M3 type 2 (as for M3 in Table 1, but with 1·15%C and 3%V) which offers a reasonably economic approach to the cutting of harder materials, also M46 (1·25C–4·0Cr–3·2V–2·0W– 8·25Mo–8·25Co) which with a higher carbon and vanadium content than M42 and hardness up to about 68 HRC suggests a further advance at the high-duty end of the range.

MANUFACTURE OF HIGH-SPEED STEELS

In recent years, developments in manufacturing techniques have been mainly aimed at minimizing the segregation of carbide which inevitably occurs during the solidification of high-speed steels. Before discussing methods of manufacture in detail, it would therefore be appropriate at this point to briefly describe the solidification and as-cast structure of high-speed steel by referring to the constitutional diagram for 18–4–1 (T1) steel shown in Fig.5.

The first phase to appear on solidification is ferrite which precipitates from the melt in the form of dendrites. The ferrite then reacts with liquid to form austenite which precipitates around the existing ferrite dendrites. As the temperature falls further a peritectic reaction occurs where carbide and austenite are precipitated from the melt and ferrite transforms to austenite. In practice, equilibrium conditions are not achieved and therefore the peritectic reaction rarely if ever goes to completion and the remaining liquid decomposes by a eutectic reaction to give austenite and carbide. Finally, the ferrite formed decomposes by a eutectoid reaction to give austenite and carbide. These various reactions can be summarized as follows:

   (i) liquid+ferrite=austenite
   (ii) liquid+ferrite=austenite+carbide (peritectic reaction)
   (iii) liquid=austenite+carbide (eutectic reaction)
   (iv) ferrite=austenite+carbide (eutectoid reaction).

The resulting as-cast structure is illustrated in Fig.6 which shows how the carbide eutectic forms networks around the grains and clusters at their boundaries. As the result of uneven cooling in the ingot, the size of the carbide networks tends to increase towards the centre, and in addition pronounced segregation is likely to occur from the last metal to solidify in the top central zone of the ingot. The distribution and form of the carbide eutectic networks can only be modified by amended casting techniques or by hot working after

solidification and in the descriptions of methods of manufacture which follow, references will therefore be made to the techniques commonly used to modify carbide segregation.

Melting

High-speed steels were originally melted by the crucible process but this has now been superseded by the high-frequency and electric-arc melting processes. A further melting process, which has received increasing attention during the last 3 or 4 years, is electroslag refining and, in view of the future potential and important advantages of this process, it will be discussed in detail separately.

As far as conventional air-melting practices are concerned, there is little to choose between high-frequency and electric-arc melting in terms of final quality, and the choice is mainly dependent on availability with electric-arc melting being usually preferred for the production of large ingots. With both processes, tungsten and molybdenum are normally added as low-carbon ferroalloys in order to facilitate solution in the melt, but in the case of chromium, high-carbon ferroalloy is used, the higher carbon limiting oxidation prior to solution. In addition to the ferroalloys, quantities of scrap are used and careful control of quality is necessary to minimize tramp elements, in particular nickel. The early addition of tungsten and the use of high melting temperatures help towards attaining a homogeneous melt.

Casting

Over the last ten years or so, particular attention has been paid to the possibilities of controlling casting conditions to suppress dendrite formation, refine primary grain size, and minimize carbide segregation. In the literature, there exists a great variety of opinions as to methods of controlling solidification, and references have been made to casting temperature, inoculation, the size and shape of ingot moulds, water-cooled and heated moulds, moulds with copper bottoms, shaking, stirring by magnetic methods, ultrasonic vibration, etc. However, many of these methods have not proved to be consistent in practice and there are also technical and economic considerations which have limited their use. Further discussions will therefore be confined to those methods which are most widely used in practice.

Control of the casting temperature to a minimum commensurate with satisfactory cleanness is commonly used to promote rapid solidification rates which tend to reduce segregation. This is contrary to certain statements in the literature claiming that carbide segregation can be suppressed by slow cooling but the fact is that unless extremely slow rates of cooling are obtained, as in the Ferrinca process,[11] then faster cooling gives the better structure. Very slow rates of cooling are achieved in the Ferrinca process by the use of exothermic moulds and although a more uniform grain size and less continuous carbide network can be obtained compared to conventionally cast ingots, the grain size is considerably coarser and the ingots have proved to be difficult to forge.

Major improvements in ingot structure can be obtained by careful attention to mould design and the use of short squat moulds as outlined by Field[12] is certainly effective in minimizing the larger segregates which tend to occur in the top central zones of the ingot. A mould with a height to width ratio of less than about 3:1 will yield an ingot which is acceptable to most consumers as regards central segregation over most of its chill length.

In addition to short stubby moulds, Field[12] also recommends the use of inoculation to improve the cast structures of high-speed steels, but he does not give details of the type of inoculant used or the method of adding it to the melt. However, there are many other references in the literature which do give details of inoculation techniques and, as evidenced by some recent papers[13–15] further development work is still being carried out. While Hoyle[16] is of the opinion

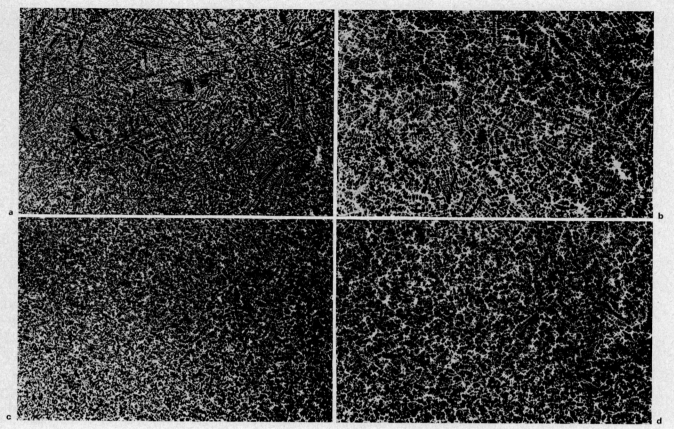

a outside, and *b* mid-radial, normal material; *c* outside, and *d* mid-radial, inoculated material

7   Effect of inoculation on ingot structure of high-speed steel   ×8

that there are sufficient nucleation centres in the melt to make any further additions superfluous, inoculation techniques are capable of reducing dendritic segregation and minimizing central segregation (*see* Fig.7) and they are therefore quite widely used. The inoculants used for high-speed steels are many and include the oxides, nitrides, and carbides of elements such as Al, Mg, Ti, Ca, Ce, Zr, V, etc., which are either used singly or in various combinations. Many of these inoculants produce harmful side effects such as inclusions and ingot-surface problems and reproducible results are often difficult to achieve without very careful control.

Hot working
Having obtained a satisfactory as-cast structure, the next stage in manufacture is the hot working of the ingot to bar or to other forms suitable for tool manufacture. In general, conventional methods of forging and rolling are used but with the larger ingots now being produced, initial forging is best carried out on rapid-acting presses equipped with double manipulators. Such equipment enables forging to be carried out quickly with few reheating operations and this helps to improve billet yields by (*a*) minimizing surface decarburization and (*b*) allowing reduction to be carried out within a relatively narrow temperature range, thus avoiding the extremes of the normal hot-working range where surface cracking problems are likely to occur. In this context, mention should be made of the work currently in hand at BISRA which is concerned with the use of infrared heaters to provide heat to the stock while it is being forged. This approach is intended to eliminate reheating operations completely and maintain the stock at a uniform temperature throughout the forging operation. Details of this technique and its considerable potential for use in the working of high-speed steels, are given in one of the papers presented at this conference.[17]

Apart from small ingots (i.e. less than 6in square) which

can be handled on bar mills, the initial hot working of high-speed steels is usually by forging and not by rolling. There are various reasons for this such as the fact that reheating operations are normally required which slow down production rates and the fact that the manipulating equipment on cogging mills is not always capable of handling the short ingots which are used to achieve satisfactory central segregation. However, the roll cogging of quite large ingots is practised, particularly with the more workable M1 and M2 grades, and it is claimed that with proper attention to roll pass design, ingot preparation, heating cycles, etc., rolling can be accomplished without reheating.[18]

Control of surface decarburization to the low levels now demanded by tool-steel users has presented, and still does present, problems during hot working and one approach is to reduce the number of reheating cycles as discussed earlier. Clearly, the time required for all reheating operations must be minimized and a process of gradual attrition has persuaded forges to reduce ingot heating times from several days in an atmosphere of flame and smoke to a few hours in atmospheres more amenable to reasonable control. While advances in furnace design and mechanical handling can bring further improvement, it seems that, for the present, a decarburized billet surface must be accepted. Various proprietary paints and coatings (usually refractory base) have been tried in an attempt to reduce scaling and decarburization but although some of these have shown promise on laboratory assessments they have not been very successful in practice as they proved susceptible to local breakdown and this has given rise to the problem of decarburization occurring in patches. At a relatively early stage in decarburization investigations, the detrimental effects of uneven billet grinding in leaving patches of decarburization were recognized. Figure 8 shows the effect which such practice can have upon the final rolled bar, no matter how careful the billet reheating practice. Means must,

8 Patchy decarburization on billet and bar

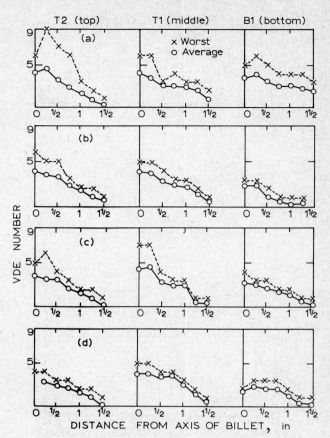

a T1 standard mould; b T1 short high-taper mould; c M2 standard mould; d M2 short high-taper mould

9 Variation in carbide grading in 3in billets from 8in ingots (tapping temperatures 1530–1540°C)

therefore, be found of even and economic removal of metal from billet surfaces and, as a result, the use of automatic grinding machines has been widely adopted. Reheating for rolling must obviously be considered and here there is excellent potential for gas-fired rapid heating which has been studied in various forms in the UK by the Gas Council, as well as by BISRA, in conjunction with Shell-Mex and BP Ltd and Birlec Ltd.[19] The savings in expensive materials which are possible by a reduction in metallic losses due to scaling and grinding to eliminate decarburization may well offset the cost of the more sophisticated heating equipment required, providing application can be made on a large enough scale. This is a process development which will clearly receive further attention in the future. Having taken precautions during hot working to minimize decarburization, it is important that these are not cancelled out by lack of proper attention to final annealing practice. Particular care is needed with small sections such as sheet and drill rod and these often warrant the use of controlled atmosphere furnaces of the bright annealing type with associated descaling facilities.

Turning now to the influence of hot-working practice on carbide segregation, this is discussed in two parts. The first deals with small section sizes (up to 3in diameter) and includes comments on the problem of assessing carbide segregation plus evidence which indicates that for certain applications quite wide variations in carbide segregation have no significant effect on cutting performance. The second part covers large section sizes (4–12in diameter) and is mainly concerned with the use of modified forging techniques designed to give optimum carbide distribution, especially in material used for the manufacture of unground hobs.

Quality problem in sizes up to 3in

As discussed earlier, after limited forging reduction, high-speed steel may still show evidence of the original carbide eutectic network. In products under 3in diameter, all customers demand freedom from such structures. Specifications call for complete absence of carbide network on a microscale at ×100 and of macrostreaks resulting from the eutectic.

From the suppliers' point of view this presents no problem, as ingots with a satisfactory distribution of carbide eutectic may be cast and given adequate reduction, generally accepted to be >94%, to be free from both network (at ×100) and streaks (×1).

Even after very high reductions from the ingot (>98%) the banding of carbides still persists on a microscale at say ×100. For some applications there is no evidence that banding has any effect on either the manufacture or the performance of the tool. This is the case with toolbits and butt-welded tools and it is not normal practice to specify any particular degree of banding for the cobalt grades generally used for such tools.

Manufacturers of drills, cutters, thread chasers, and cold-extrusion tooling invariably impose some standard for carbide banding. These applications have in common the need for the tool to withstand tensile stresses or impact loading either during manufacture or service. Severe carbide banding might be expected to reduce the toughness characteristics of high-speed steel and so be detrimental for such applications. The manufacturers have therefore imposed 'house specifications' based on either longitudinal or transverse examination, but the UK has still to set up a national standard for carbide assessment. This deficiency requires action and is now receiving consideration by the appropriate committee of the British Standards Institution.

The best existing method of assessing carbide segregation was developed in Germany,[20] and essentially measures the overall width of the carbide banding at a standard magnification of ×100 and distinguishes between light and dense banding. Only stringers whose length is at least four times their thickness are considered and if the distance between adjacent stringers is not greater than the thickness of the smaller stringer, the pair are evaluated as one stringer. The resulting VDE number increases with increased band width.

This method has been applied by users mainly by specifying a maximum acceptable grading. It is customary to examine the entire cross-section of a longitudinal sample, but at any given radius the number of fields examined is small. The problems created by this approach are shown by studying the carbide rating for the whole of the product of one ingot, some typical examples being shown in Fig.9.

In each case the highest VDE rating found by examining 20 fields at the indicated position is compared with the mean VDE rating of the 20 fields. The examples shown are typical of a study made of the product of 8in ingots of varying geometry, cast under systematically varied conditions of casting temperature and nozzle size in both T1 and M2 grades. The noteworthy features are the difference between the worst and

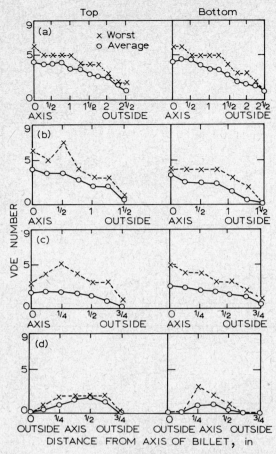

a 5in square billet; *b* 3in square billet; *c* 1⅜in square billet;
d ¾in diameter bar

10   Carbide grading at various stages in reduction of a 14½in ingot of T1 high-speed steel

the mean VDE ratings and the somewhat random occurrence of the worst fields. Quality control based on the worst carbide rating found in the limited number of fields normally examined is therefore likely to be unreliable.

The mean VDE rating was surprisingly consistent throughout the product, particularly for the M2 quality, even when cast in standard moulds under varying casting conditions. At the outside of the bars the worst ratings were low (< 1 for M2).

It is also interesting to note the effect on the VDE rating of increasing forging reductions carried out on a bar of T1-type high-speed steel made from a 14½in square ingot (Fig. 10). Although, as would be anticipated, the average carbide quality is progressively improved, e.g. decreasing from 4½–1½ at the centre of the product, fluctuations in the worst rating were found and even at 1⅜in square (99% reduction) the worst fields were grade 5 and no better than those at 5in square (88% reduction).

It is noteworthy that VDE ratings are meaningless if the structure still contains remains of the eutectic network. This is normal on commercial materials above about 4in diameter and the 5in square billet referred to in Fig.10 did in fact contain network.

From the steelmakers' point of view, effort should therefore be directed at avoiding the occurrence of those random areas of high VDE rating, and as shown in the examples in Fig.9 the use of M2 instead of T1, and the use of short ingots with a low length to width ratio and with high tapers of at least 1in/ft helps.

From the point of view of specifying carbide banding, the user must therefore exercise caution in specifying only worst field VDE gradings. To be meaningful, very extensive

Table 6   Single severity tests on ½in drills of M2

Drill	No. of holes to failure	Mode of failure	Hardness, HV, on body clearance 30 kg			
T2	27	Corners worn	842,	876	780,	775
T3*	2	Corners worn	869,	876	883,	876
T4	104	Corners worn	823,	862	792,	810
T5	8	Corners worn	856,	862	876,	869
T7	16	Corners and lips worn	869,	862	883,	876
T8	20	Corners worn	836,	856	817,	856
T9	8	Corners worn	836,	869	792,	842
T10	55	Corners worn	876,	890	856,	856
T11*	24	Corners worn	849,	849	810,	842
T12	70	Corners worn	869,	883	862,	856
T13	59	Corners worn, lips chipped	829,	849	757,	804
T14	23	Corners worn, lips chipped	862,	862	883,	883
T15	17	Corners worn	817,	842	836,	862
T16	18	Corners worn, lips chipped	836,	810	849,	862
B1	20	Corners worn	804,	842	780,	792
B2	103	Corners worn, lips chipped	869,	869	862,	876
B3	21	Corners and lips worn	817,	842	849,	856
B5	18	Corners worn	823,	836	786,	829
B6	23	Corners worn	829,	836	836,	849
B7	116	Corners worn	862,	862	869,	869
B8	6	Corners worn	849,	862	849,	862
B10	13	Corners worn	837,	842	804,	817
B11	14	Corners worn	798,	810	849,	869
B12	6	Corners worn	780,	856	856,	856
B13	13	Corners worn	836,	823	798,	823
B14	76	Corners worn	862,	856	876,	862
B15	20	Corners worn	849,	842	810,	842
B16	33	Corners worn	836,	856	876,	862

Drills preconditioned over 10 holes at 9.6 in/min penetration; tests to failure at 13.1in/min penetration; mean life for T drills 32.2 holes, VDE, rating 5d; mean life for B drills 34.4 holes VDE, rating 2L
* These two drills chipped on the lips during the conditioning run, and were therefore repointed and reconditioned

sampling is essential, and a high cost will result from both testing and the process route required.

Unfortunately macroexamination, which permits much more representative sampling to be readily carried out, cannot detect even severe carbide banding equivalent to VDE6. The specifications imposed by drill manufacturers and other demanding users normally call for carbide banding less than VDE3 for sizes below 2in and from VDE3–4 for sizes between 2 and 3in and such quality requirements dictate the use of low-yield high-taper short ingots which cost more to process than conventional ingots.

It is pertinent at this stage to consider to what extent the users demand for minimum carbide banding is based on evidence that carbide banding affects service performance. Although there is a widespread belief among the users that this is the case, documented evidence is rare and does not always support the belief.

A BISRA collaborative committee[21] studied the performance of ½in drills made from ⅝in diameter bars of M2 produced by normal commercial practice. Both single severity level (Table 6) and multiple severity level (Table 7) tests were carried out. The drills designated T were produced from the top of a 3¾in square ingot and had a VDE rating of 5d. Those designated B were from the bottom of a 9in square ingot and had a VDE rating of 2L.

Table 7   Multiple severity level tests on ½in drills of M2

Level no:	1	2	3
Rev/min:	541	541	541
Penetration:	8.6in/min	10.6in/min	13.1in/min
B1	OK 10 holes	OK 10 holes	Failed 1 hole
B5	Failed 7 holes		
B6	OK 10 holes	OK 10 holes	Failed 4 holes
B7	OK 10 holes	OK 10 holes	Failed 2 holes
B12	OK 10 holes	OK 10 holes	Failed 3 holes
B13	OK 10 holes	OK 10 holes	Failed 1 hole
B14	OK 10 holes	OK 10 holes	Failed 2 holes
T2	OK 10 holes	OK 10 holes	Failed 3 holes
T4	OK 10 holes	OK 10 holes	Failed 2 holes
T5	OK 10 holes	OK 10 holes	Failed 2 holes
T7	OK 10 holes	OK 10 holes	Failed 2 holes
T13	OK 10 holes	OK 10 holes	Failed 1 hole
T14	OK 10 holes	OK 10 holes	Failed 2 holes
T16	OK 10 holes	OK 10 holes	Failed 3 holes

T drills VDE rating 5d; B drills VDE rating 2L

a 4in diameter; b 6in diameter; c 8in diameter; d 10in diameter; e 12in diameter

11   Typical carbide distribution at mid-radial positions of M2 high-speed steel bars sectioned longitudinally   ×100

For each group of drills the correlation between drill life and hardness was examined and a slight dependence (at the 1% significance level) found in the case of the B drills only.

The data did not provide any evidence of a difference in drill life due to carbide segregation, the mean lives being 32·2 holes for T drills and 34·4 holes for B drills in the single severity level tests. The multiple severity level tests confirmed that there was no difference in quality between the samples.

The importance of determining the true significance of carbide rating of high-speed steel with respect to service performance cannot be overstated. Unnecessarily severe specifications will cost the user money and achieve nothing.

Quality problem in sizes over 3in

The majority of high-speed steel bar 3in diameter and larger is used for tools where the axis may be exposed by machining and it is therefore essential that material in this size range should be free from gross axial carbide segregates. These have relatively lower melting points than the surrounding metal and readily overheat even at the lower ranges of hardening temperature and the use of austenitizing temperatures at the top side of the range commonly results in incipient

fusion. Bearing this in mind it is obviously necessary to control the manufacture of high-speed steels so that the incidence of central carbide segregation is reduced to an acceptable minimum level and this may be achieved as mentioned previously by the utilization of numerous factors, the principal ones being the use of low casting temperatures, striking a reasonable balance between ingot length, cross section and taper, and finally a limitation with regard to mass. With regard to the latter factor, satisfactory ingots relatively free from axial carbide segregates can only be produced up to a certain size since, with increasing mass, it becomes impossible to restrict segregation to a reasonable commercial acceptance level. This size limitation would be of the order of 16in square for molybdenum high-speed steels but in the tungsten types, due to the greater proportion of carbide formed, the maximum ingot size would be restricted to approximately 12in square section.

It is universally accepted that the as-cast carbide eutectic can only be broken down by mechanical means and the standard methods are press or hammer forging and rolling to ensure that the material receives a reduction of at least 90% from ingot to bar. To apply this rule to the letter means that it is only possible to produce bars of up to 6in

12   **Typical mid-radial microstructure of 10in diameter T1 high-speed steel bar sectioned longitudinally; for comparison with Fig.11d**   × 100

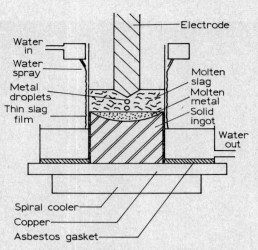

13   **Diagram showing principle of the ESR process**

diameter in molybdenum high-speed steels and 4¼in diameter in the tungsten types from normal commercially acceptable ingots, and that larger bar sizes will inevitably contain a marked cellular carbide network becoming more coarse with increasing size of bar. With the mass-effect limitations previously described taken into consideration, it is apparent that to produce bars greater in section from conventional ingots, advantage of special forging techniques must be taken. Several years ago large high-speed steel bar material with apparently excellent metallurgical characteristics was available. As is widely known, these were manufactured by upsetting ingots and forging to bar at right angles to the original metallurgical axis resulting in bars with exceptional carbide distributions at both ends but the middle of the bar contained off-centre carbide segregates, the severity of which was dependent on the degree of segregation present in the top portion of the ingot. The fact that this carbide segregation lay on one side of the bar was responsible for high incalculable distortion and severe localized overheating in the resultant tools produced. However, the idea had some merit and bars in the greater size range can now be produced by upsetting conventional ingots up to 50% and then drawing back to the original ingot size prior to forging, normally parallel to the metallurgical axis. This is obviously expensive but nevertheless does substantially improve the degree of primary carbide eutectic breakdown which could not otherwise be applied to large bars, and also consolidates the ingot axis to yield sound stock. It must be appreciated however that in all large high-speed steel bars, the primary cellularity inherent from the ingot will persist although the individual skeletal eutectic pattern should be broken down to the spheroidal type required. Figure 11 is typical of the carbide distribution which may be expected at mid-radial positions from bar stock over the 4–12in diameter range and clearly illustrates the increasing severity of cellularity which can be expected with increase in section over 4in diameter. In addition, Fig.12 has been included to demonstrate the appreciable difference between microstructures of 10in diameter T1 and M2, but it must be noted that these bars have been produced from the largest ingot sizes possible compatible with maximum control on central carbide segregation.

Up to the present date, there are no national or international standards for carbide segregation which take into consideration macrosegregation or degree of cellularity in the larger bar size range over 4in diameter section. Neither are

there standards which differentiate between the molybdenum-based or tungsten-based high-speed steels and anyone familiar with the standards achievable in these types will appreciate that there is a considerable difference. It is therefore likely that standards for the larger high-speed steel bar material will remain, as before, a matter of mutual agreement between supplier and consumer.

One of the major outlets for bar stock in excess of 3in diameter is for hobs and other tools where the teeth or cutting area are coincident with the immediate peripheral zone of the bar. In the past most of the major users have specified that such tools should be manufactured from individually forged blanks but, with few exceptions, most now realize that the utilization of large bar stock will provide material with the finest possible carbide distribution at the most critical position and this is particularly important in the larger size range.

The manufacture of cutters and hobs in the medium size range from 3–5in diameter bar has been fairly standard practice for some time and is increasing. In certain types of hobs it is essential that movement in heat treatment should be predictable, therefore, since volume change is appreciably affected by carbide distribution, bars from which a large number of hobs are to be produced require a good carbide distribution and a high degree of homogeneity throughout the bar length. With the development of the free-machining high-speed steels some years ago, and with specific reference to the M2 type, it became possible to rough machine hobs, apply a modified heat treatment to stiffen the material to hardnesses of the order of 38–40 HRC and finish machine to size, thus obviating the final grinding operation. This, to a large extent, highlighted the importance of having material of high, consistent quality. As far as is known there is no standard test procedure to accurately determine the degree of movement resulting from heat treatment but many consumers employ their own specialized techniques in approaching this problem. While high-quality stock is essential, there is no doubt that there are several other factors, such as adequate relief of machining stresses and very closely standardized preheating and austenitizing temperatures and times, which are probably of greater importance in controlling distortion.

Sheets in high-speed steels

The main uses for high-speed steel sheets are in the manufacture of hand and power hacksaw blades and circular slitting cutters or machine knives, the grades mostly required being 14%W (often with 1%V), T1, and M2. As in the case of tools made from bar the trend has been towards M2 during recent years, though the need for careful adaptation of heat-a

b

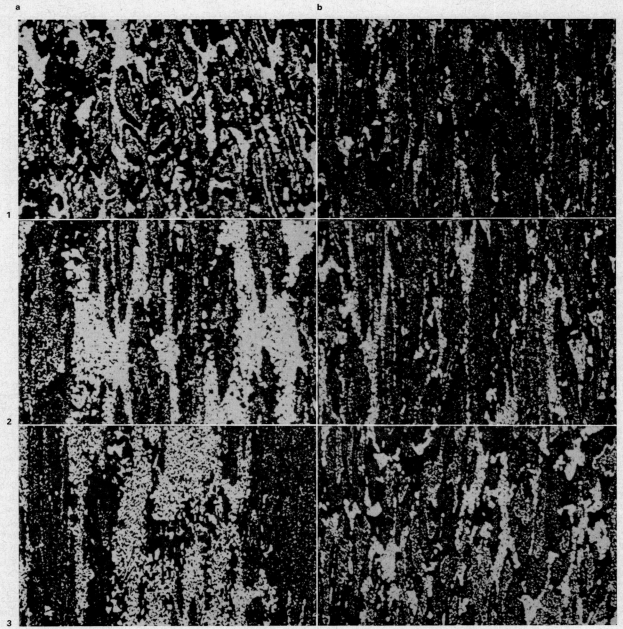

1

2

3

a air-melted; *b* ESR; 1, 2, and 3 represent longitudinal sections from outside, mid-radial, and centre positions respectively

14 T1 high-speed steel 7¼in diameter   ×100

treatment techniques has retarded the rate of change in some instances.

Quality requirements are similar to those applied to bar except that control of surface decarburization is more critical since no surface machining is carried out during the manufacture of hacksaw blades. The heavy reduction in cross-section in processing from ingot to sheet generally takes care of carbide breakdown requirements though the need to avoid the laminating effects of unduly heavy carbide bands or continuous non-metallic inclusion stringers demands similar careful attention to technique during steelmaking and casting as has been described relative to bar production.

A significant manufacturing development has resulted in the recent availability of high-speed steel sheets with a 'cold-planished' finish. This offers distinct advantages to the user in terms of considerably improved sheet flatness, virtual absence of surface scale and decarburization levels about half those normally obtained in the black-rolled product, e.g. 0·062in thick: 0·001in maximum partial decarburiza-

tion *v.* 0·002in maximum decarburization for black-rolled material. Improved tool manufacturing efficiency may be achieved by the ready application of the sheets to automatic machine feeding or, in the case of slitting cutter and machine knife production, the elimination of expensive smithing operations for the purpose of flattening prior to surface grinding. Minimum surface scale and decarburization offer a potential improvement in performance in the case of hacksaw blades.

ELECTROSLAG-REFINED HIGH-SPEED STEELS

It must be plain from the foregoing references to additional processes and increased inspection involved in meeting the quality requirements specified by a rapidly growing number of high-speed steel users, that there is an increased burden upon the technical and production resources of the steelmakers. The increasing use of stronger materials of construction, the tendency towards higher machining rates and the use of automated machines in the attainment of more

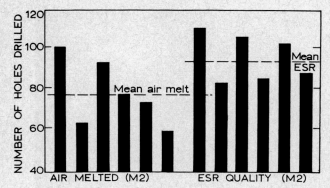

15 Comparative life of ESR and air-melted M2 high-speed steel drills, $\frac{13}{32}$in diameter, cutting 50 tons/in²; En9 blind holes $\frac{13}{32}$in deep at 732 rev/min (90 ft/min) and 10in/min feed increasing by 1in/min after every 20 holes

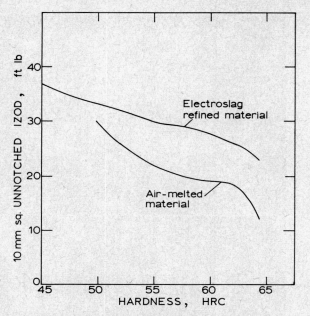

16 Comparative toughness of ESR and air-melted T1 high-speed steel

economic production in the engineering industries can only throw increasingly severe demands upon the tools used. Consistently high tool performance with an absolute minimum of premature failures must be the objective. It seems, therefore, that the quality standards which hitherto have been considered as special will, in the near future, become universal and an overall improvement in quality will be required from the steelmakers.

It is with these considerations in mind that several British steelmakers are now using the electroslag- (or electroflux-) refining process for the production of high-speed steels in order to take advantage of the considerable improvements that can be gained from this process in terms of final product quality and manufacturing efficiency.

The electroslag-refining (ESR) process is indicated in diagrammatic form in Fig.13. The process consists of melting and refining a consumable electrode of suitable composition through a molten slag, the heat for the process being generated by passing an alternating current between the mould base and the electrode through the slag. When the temperature of the slag pool rises above the melting point of the electrode, droplets of metal fall through the hot slag thereby being refined before collecting in a molten pool below the slag to solidify in the mould. As can be seen from the diagram, a layer of slag solidifies round the outside of the ingot and this helps to maintain unidirectional freezing from the base and also gives a very smooth ingot skin. The slags used for high-speed steels are usually of the CaF_2 plus Al_2O_3 type with CaO added where desulphurization is required.

Since freezing occurs from the bottom upwards, and only a comparatively small amount of metal is liquid at any one time, there is usually no macrosegregation of the type encountered in a conventionally cast ingot, and ingots produced by this process are extremely uniform in composition. It is possible therefore, with proper control of operational parameters, to obtain a uniform distribution of carbide throughout ESR ingots of high-speed steel and thus avoid the

Table 8   Drill test results, $\frac{1}{2}$in diameter TS drills

		No. of holes drilled (6 drills)	Average no. of holes drilled	
1st test bar	Electroslag top	126, 130, 120, 137, 138, 122	129	} 126
1st test bar	Electroslag btm	116, 121, 136, 115, 127, 129	124	
1st test bar	Air-melt top	129, 128, 127, 125, 114, 130	125	} 122
1st test bar	Air-melt btm	105, 112, 134, 121, 128, 116	119	
2nd test bar	Electroslag top	140, 121, 139, 153, 157, 130	140	} 138
2nd test bar	Electroslag btm	156, 137, 130, 159, 113, 127	137	
2nd test bar	Air-melt top	151, 129, 130, 122, 123, 117	129	} 128
2nd test bar	Air-melt btm	127, 140, 98, 137, 127, 131	128	

Test procedure: Standard taper shank drills $\frac{1}{2}$in diameter were manufactured from $\frac{13}{16}$in diameter bars and subjected to step-up drilling tests in BS drill bar using 'Dixol' soluble oil in water (1:10) as coolant

problems of top end central segregation and minimize the differences in segregate band widths between centre and outside that obtain with conventional ingots. The benefit structurally is most obvious in the case of the larger diameter bars, and in Fig.14 a longitudinal section of a $7\frac{1}{4}$in diameter T1 high-speed steel bar from an 18in diameter ESR ingot is compared with that of a similar bar produced from a 16in square ingot in air-melted steel. The fact that the steel is T1 type and the air-melt ingot was of a size which has previously been said to be particularly prone to gross segregation, means that the comparison shown in Fig.14 is rather exaggerated and a similar comparison for M2 steel would not be expected to show the same order of difference in structure. Nevertheless, it can readily be appreciated that it is certainly easier to set standards for carbide segregation for ESR than with air-melted material. In general, the grain size, the dendritic structure and the size of the carbide networks in ESR high-speed steel ingots is not significantly different from that obtained in the outer regions of similar-sized air-melted ingots, and it is still necessary to apply about 90% reduction to ESR ingots in order to break up the carbide networks. However, it has been demonstrated by BISRA,[22] that it is possible to refine the grain size of ESR ingots by adding inoculants via the slag, and if such a technique could be made to work with high-speed steels, then further significant improvements in structure might result. Other methods of refining the structure of ESR ingots, e.g. ultrasonic vibration, reciprocating moulds, magnetic stirring, etc., are also being investigated and with new developments of this type plus further operational experience it is not unreasonable to expect still further improvements in the structural quality of ESR high-speed steels.

The cleanness of ESR high-speed steels is of a high standard, and it is unusual to find any non-metallic inclusions exceeding about 0·004in in length. The cleanness is considerably superior to that of air-melted steel and compares very well with vacuum arc remelted (VAR) material. This is to some extent borne out by the fact that on rolling contact fatigue tests where inclusion size and frequency are known to have a significant effect on the results, ESR T1 steel has given lives at least as good as those obtained with VAR material and as a consequence is now being specified as an alternative to VAR material for aircraft-engine bearings.

An example of what the ESR process may mean to cutting efficiency is shown in Fig.15.[6] Here M2 quality twist drills made from ESR steel showed an average improvement of 23% over similar drills made from air-melted steel. In these tests, the greater consistency of performance and the absence of chipping with the ESR drills suggested that more reliable attainment of full cutting potentiality would be realized when ESR quality was used. However, the results of similar tests, given in Table 8, indicate that the improvement in cutting efficiency of ESR over air-melted drills is often only marginal and would not justify the extra cost for ESR quality. At the present time, it is perhaps true to say that the benefits to be gained by using ESR material for small-section cutting tools are probably marginal and confined to reducing premature failures such as can occur with air-melted tools due to the presence of the occasional large inclusion or carbide segregate. On the other hand, for large-section tools the improved carbide distribution in ESR material would be expected to give significant improvements in service performance compared with air-melted steel, and it would be interesting to see cutting tests carried out with, for example, large-size hobs to determine whether this is true or not. In this context, it is worthwhile pointing out that the more uniform carbide distribution found in ESR steel compared to air melt, is conducive towards more consistent dimensional changes during the heat treatment of unground hobs.

An attempt has also been made to determine the effects of ESR on the toughness of high-speed steel and results obtained for ESR T1 material are shown in Fig.16.[6] These tests were made on unnotched 10 mm square Izod specimens machined from $\frac{5}{8}$ in square bars in each case. It will be seen that the ESR steel is superior in toughness to the air-melted steel at all hardness levels, with a difference of about 10 ft lb at normal working hardness values.

Clearly the provision of ingots having a surface requiring no conditioning, having improved and more consistent forgeability, and possessing a consistently high standard of structural refinement along the whole length, must result in increased yields and lower rejection rates being obtained for ESR compared with air-melted steel. However, these improvements in manufacturing efficiency only partially offset the cost of providing the ingots and therefore ESR quality products cost more to produce than normal products. The increased cost is the only factor likely to limit the application of the ESR process for the production of high-speed steels for cutting tools, and while costs will be reduced as higher productivity equipment attains full capacity and still further quality improvements will come from additional developments, the future of the process will depend on the full exploitation by the steelmaker of the in-process benefits which he can derive and on the user being prepared to acknowledge that, having specified higher quality he must pay something extra for it.

SUMMARY

Sufficient has been said to indicate the many problems which exist both in alloying relative to desired cutting performance and in manufacturing relative to quality requirements in such inherently complex material as high-speed steel. As in most industrial situations, economic considerations provide a major limiting factor to what can realistically be provided. It is important that users, therefore, should guard against overspecification both with regard to the type of steel used and the quality levels required.

The comments of users will be especially welcome on the difficult subject of relative cutting performance of the various high-speed steel grades. Although the issue is by no means clearcut, an attempt has been made in this paper to provide guidelines to selection on the basis of properties that are readily determined in the laboratory. Other properties such as surface frictional characteristics may need to be considered in the search for optimum compositions. In order to achieve a further reduction in the number of grades produced as an aid to more efficient production and supply, and to assess the true place of the newer compositions mentioned in the text, closer collaboration between manufacturer, tool producer, and user is undoubtedly required.

The major quality problem discussed, that of carbide segregation, is related to an adverse structural condition inherent in the cast ingot, which, though reasonably controllable by the developments in conventional casting techniques described, and distinctly improved in terms of consistency and freedom from gross segregates by ESR, remains a constant challenge to the manufacturer, especially in larger bar sizes. Further development of the ESR process holds promise of greater success, but considerable potential is shown by the powder-metallurgy route where extremely fine carbides are formed in the powder in the course of atomization of a molten-metal stream. The development of such a procedure allied with compacting techniques designed to achieve maximum density and reliability is, therefore, receiving considerable attention at the present time especially in the USA and is the subject of another paper presented at this conference.[23] Apart from the many technological problems which are involved, there would appear to be a major economic obstacle to overcome before the process can be applied in production to all high-speed steel grades.

The greater latitude for alloying at the higher levels which may be achieved in a powder-metallurgy product offers a further potential benefit in the case of high-performance tooling, and it is in this area that the first applications seem likely to be achieved.

Further movement towards the use of automated high-output machine tools with the associated heavy cost penalties resulting from early tool failures, will tend to set consistently high tool performance at a premium and may have a profound effect on the viability of the more advanced high-speed steel manufacturing techniques.

REFERENCES

1 G. STEVEN et al.: Trans. ASM, 1964, **57,** Part IV, 925–948
2 P. JUBB: 'A British Standard for tool steels', this volume
3 F. A. KIRK AND N. BURTON: Iron Steel, 1968, Special issue, 53–55
4 G. HOYLE AND E. INESON: JISI, 1959, **191,** 44–55
5 G. STEVEN AND J. R. HANDYSIDE: Proc. ASTM, 1963, **73,** 1122
6 F. A. KIRK: Steel Times, 1969, **197,** 39
7 G. STEVEN AND C. P. MCSHANE: Cutting Tool Engineering, Oct., 1967
8 T. MURAKAMI AND A. HATTA: Sci. Rep. Tôhoku Univ., Honda anniversary vol., 1936, 882
9 K. KUO: JISI, 1955, **181,** 128
10 H. J. GOLDSCHMIDT: ibid., 1957, **186,** 68
11 Inca Steel Co., British Patent no.879, 670
12 J. FIELD: Iron Age, 1959, **12,** 42–43
13 R. MITSCHE AND E. KUDIELKA: Radex Runds. 1967, **1,** 407–413
14 E. BERNART AND V. DLOUCHY: Hutnik (Prague), 1968, **18,** 4, 165–167
15 M. PETZ et al.: Berg- Hütten. Monatsh., 1968, **113,** 3, 135–141
16 G. HOYLE: Met. Rev., 1964, **9,** 33, 49–89
17 A. TOMLINSON AND D. E. BEARD: 'Controlled forging of tool steels', this volume
18 E. BONDI AND G. TIMO: Met. Ital., 1963, **55,** 4, 159–162
19 J. PERRY et al.: Met. Forming, 1967, **34,** nos.6, 7, and 8
20 VDE test sheet no.1615, 1959
21 J. R. POWELL: BISRA rep. MG/AL/11/63
22 W. E. DUCKWORTH: Steel Times Ann. Rev., 1968, 183–194
23 E. J. DULIS AND T. A. NEUMEYER: 'Particle metallurgy high-speed tool steel', this volume

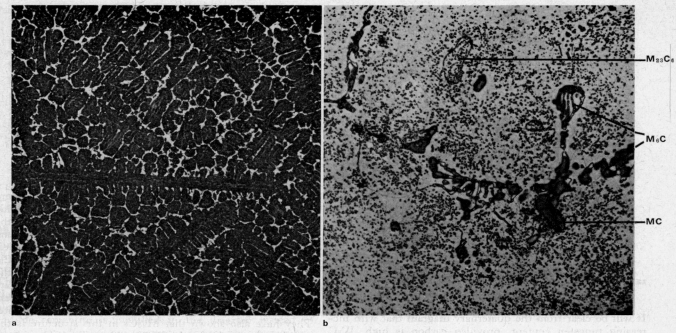

a etched in Steads' reagent $\times 75$; b first etched in $4\%NaOH$, saturated in $KMnO_4$, and then electroetched in $1\%Cr_2O_3$ $\times 750$

15 Structure of as-cast high-speed steel

HEAT TREATMENT

Heat treatment in high-speed steels consists of an austenitizing treatment followed by oil quenching or air cooling. Because of the high hardenability of these steels. air cooling is often sufficient for obtaining a martensitic structure. Some austenite is, however, retained, whatever the method of cooling. Single or multiple tempering is then carried out.

Isothermal treatment giving rise to bainitic structure is not popular because of the very long time necessary for the transformation. Moreover a maximum of 70% of austenite transforms during isothermal cooling and multiple treatment is necessary for further transformation.[3]

Hardening temperature

Hardening temperature is an important parameter in the heat treatment of tool steels. The idea is to get as much of the carbides in solution without appreciably coarsening the grain size. Several workers[15,16,29] have studied the extent of solution of carbides during austenitizing. The $M_{23}C_6$ and M_7C_3 carbides go into solution completely during commercial austenitizing treatments, while the others go into solution in different degrees, depending on austenitizing temperature

and time. Figure 18 shows the relation between austenitizing time and temperature and the volume fraction of undissolved carbide in 18–4–1 tool steel. The grain size, as would be expected, increases with increasing austenitizing time and temperature and the relationship is given in Fig.19 (see Appendix).

The as-hardened hardness increases with increasing austenitizing temperature to a maximum and then decreases slowly as the retained austenite content increases. When the temperature is sufficiently high for liquidation to occur at the grain boundaries, the hardness drops drastically. Fig.20a shows the relationship between austenitizing temperature and hardness as well as retained austenite content, grain size, and hardness after refrigeration in M2 type tool steel, and Fig.20b shows the relationship between the temperature and hardness in M50 tool steel. It will be clear from Fig.20a that increase in hardness due to refrigeration is high when the retained austenite content is high. Also note that the retained austenite content (measured in terms of austenite–ferrite ratio) of the sample hardened at $1230°C$ is less than that at $1210°C$. Incipient liquation at grain boundaries was observed at $1230°C$ and the lower austenite content may be due to some ferrite retention.

Matrix composition on hardening and its effect

The matrix composition changes as the carbides go into solution; Fig.10, due to Goldschmidt, shows the variation of matrix composition of 18–4–1 high-speed steel. Clearly the composition of the matrix in the hardened condition is much different from that in the annealed condition, the difference being mainly due to increase in carbon and tungsten content. Notice that there is little solution of carbides when the matrix is ferritic as would be expected because of the low solubility of carbon in ferrite.

Tungsten and molybdenum, because of their large atomic diameter are expected to give significant solid solution hardening. Carbon is known to increase the hardness of martensite, but increasing carbon in the matrix leads to increasing amount of retained austenite, and lowering of solidus temperature. An interesting example of this phenomenon is seen in carbon-rich segregation (spot segregates) sometimes seen in vacuum-arc remelted tool steels. Figure 21 shows one such segregate in a fully treated 18–4–1 steel. Notice the liquation

16 Typical structure of high-speed steel, found in commercial drill rods; central region of 0·3in diameter rod $\times 100$

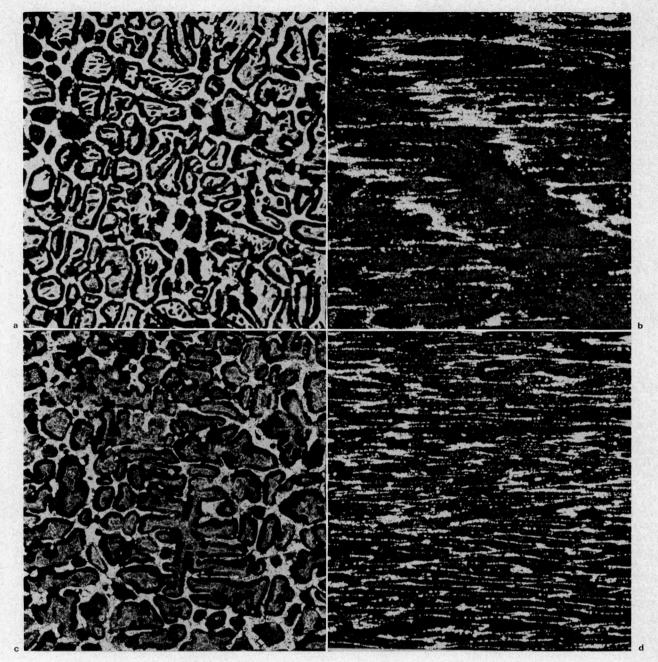

a and *c* as-cast; *b* and *d* as-forged; *a* and *b* coarse cast structure giving coarse forged structure; *c* and *d* fine cast structure giving fine forged structure

17 Relation between as-cast and as-forged structures of 18–4–1 high-speed steels   ×110

at grain boundaries and the large amount of retained austenite in the segregated area while the rest of the structure is typical of normal high-speed steel.

Tempering

The aim of the tempering operations is to obtain a structure containing tempered martensite and undissolved carbides only. However, the hardened structure contains a significant amount of retained austenite. This austenite decomposes during tempering, although there is some controversy regarding the stage at which the decomposition takes place. The secondary hardening precipitate is also a matter of some dispute. It is intended, therefore, to give a brief review of the literature on this aspect and then describe the work done at the author's laboratory.

REVIEW ON PRECIPITATION IN MARTENSITE

High-speed steels rely on the stability of the precipitates formed during tempering for their strength and red hardness. Kuo[30] showed, by X-ray diffraction of extracted carbides, that W_2C is the secondary hardening precipitate in 18–4–1 tool steel. However, White and Honeycombe,[31] working on the same composition, failed to detect W_2C at any stage of heat treatment. They concluded from electron diffraction of carbides on extraction replicas that $M_{23}C_6$ is the secondary hardening precipitate. White and Honeycombe, however, used ring patterns rather than single crystal patterns and, in these steels where more than one carbide can precipitate out, this is not the ideal method of identification. It is also difficult to see how $M_{23}C_6$ which is fcc with a very large parameter can lead to secondary hardening in bcc ferrite, especially

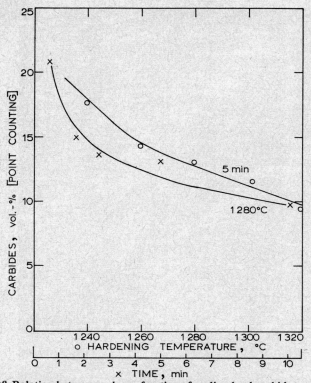

18 Relation between volume fraction of undissolved carbides and hardening time and temperature, 18–4–1 high-speed steel

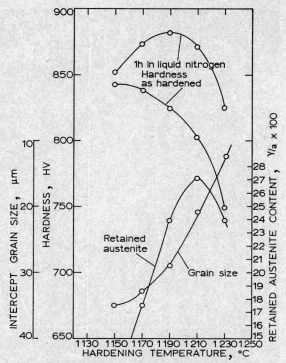

20a Relation between hardness, grain size, and retained austenite and hardening temperature in M2 high-speed steel

because the parameters are not compatible. $M_{23}C_6$ also coarsens rapidly in ferrite.[33] Colombier and Leveque in an 18W–4Cr steel (18–4–1 without vanadium) have identified W_2C by electron diffraction from single crystals of W_2C in thin foils and attribute this carbide to secondary hardening. V_4C_3 has been shown to be the cause of secondary hardening in many steels containing vanadium.[9,34,36] Payson[9] has also shown that Mo_2C or W_2C gives rise to secondary hardening in simple molybdenum or simple tungsten steels, but when vanadium is present above a minimum amount (0·5%) V_4C_3 gives rise to secondary hardening. He also showed that the iron carbide, Fe_3C, may be present along with the secondary hardening alloy carbides over some or all the tempering range up to or beyond the secondary hardness peak in the tempering curve.

Thus, there is still some doubt regarding the secondary hardening precipitate in these steels, and specific work on a few steels was carried out to explain the controversy.

REVIEW ON DECOMPOSITION OF RETAINED AUSTENITE

It was mentioned earlier that the hardened steel consists of martensite, undissolved carbides, and retained austenite. All or most of the retained austenite decomposes during the

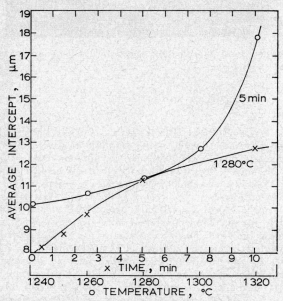

19 Relation between grain size and hardening time and temperature, 18–4–1 high-speed steel

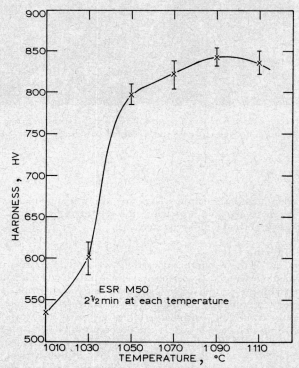

20b Relation between hardness and hardening temperature in M50 high-speed steel

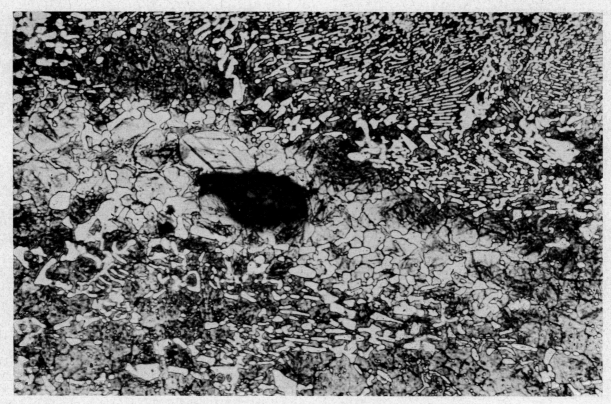

21 Spot segregate in vacuum-arc remelted 18–4–1 high-speed steel; etchant Glycia-regia   ×400

multiple tempering processes. White and Honeycombe[31] detected $M_{23}C_6$ precipitates in the austenite region tempered in the range 300–400°C, and this has been considered as an evidence to suggest that decomposition of austenite does not contribute to secondary hardening. Payson[9] considers that since there is no discontinuity in the master tempering curves for tool steels, decomposition of austenite cannot take a significant part in secondary hardening, otherwise an abrupt change in the slope of the hardness curve would be expected. It appears, therefore, that it is accepted that decomposition of austenite does not give rise to secondary hardening, although at which stage it decomposes, and its effect on the tempering curve is not well known.

PRESENT WORK ON TEMPERING HIGH-SPEED STEELS
Tempering curves
Figures 22–24 show tempering curves for 18–4–1 (W–Cr–V) steel (T1), 18–4–0 which is 18–4–1 without vanadium,

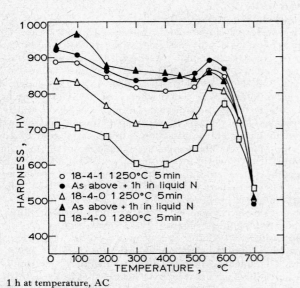

1 h at temperature, AC

22 Tempering curves for 18–4–1 and 18–4–0 steels

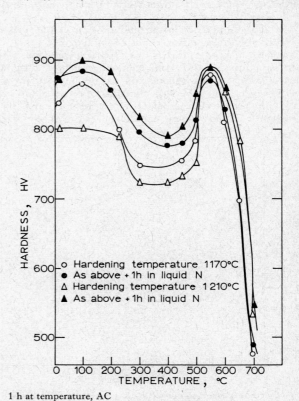

1 h at temperature, AC

23 Tempering curves for M2 high-speed steel

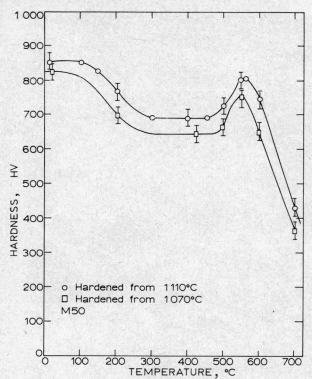

1 h at temperature, AC

24   Tempering curves for M50 high-speed steel

6–5–4–2 (W–Mo–Cr–V) steel (M2), and 4–4–1 (Mo–Cr–V) steel (M50) and Fig.25 shows isothermal tempering characteristics of 18–4–1 at 560°C. The tempering curves for hardened and refrigerated materials have been included in a few cases for showing the effect of retained austenite.

Effect of hardening temperature on tempering characteristics

Hardening temperature affects the tempering characteristics in two ways:

(i) by the difference in the amount of austenite retained
(ii) by the difference in composition of the matrix.

It has been mentioned earlier that the hardness of the hardened structure decreases with increasing retained austenite. This effect is usually retained on tempering. The 18–4–0 steel is expected to have large amounts of retained austenite (54% in the steels hardened from 1250°C and 65% in those hardened from 1280°C) because of the large

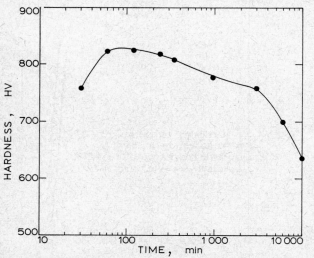

25   Isothermal tempering curve for 18–4–1 high-speed steel at 560°C; hardening treatment: 2½ min at 1250°C

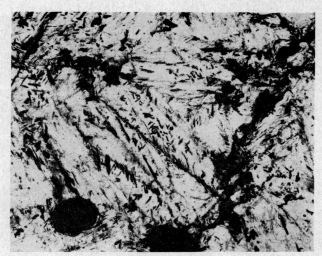

26   Structure of 18–4–1 high-speed steel after tempering for 1 h at 560°C; electron micrograph showing large undissolved carbides, rod-like cementite precipitates and darkening of martensite needles × 10 000

amount of carbon available to the matrix (no carbon being tied up with difficult to dissolve V_4C_3). Therefore the effect of hardening temperature is rather exaggerated in this case, the samples hardened from 1280°C having much lower hardness than those hardened from 1250°C, and the tempering curve for the samples hardened from 1280°C is correspondingly lower than those from 1250°C as shown in Fig.22. In the case of M50 with the hardening temperatures used, samples hardened from the higher temperature have higher hardness and the effect on the tempering curve is very similar as shown in Fig.24. In the case of M2 steel the samples hardened from 1170°C produced a tempering curve higher in hardness than that obtained from 1210°C except at and above 560°C. When the hardened samples from these two temperatures are refrigerated, the position of the tempering curves were reversed, i.e. the one corresponding to 1210°C is higher than that to 1170°C. The retained austenite content of the refrigerated samples for 1210°C and 1170°C was 14·2 and 9·1 respectively, although the ratio before refrigeration was 17·5 and 27 respectively. This difference in the decomposition characteristics of austenite on cooling is not understood.

Metallography and identification of carbide on martensite

T1 steel 18–4–1 (W–Cr–V)

No discrete precipitate could be seen up to about 400°C in the replicas of any of these steels, but the surface appeared to be cross-hatched and presence of hexagonal ε-carbide is suspected.[37] At and above 400°C cementite is present in 18–4–1 in the form of short needles. Figure 26 shows the structure after 1 h at 560°C when peak hardness is reached. No other freshly precipitated carbides can be identified at this stage although a diffused ring pattern could often be seen in selected area diffraction patterns. A selected area diffraction pattern obtained from a sample tempered for 2 h at 560°C of a cementite needle is shown in Fig.27a. The diffraction is streaked in the <001> direction. It is well known that the secondary hardening carbides, when they impart maximum hardness are often too small to identify and it is easier to identify them when the hardness starts dropping at higher temperature or after longer tempering at the same temperature. The latter method was used for identifying the secondary hardening precipitate of 18–4–1. An fcc phase of lattice parameter 4·17Å was detected first after 5½ h from ring patterns. This parameter is very near to that of V_4C_3 (4·16Å) and it is concluded that the secondary hardening precipitate

a after 2 h showing presence of cementite; *b* after 5½ h showing two single crystals of cementite and ring pattern from V_4C_3 precipitate; *c* after 175 h showing arced diffraction rings from V_4C_3 precipitate; diffraction patterns from two cementite single crystals can also be seen

27 Selected area diffraction patterns from 18–4–1 high-speed steel after different times of tempering at 560°C

in this steel is V_4C_3. Figure 27*b* shows a selected area diffraction pattern from a small area from the steel after 5½ h tempering. The area contained one precipitate of cementite which has given the single crystal pattern, streaked in the $<001>$ direction. The cementite persisted even after 175 h tempering at 560°C, although the amount progressively decreased as shown in Fig.28. Ring patterns of V_4C_3 at this stage showed distinct arcing (Fig.27*c*) indicating Widmanstätten precipitation. After 25 h at 560°C, distinct platelike precipitates of V_4C_3 can be seen and single crystal diffraction patterns from this phase can be obtained as shown in Fig.29. Figure 30 is a dark field electron micrograph using a [100] V_4C_3 zone reflection. It was observed that many of the V_4C_3 precipitates nucleated and grew at the surface of cementite. This phenomenon has been previously reported by Davenport in some steels hardened by V_4C_3 precipitates. It should be pointed out here that it was possible to identify V_4C_3 crystals at the grain boundaries after 2 h tempering at 560°C.

18–4–0 STEEL

In the 18–4–0 steel, cementite was observed in the same way as in 18–4–1 and diffraction patterns containing the $<001>$ direction were streaked. Figure 31 shows the structure after tempering for one hour at 550°C (tempered to peak hardness). Notice the following aspects:

(i)   there are two types of martensite, tempered and untempered; the untempered martensite must have formed during cooling down after tempering

(ii) there is no precipitate in retained austenite

(iii) the cementite precipitates in tempered martensite are parallel and probably have formed on twin boundaries

(iv) there is a cloud-like precipitate (probably a product of etching) on the tempered martensite, which could not be identified and will be discussed later.

M2 STEEL 6–5–4–2 (W–Mo–Cr–V)

The results from M2 steel are very similar to that from

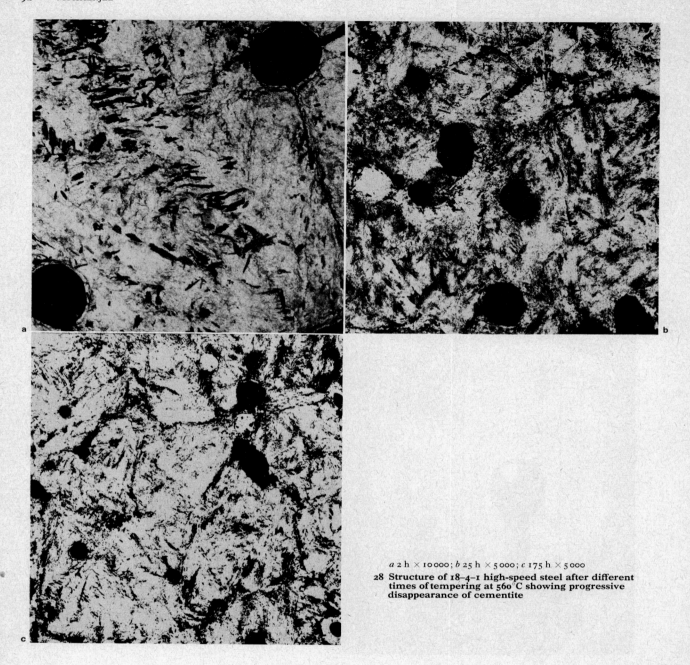

a 2 h × 10000; b 25 h × 5000; c 175 h × 5000

28 **Structure of 18–4–1 high-speed steel after different times of tempering at 560°C showing progressive disappearance of cementite**

18–4–1 steel. Again V₄C₃ appeared to be the precipitate giving rise to secondary hardening, although a large amount of alloyed cementite is present when peak hardness is reached. In this case V_4C_3 was identified after tempering for 1 h at 650°C. At this stage there was very little cementite present. The structures after tempering at different temperatures are shown in Fig.32. The secondary hardening carbide, alloyed V_4C_3, have a parameter of 4·12Å. It is difficult to explain how a parameter of 4·12Å is possible for V_4C_3 (4·16Å) in this steel containing large amounts of tungsten and molybdenum which are expected to expand the lattice. But it has been shown that the parameter of V_4C_3 decreases when chromium and/or nitrogen are present in the lattice;[39,40] the chromium present in the steel and the nitrogen, if present, probably go into the V_4C_3 lattice, thus explaining the smaller parameter.

M50 STEEL 4–4–1 (Mo–Cr–V)

The cementite precipitates in M50 tool steel (Figs. 33–36),

are invariably on twin boundaries. For example, Fig.33 shows the structure at peak hardness, and the similarity between this structure and that produced by Kelly and Nutting[41] showing precipitation of Fe_3C along twin boundaries of high-carbon martensite is striking. V_4C_3 was detected after 2 h at 600°C. The characteristically needle-like Mo_2C carbides were never observed (Fig.34) and as such V_4C_3 can be taken as the secondary hardening carbides in this steel. Selected area diffraction patterns of a number of cementite crystals in a zone always showed arcing, indicating Widmanstätten precipitation. This is shown in Fig.35a and b. It can be shown that the cementite crystals are elongated in the $<010>$ direction. Figure 36a and b shows the structure at different tempering temperatures. One significant feature in Fig.36b is the presence of white patches in between needles of tempered martensite. These are areas of retained austenite which may or may not have been transformed into martensite during cooling down after tempering. This micrograph is important in the sense that it shows no precipitation in the retained austenite region.

29 Single crystal pattern from V₄C₃ after 5½ h tempering at 560°C, 18–4–1 steel

31 Structure of 18–4–0 steel after tempering for 1 h at 550°C showing undissolved carbides, retained austenite, tempered martensite, and freshly formed martensite

Decomposition of retained austenite

Comparison of the tempering curves of refrigerated and unrefrigerated materials will show (Figs.22 and 24a) greater secondary hardening intensity in the materials that were not refrigerated. The peak hardness values achieved, however, are not very different. The hardnesses of the refrigerated samples are high before tempering, because of the larger amount of martensite. Increase in hardness on tempering in this case is mainly due to precipitation of secondary hardening carbide. But in the unrefrigerated samples which contain large amounts of retained austenite (20–25%) the increase in hardness is due to two major causes:

(i) due to secondary hardening carbide precipitates on existing martensite
(ii) due to transformation of retained austenite into secondary martensite.

30 Dark field electron micrograph of 18–4–1 steel after 5½ h tempering at 560°C using (100) V₄C₃ zone reflection ×10 000

It is clear from Fig.31 that the freshly formed martensite has no precipitate, and it can be safely concluded that this martensite has formed during cooling down after tempering. It appears, therefore, that although retained austenite does not contribute to secondary hardening at the tempering temperature, its decomposition into martensite after tempering results in considerable increase in hardness at room temperature.

Secondary hardening carbides

The presence of fine cementite even after long hours of tempering shows its surprising stability. The high stability of cementite in 3% tungsten steel has been reported by Pickering.[36] Durnin and Mukherjee[42] commented on the streaked diffraction patterns of cementite in alloy steels. It appears that cementite precipitates have thin regions of faulting in <001> direction. This may be due to the presence of much larger atoms of molybdenum or tungsten in the cementite lattice leading to considerable lattice strain which results in systematic or random faulting. However, the possibility of some ordering in the structure cannot be ruled out.

Although cementite could be seen in large amounts in all the steels examined at peak hardness, it is difficult to see how this carbide can give secondary hardening, especially when the size is fairly large (200Å). In the steel containing 18W–4Cr, one would expect precipitation of W₂C type carbide as has been observed by Colombier and Leveque[32] in thin foils. It is possible that W₂C dissolves in the etchants used (HNO₃ and HF in ethanol) giving tungstic acid (H₂WO₄) and probably some homopoly or heteropoly acids.[43] Therefore, the absence of W₂C in replicas made from this steel should not be taken as a conclusive proof of its absence. Pickering[36] has shown that in 3% tungsten steel, retarded softening rather than secondary hardening is observed during tempering, coincident with the precipitation of cloud-like W₂C. However, at higher temperatures M₂₃C₆ and M₆C carbides grow at the expense of W₂C carbides.

The secondary hardening intensity in the 18–4–0 steel is smaller than that in 18–4–1, when the refrigerated samples of steels are considered. It is easier to compare the secondary hardening intensities of the carbides in this condition as the complexity due to transformation of retained austenite is less important. The higher hardening intensity in 18–4–1 steel is due to the precipitation of V₄C₃ in contrast with that

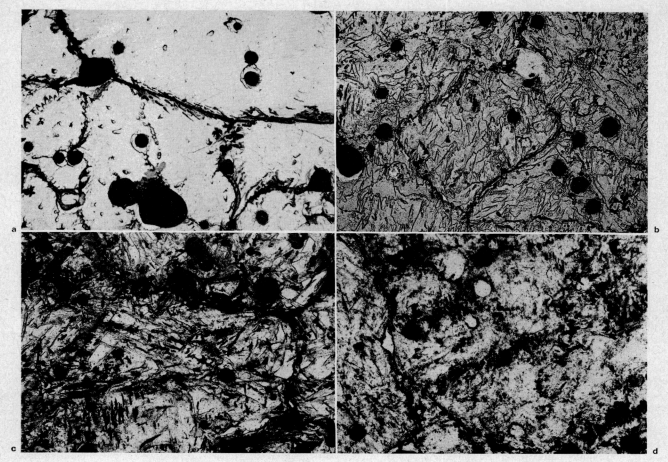

a as-hardened showing matrix and undissolved carbide; *b* tempered for 1 h at 400°C showing practically no precipitation of cementite; *c* tempered for 1 h at 560°C showing precipitation of cementite; *d* tempered for 1 h at 650°C showing significant dissolution of cementite; V_4C_3 can be identified at this stage but they are too small to be resolved at this magnification

32   Structure of M2 high-speed steel; electron micrographs from extraction replicas   ×5 000

33   Structure of M50 high-speed steel after 1 h tempering at 560°C showing precipitation on twin boundaries   ×6 000

due to W_2C in the 18–4–0 steel. Payson concluded that when the vanadium content is greater than 0·5%, V_4C_3 can be the hardening precipitate. In the steels examined, V_4C_3 was found to precipitate on the cementite as well as directly in the matrix. It was not attempted to determine the orientation relationship between cementite and V_4C_3 but this has been done by Davenport.[38]

SUMMARY

It is now possible to give a coherent and consistent picture of the full heat-treatment cycle of high-speed steels.

Although there are many specifications for high-speed tool steels with widely varying compositions, the physical metallurgy of these steels is very similar. The different analyses necessitate different temperatures for heat treatment and result in different properties of the end product, but almost all the end products consist of tempered martensite with variable amounts of M_6C and V_4C_3 carbides. Annealed steels when heated for austenitizing do not change very much in structure until AC_1 temperature is reached. When ferrite is transformed into austenite, $M_{23}C_6$ or M_7C_3 type carbides go into solution first, complete solution occurs at about 1 100°C. At higher temperatures M_6C and $V_4C/_3VC$ type carbides go into solution. The extent depends on time and temperature of austenitizing.

Hardened steels contain martensite, retained austenite,

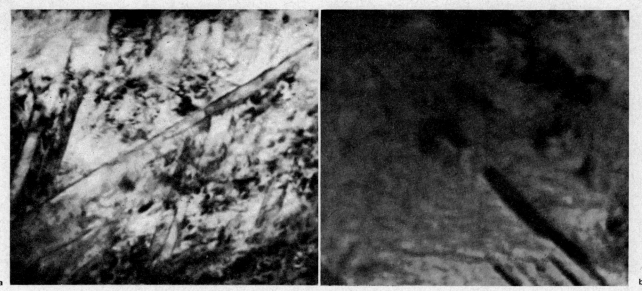

a showing tempered twinned martensite × 200000; *b* cementite precipitates can be observed but no rod-like Mo₂C precipitates × 250000

34 Thin foil electron micrograph of M50 high-speed steel after tempering for 1 h at 600°C

and undissolved carbides. The retained austenite content increases with increasing carbon in solution.

Tempering reactions follow the usual route: hexagonal ϵ-carbide→cementite→W₂C/V₄C₃.

The retained austenite decomposes during cooling down after tempering, presumably due to stress relief; retempering is necessary as the secondary martensite after first tempering is untempered. The number of tempering treatments depends on the extent of austenite decomposition in each tempering process.

It is difficult to specify what the ideal structure of a high-speed steel should look like as this will depend on the purpose for which it is used, e.g. a larger volume fraction of vanadium carbide will be necessary for high wear resistance. However, it is obvious that a uniform distribution of primary carbides is desirable from the mechanical properties point of view. It is also advantageous metallurgically as the retained austenite content in high-speed tools has always been seen to be higher around the carbide bands than that away from it. For a par-

ticular steel, a tempered martensitic structure containing a uniform distribution of carbides, fine grain size, and no retained austenite is expected to give the best properties.

CONCLUSIONS

1 Carbides present in annealed high-speed steels are usually M₆C, V₄C/₃VC and M₂₃C₆. In high-carbon steels, however, M₂₃C₆ is replaced by M₇C₃.

2 Nitrogen stabilizes M₆C and V₄C₃/VC type phases producing the corresponding carbonitrides.

3 Hardened steels contain M₆C and V₄C₃/VC type carbides with martensite and retained austenite.

4 Retained austenite content increases with increasing carbon in solution.

5 V₄C₃ is the secondary hardening precipitate in the steels investigated although the possibility of simultaneous presence of some W₂C cannot be excluded.

6 The increase in room-temperature hardness after tempering is due to precipitation of secondary hardening carbides

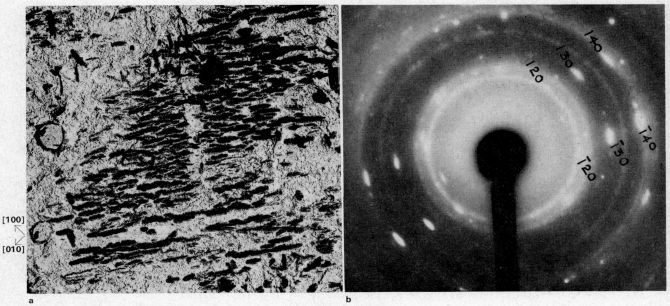

[100]
[010]

35*a* Cementite crystal colony in M50 high-speed steel after tempering for 1 h at 560°C
35*b* Selected area diffraction pattern from 35*a*

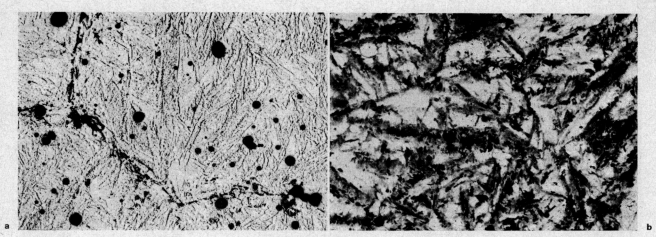

a 400°C showing very little precipitation inside the grains; b 600°C showing tempered martensite and retained austenite (white patches); undissolved carbides have not been extracted

36 Structure of M50 high-speed steel after tempering for 1 h at different temperatures (*see also* Figs. 1, 2, 3, and 33) ×8000

in the martensite and transformation of retained austenite into martensite.

7 No precipitates were observed in the retained austenite at any stage of tempering.

ACKNOWLEDGMENTS

The author is grateful to Mr W. H. Bailey, Technical Manager, Special Steels Division, British Steel Corporation, for permission to publish this work. Thanks are due to Dr K. A. Ridal for his continued interest in this work and for helpful discussions, and to many colleagues, particularly Mr J. Durnin, Mr M. Perne, and Mr K. Wragg for assistance in preparation of the paper.

REFERENCES

1 G. A. ROBERTS et al.: 'Tool steels', 1962, Ohio, ASM
2 P. PAYSON: 'Metallurgy of tool steels'; 1962, New York, London, J. Wiley
3 G. HOYLE: Met. Rev., 1964, 9, 49
4 Study of the metallurgical properties that are necessary for satisfactory bearing performance and the development of improved bearing alloys for service up to 1000°F, WADC Tech. Re. Nov. 1957; 57–343, Part I G. K. BHAT AND A. F. NEHRENBERG, WADC Tech. Re. Nov. 1957, 57–343; Part II T. V. PHILIP et al.: ibid., Oct., 1958
5 G. STEVEN et al.: Trans. ASM, 1964, 57, 925
6 D. J. BLICKWEDE et al.: Trans. ASM, 1950, 42, 1161
7 L. L. BURNS et al.: Trans. ASM, 1938, 26, 1
8 R. W. BALLUFFI et al.: Trans. ASM, 1951, 43, 497
9 P. PAYSON: Trans. ASM, 1959, 51, 60
10 V. K. CHANDHOK et al.: Trans. ASM, 1963, 56, 677
11 H. J. GOLDSCHMIDT: 'Interstitial alloys'; 1967, London, Butterworths
12 H. KRAINER: Arch. Eisenh, 1950, 21, 39
13 S. W. K. SHAW AND A. G. QUARRELL: JISI, 1956, 184, 262
14 L. COLOMBIER et al.: Rev. Met., 1967, 987; Eng. tr. BISIT 6187
15 D. J. BLICKWEDE AND M. COHEN: Trans. AIME, 1949, 185, 578
16 F. KAYSER AND M. COHEN: Met. Prog., 1952, 79
17 T. K. JONES AND T. MUKHERJEE: JISI, 1970, 208, 90–92
18 J. H. WOODHEAD AND A. G. QUARRELL: JISI, 1965, 203, 605
19 G. STEVEN et al.: Trans. ASM, 1969, 62, 180
20 A. RANDAK AND J. KURZEJA: Stahl Eisen, 1966, 86, 1017
21 H. SPITZER AND P. W. BARDENHEUER: Paper presented in 2nd international symposium on electroslag technology, 1969, Pittsburgh
22 H. J. GOLDSCHMIDT: JISI, 1957, 186, 68
23 T. MURAKAMI AND A. HATTA: Sci. Rep. Tôhoku Univ., 1936, Honda anniversary vol. 882
24 K. KUO: JISI, 1955, 181, 128
25 H. J. GOLDSCHMIDT: JISI, 1957, 186, 79
26 H. J. GOLDSCHMIDT: JISI, 1952, 170, 189
27 R. SMITH: ISI Spec. Rep. 64, 1959, 307
28 K. W. ANDREWS: Private communication
29 T. MALKIEWICZ et al.: JISI, 1959, 192, 25
30 K. KUO: JISI, 1953, 174, 223
31 C. H. WHITE AND R. W. K. HONEYCOMBE: JISI, 1961, 197, 21
32 L. COLOMBIER AND R. LEVEQUE: Mém. Sci Rev. Mét., 1968, 65, 229; Eng. tr. BISIT 6791
33 T. MUKHERJEE et al.: JISI, 1969, 207, 621
34 A. K. SEAL AND R. W. K. HONEYCOMBE: JISI, 1958, 188, 9
35 R. G. BAKER AND J. NUTTING: ISI Spec. Rep. 64, 1959, 1
36 F. B. PICKERING: ISI Spec. Rep. 64, 1959, 23
37 J. GORDINE AND I. CODD: JISI, 207, 461
38 A. T. DAVENPORT: PhD thesis Sheffield/Cambridge, (1968)
39 H. KRAINER: Arch. Eisenh, 1950, 21, 33
40 P. DUWEZ AND F. ODELL: J. Electrochem. Soc., 1950, 97, 299
41 P. M. KELLY AND J. NUTTING: Proc. Roy. Soc., 1960, A259, 45
42 J. DURNIN AND T. MUKHERJEE: JISI, 1969, 207, 369
43 A. PAUL: (Dept. of glass tech., Sheff. Univ.) private communication

APPENDIX

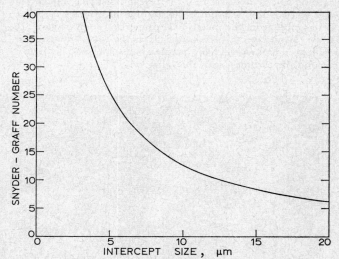

A1 Snyder-Graff number *v.* intercept grain size

Grain size in high-speed steels is usually expressed in terms of Snyder-Graff number, which is the number of grains in a length of 5×10^{-3}in. Figure A1 shows the relationship between Snyder-Graff number and intercept grain size in μm.

Controlled forging of tool steels

A. Tomlinson and D. E. Beard

In many present-day forging procedures it is not possible to complete all the required shaping in one working period due to cooling of the stock. Consequently, it is necessary to return the piece to the furnace for reheating before further work can be done. The necessity of reheating gives rise to a jobbing type of production and has a substantial bearing on productivity levels in forges. The situation is particularly relevant to materials of low workability such as high-speed tool steels because of the narrow range of working temperatures employed to minimize cracking during working. Substantial loss in metal due to oxide scaling and decarburization is a further result of the continual reheating required for high-speed steels.

To improve the working of these difficult materials the possibility of supplying heat during forging has been investigated. The investigation has progressed through a number of stages. As a first step various methods of supplying heat have been studied. This indicated that the only real practical solution was offered by an electrical method employing infra-red heating elements. Based on this technique, a prototype heating installation has been developed on the BISRA experimental forge. The primary aims of the equipment development were, firstly to evaluate improvements in processing leading to time and cost savings, and secondly, to establish data and practical experience for specifying design criteria for industrial installations. Following this stage the aim was to employ the experimental equipment as a research tool in establishing thermomechanical procedures for promoting better mechanical properties.

HEAT-TRANSFER CALCULATIONS
In order to specify any form of heating for application during forging it is necessary to know how much heat is required. It is a fairly straightforward calculation to determine the rate of loss of heat from the surface of a workpiece at a particular forging temperature, and standard equations are available.[1] To supply heat at a rate equal to the heat loss would maintain the forging indefinitely at that temperature.

Table 1 shows the rate of heat loss at various temperatures. The values given are the sum of the radiative and convective heat losses. The former is given by the Stefan–Boltzmann Law:

$$E = \sigma \epsilon (T_F{}^4 - T_S{}^4) \text{ BTU ft}^{-2} \text{ h}^{-1}$$

where σ = Stefan's constant $1 \cdot 73 \times 10^{-9}$ BTU ft^{-2} °K^{-1} h^{-1}
ϵ = emissivity

T_F = ingot temperature and T_S the temperature of the surroundings.

Convection is relatively small at these temperatures and is given by

$$E = 0 \cdot 80(T_F - T_S)^{5/4} \text{ BTU ft}^{-2} \text{ h}^{-1}$$

One of the features evident from Table 1 is the high power necessary to maintain temperatures of the order of 1 200°C. Fortunately, for all practical applications, it is unnecessary to maintain such high temperatures. A rational approach, as a first consideration, appears to be to heat the stock in the furnace in the usual way to maximum forging temperature where the low yield stress allows maximum section to be worked, and to supply a moderate heat input to retard cooling sufficiently to allow forging to be completed in one period.

Based on the equations defining rate of heat loss a numerical method[2] has been employed to calculate temperature distributions during cooling in air for various sizes of stock. The method has subsequently been extended to take into account the effects of supplying heat to the surface of the stock in order to retard cooling. The rates of heat supply chosen were 5, 10, and 15 kW/ft². Figure 1 depicts a typical result and shows the cooling rates calculated for the surface of a 6in square bar. Similar calculations have been made for a range of stock sizes.

INFRA-RED HEATING EQUIPMENT
The results of the heat calculations provide an indication of the amount of heat required for various forging sizes and enable the suitability of heating methods to be assessed.

A variety of heating methods has been considered. These include: exothermic coatings applied to the stock surface, methods based on the use of combustible fuels, high energy beam techniques, and a range of electrical methods comprising induction, direct resistance, and radiant heating methods. Apart from the radiant heating method employing infra-red heating elements, all the methods examined were found to have one or more disadvantages rendering them unsuitable for applying heat during forging. A full review of the considerations leading to this conclusion is outside the scope of the present paper and this aspect is not, therefore, reported in detail.

Hitherto the main reason why radiant heating has not been employed for heating steel to hot-working temperatures has been the lack of suitable heating elements. As the heat-transfer process is entirely by radiation it is necessary to have a source of temperature considerably in excess of the desired stock temperature in order that high heat-transfer rates may be obtained. Heating elements capable of meeting these requirements were introduced by electrical manufacturing

Cooling of the stock during working presents problems in various hot-working processes, but particularly so in the case of forging where many products require to be reheated at intervals during the shaping procedure. Infra-red heating equipment is being developed for applying heat during forging. This should lead to improved productivity and cost savings for a variety of forged products but the technique has immediate implications for the working of high-alloy tool steels. A prototype heating rig of 200 kW rating has been installed on the BISRA experimental forge from which data and practical experience are being obtained for specifying design criteria for industrial installations. Forging trials with the experimental equipment on high-speed steel ingots have shown that reheatings can be drastically reduced because of the extended working times. Additionally, the virtual elimination of cracking even in the most difficult tool steels, and reductions in oxide scaling and decarburization result in increased yield of material with substantial potential for cost savings. Future programmes include a comprehensive investigation to develop improved properties and performance for tool steels.

669.14.018.252.3 : 621.735.3

The authors are with BISRA–The Corporate Laboratories of the BSC

Table 1   Rates of heat loss

Temperature, °C	kW/ft²	BTU/ft²/h
1 200	16·5	56 400
1 100	12·9	43 800
1 000	9·6	32 700
900	7·3	24 800
800	5·3	18 200

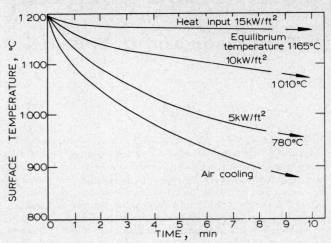

6in square bar

1   Rate of surface cooling

companies in the late 1950s. The heaters were originally of 1 kW power but development has proceeded rapidly and elements are currently available up to 20 kW with a guaranteed life of 4 000 h. A particular feature of the heaters is that of immediate power availability, virtually full power being achieved within a matter of seconds.

Figure 2 illustrates the make-up of a typical heating unit which has been developed to house the infra-red elements.[3] The heater itself comprises a spirally wound tungsten element enclosed in a quartz envelope. This is backed up by a water-cooled reflector of anodized super purity aluminium which may be of parabolic or elliptical profile to direct a parallel or focused beam of radiation onto the stock. The reflector may be fitted with a quartz window in front of the element to prevent deterioration of the anodized surface of the reflector. Element working temperatures are currently in the region of 2 500°K to 3 000°K and considerable glare is emitted, but for those applications where it is considered necessary this can be reduced by coating the quartz window to filter out the visible light.

For the purpose of applying infra-red heating to forging procedures the intention was to employ a number of heater units mounted on the forging equipment. To evaluate the technique and to obtain design data and operating experience, a prototype heating rig was designed, manufactured, and installed on the experimental forge in the metalworking laboratory in Sheffield. For industrial installations it is probable that 20 kW units would be employed, but because of space limitations on the 2 MN (200 tonf) laboratory press 4 kW parabolic units were used for the prototype. A view of the heating equipment after installation is shown in Fig.3. The heater units are mounted in the form of arcs to direct the maximum amount of radiation onto the forge stock. There are six heaters in each arc and the rig comprises six arcs, making up a total power of 144 kW. For experimental flexibility, the equipment can be overvolted from a step-up transformer providing power ratings of 180 kW and 216 kW. To follow the traversing movements of the workpiece during forging

the heater assemblies are capable of moving within the press columns and are motor operated. Electricity and water supplies are brought in by overhead cables and pipes.

Using the parabolic units in an arcuate configuration results in high efficiency levels being maintained for a wide range of workpiece sizes. For furnaces designed for primary heating of billets to hot working temperatures (~1 250°C) full enclosure is possible, and overall efficiencies between 40% and 50% are achieved. Tests with the on-forge heating rig, where full enclosure is not possible, indicated an efficiency of around 30%. This is more than sufficient to offset the high cost of electricity when compared with fuel-fired furnaces employed for reheating.

FORGING TRIALS

Forging trials using the experimental infra-red heating rig have been carried out on various high-speed steel compositions as shown in Table 2. Most of the trials were carried out on 4in square ingots and these were reduced 4:1 in area to produce 2in square billets. In some trials a portion of the billet was further reduced to about 1·6in square to examine the effect of different reductions.

Heating for forging was carried out in two stages; by preheating in a gas furnace to almost 750°C, followed by transfer to a second furnace at the specified forging temperature. For

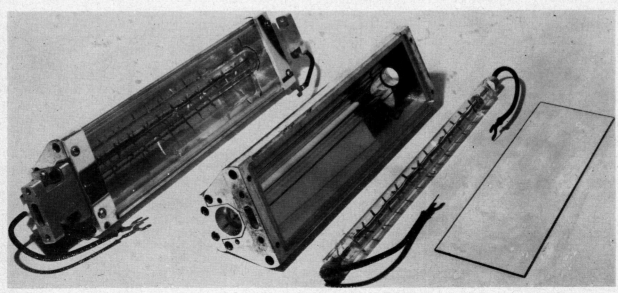

2   Make-up of infra-red heater unit

3 Prototype infra-red heating rig

trials where accurate control of temperature was critical, a thermocouple was embedded below the ingot surface. Minimal soaking was allowed before commencing forging. Temperatures and times were recorded at the start and finish of forging using a total radiation pyrometer and in some cases additional readings were taken while forging was progressing. Preheated forging tools with a nickel-base alloy cap were employed to reduce the considerable cooling effect in the stock imparted by the tools when forging. The billet was slow cooled on completion of forging to prevent thermal cracking.

Table 2 Compositions of high-speed steels

	C	Cr	V	W	Mo	Co
M2	0·85	4·0	2·0	6·0	5·0	..
M4	1·30	4·0	4·0	5·5	4·5	..
M15	1·5	4·0	5·0	6·5	3·5	5·0
T1	0·75	4·0	1·0	18·0	..	..
T6	0·8	4·5	1·5	20·0	..	12·0
T42	1·30	4·0	3·0	9·0	3·0	9·5

After forging the extent of any surface cracking that had resulted from the forging operation was noted. The amount of oxide scaling that had occurred during heating and working was obtained by weighing the ingot prior to forging and the billet after forging. Spectrographic techniques were used to measure the depth of decarburization.

A summary of the results of the trials is given in Table 3. In trials 1–3, 4in ingots in M2 composition were forged to 2in billet over the normal range of working temperature, 1175°C down to 900°C. The press penetrations were typical of those used for cogging carbon steels, i.e. 15–20%, and the 4:1 reduction from ingot to billet was effected in one working period. The ingots worked down satisfactorily and only occasional minor cracking occurred along the corners of the billets. The loss due to oxide scaling varied between 1% and 2% of the original ingot weight. The spectrographic measurements of surface decarburization showed that full core carbon was attained at depth between 0·025in and 0·030in and 90% of the core carbon level was reached at 0·018in/0·019in. The metal losses resulting from scaling and decarburization are less than would be expected for conventional working, and this is further reflected in the yields of 75–81% obtained after grinding the billets. The loss of metal was accounted for, in general terms, by 1–2% scale, 8–14% top and bottom discard, and 6–10% by removal in grinding. A similar result to the above 4in ingots was obtained from a 6in ingot forged to 4in billet (trial 4) but in view of the larger finishing size the scaling and decarburization levels resulted in a lower percentage metal loss.

Further forging trials have been taken to evaluate the response for M2 composition to forging at lower temperatures. In one trial an ingot was forged between 1000°C and 900°C. The ingot was heated in the normal manner to 1175°C and

Table 3 Summary of infra-red forging trials

Ref. no.	Composition	Furnace	Surface at start forge	Surface at finish forge	Time for forging, min	Forging response	Oxide scale loss, %	Decarburization Depth to 90% core carbon, in	Equivalent metal loss, %	Depth to full core carbon, in	Equivalent metal loss, %	Scale and decarb. loss, %	Ingot to ground billet yield, %
1	M2	1175	1060	975	10·00		2·0	Not tested		..	..	..	75
2	M2	1175	1060	930	10·10	Occasional minor cracks	1·25	0·019	3·8	0·025	5·0	6·7	80
3	M2	1175	1080	850	12·00		1·0	0·018	3·6	0·030	6·0	7·0	81
4	M2	1175	1086	912	20·15 (NB 6in ingot)	Occasional minor cracks	1·0	0·019	1·9	0·025	2·5	3·5	..
5	M2	1175	1000	900	16·15	No cracking	..	0·013	2·6	0·038	7·6	..	..
6	M2	820	835	920	15·00	Moderate cracking During final passes	<0·4	Examination suggests slight carburization of surface					
7	M2	800	800	800	6·30	Moderate cracking During final passes	<0·3	No decarburization detected				<0·3	..
8	M4	1175	1075	825	17·00	Minor cracks	2·0	No decarburization detected				2·0	80
9	M4	1175	1050	850	10·20	Minor cracks	1·75	Not tested				..	86
10	M15 (Substandard ingots)	1175	1040	895	13·00	Occasional moderate to large cracks	1·85	Not tested				..	..
11	M15 (Substandard ingots)	1180	~1100	~950	11·30	Segregated areas of large cracks	..	Not tested				..	..
12	T1	1175	~1100	920	6·10	No cracking	1·4	0·014	2·8	0·023	4·6	6·0	..
13	T6	1175	1075	890	12·20	Minor cracking	2·25	Not tested				..	81
14	T6	1175	1070	900	7·00	Minor cracking	0·6	0·006	1·2	0·034	6·8	7·4	..
15	T42	1175	1040	945	13·10	Minor cracking	4·5	Not tested				..	82
16	T42	1175	1070	910	12·30	Minor cracking	1·75	Not tested				..	78

a M2 forged isothermally at 800°C (2in square); *b* T6 forged between 1070 and 900°C (2in square)

4   High-speed billets, infra-red forged

then cooled as quickly as possible to 1000°C before commencing forging. By working in the lower temperature range it was hoped to improve structural refinement. Press penetrations between 7% and 10% were used and a crack-free billet was produced. In other trials, M2 ingots were forged with the material mainly in the α condition. Wallquist[4] has shown that high-speed steel of the 18%W type exhibits reduced strength at 800°C, associated with the high-temperature α condition. The trials were carried out on the assumption that the reduced strength would be accompanied by an increase in ductility. In carrying out these tests great care was taken to ensure that during heating the ingot did not rise above 825°C. Press penetrations were kept below 10% because of the limited ductility of the material in this temperature range. In the first trial, nominally at 800°C the temperature rose from 835°C to 920°C during forging because of difficulties encountered in regulating the heater power. The stock worked down satisfactorily for most of the schedule but during the last few passes some cracking occurred along the billet corners. In the second trial at 800°C the temperature was maintained constant. The response of the material to forging at this temperature was satisfactory but some cracking occurred on two opposite corners of the billet during the later stages of working, and this is shown in Fig.4. Very little oxide scale was produced, most of it being of a powdery nature, and reference to Fig.4 illustrates the good surface finish which resulted. It was not possible to detect any surface decarburization on either of the trials at 800°C and in one billet examination suggested some slight carburization of the surface.

Examination of the standard 18%W high-speed steel composition has been confined to one trial only. The ingot was forged in one working period and the billet produced was completely free from cracking. Oxide scaling and decarburization followed the improved trend indicated for the M2 trials.

Trials have been carried out on several cobalt-bearing high-speed steels as indicated in Table 2. Normal forging temperatures for these materials lie in the range 1175°C to 900°C and these were adhered to in the trials. Also, in view of the intractable nature of these alloys, press penetrations were maintained between 10% and 15%. Using the infra-red heating rig reductions from the 4in ingot size to 2in billet were achieved in one working period. For the M4, T6, and T42 ingots only minor cracking occurred and Fig.4 shows the appearance of one of the billets in T6 composition (12% cobalt). After grinding the billets the overall yields from ingot weight to dressed billet weight ranged between 78% and 86% and were considered particularly good when compared with industrial practice.

The ingots in M15 quality which were forged had been scrapped from production because of inferior surface quality.

These were forged using infra-red heating with generally satisfactory results. Some cracking occurred at the location of the surface defects but the forged billets were acceptable for further processing.

The figures for oxide scale loss in the cobalt bearing compositions are similar to those obtained in the trials on M2 composition and show an improvement over industrial processing conditions. Only limited examination for decarburization was possible because of production commitments but the result obtained for T6 composition compares with the figures obtained for M2 composition.

DISCUSSION

In the trials on the experimental forge, using infra-red heating equipment, working times for high-speed tool steels were considerably extended. Forging reductions from ingot to billet of 4:1 and more were found to be quite feasible in a single working period. Additionally, substantial gains in metal yield were indicated due to reduced cracking, oxide scaling and decarburization.

For the standard Mo-based high-speed composition (M2) trials were taken over different ranges of working temperature with satisfactory results. At the higher working temperatures this composition could be forged with quite heavy press penetrations without fear of cracking but towards the lower end of the temperature range slight cracking tended to occur in some of the trials. This was of a minor nature and should be obviated by further work. The isothermal working trials at 800°C gave particularly interesting results. The reduced strength and increased ductility associated with the high-temperature α condition, as compared with the γ form at the same temperature, provided a successful forging response. Using extended working schedules with small press penetrations the cracking which occurred was of only a minor nature.

Trials of the standard 18%W composition and the cobalt bearing high-speed steels reflected similar improvements to those brought about for M2. Substantial forging reductions were effected in one working period, but as might be expected more careful treatment of the cobalt steels, in the way of smaller press penetrations, was necessary to minimize the tendency for cracking. One of the best results was that shown in Fig.4 for T6 composition containing 12%Co, which is considered to be the most intractable of the range of high-speed steels manufactured. The satisfactory reclamation of the M15 ingots (5%Co–5%V) which were to be scrapped because of surface defects is a further pointer to the efficacy of on-forge heating.

Further work still remains to be done to ensure complete elimination of cracking during primary working of high-speed steels, but attention to the forging technique should provide for this. Modifications to the shaping geometry which will be examined include the use of alternative tool shapes, reduced penetrations in the latter stages of working, and periodic reductions along the corners of the stock to reduce the high local strain in these areas during subsequent working.

Oxide scaling and decarburization figures recorded in the trials show some variability but are lower than the levels normally obtained in industrial production, presumably because of reduced high-temperature exposure resulting from single heat working. Values resulting from working at conventional temperatures were probably about half the corresponding industrial levels, and might well have been better with optimum furnace heating conditions. The drastic reduction in scaling and decarburization in the 800°C isothermal trials is worthy of note. Scaling amounted to less than 0·3% and was in powder rather than flake form. There was no evidence of any surface decarburization. Metallurgical investigation of the billets produced in the trials is proceeding presently.

CONCLUSIONS

Trials on the experimental forge have shown that infra-red heating during working is effective in extending working times for high-speed tool steels and that large forging reductions are feasible in one working period. The conclusion for industrial practice is that, with a heating rig of appropriate power rating, working of these difficult materials could be continued for as long as production requirements dictate. Design principles and operating data are presently being derived from the pilot-plant equipment for specifying industrial installations. In the long term the development could well change the whole pattern of working in forges by replacing existing jobbing-type production with continuous-flow production.

The investigation to date shows that substantial cost savings for these high value tool steels accrue from the use of infra-red heating, through reductions in cracking, oxide scaling, and decarburization. Current work is aimed at optimizing the improvements. In this context, working the standard Mo-based composition at around 800°C, in the 2 phase condition, is a promising line of investigation which is being pursued.

The infra-red equipment will be employed in future work as a base for development of thermomechanical treatments and a comprehensive programme has been formulated for improving the properties and performance of high-speed steels.

ACKNOWLEDGMENTS

The authors wish to acknowledge the help received from their colleagues in this work, particularly the contribution of Mr R. I. Bly, who was responsible for the equipment development.

They also thank the members of the Forging Committee of BISRA for their interest and support and the tool-steel manufacturing companies who kindly provided materials for investigation.

REFERENCES

1 H. M. SPIERS: Technical data on fuel
2 N. V. HUNT: *BHP Techn. Bul.*, 1958, **2**
3 W. R. LAWS: *Steel Times*, 1969, Feb., 26
4 G. WALLQUIST: *Jernkon Annaler*, 1961, **145**, 5, 231–254, Eng. trs. BISIT5810

A British Standard for tool steels

P. Jubb

NEED FOR A STANDARD

For a number of years it was the practice in this country to sell tool steel under makers' brand names, in many cases without disclosing the chemical composition. In defence of this practice it was claimed, with some justification, that the quality of the steel depended on the maker's skill, on the care with which manufacturing variables were controlled, and not merely on composition. Nevertheless there were many disadvantages to the system, particularly for the user of tool steel. If the composition was not known, changing to another source of supply could be difficult, if the original brand of steel became unobtainable or proved unsatisfactory for any reason. Moreover, if such a change were made to an allegedly similar steel, minor differences in composition could make the response to heat treatment different with a consequent difference in service performance. Each maker had his own ideas of the most suitable composition for a given type of steel, and an exact equivalent might not be available. It is not surprising that once users became accustomed to a particular brand they were reluctant to change. Although a change might ultimately lead to improved performance, experimental work to prove its suitability could be expensive.

With the advent of a more scientific approach to the use of tool steels, many of the larger users preferred to buy steel of known composition and, in the absence of a recognized standard, to draw up specifications of their own. Unfortunately the compositions specified differed from one another, so it has frequently been necessary for manufacturers to melt specially and keep separate stocks for individual customers. There has also been a tendency for what many people consider to be an unnecessarily large number of compositions of tool steel to be sold, differing from one another in the quantity of major alloying elements present and in the variety of minor elements added. This practice also adds to the cost of production by making excessively large tool-steel stocks necessary.

As a result of these considerations a demand arose, particularly from users, for a British Standard. The principal advantages hoped for were that uniformity of composition and quality would ensure consistent response to heat treatment and satisfactory performance in service, that the number of types of steel would be reduced enabling manufacturer and user to reduce their stocks, and that purchasing would be simplified.

There are of course disadvantages in working to a standard. The user may be forced to accept a composition of steel which is not exactly suited to his particular application. The manufacturer may on the other hand have unaccustomed restrictions imposed upon him which could add to the cost of production. He will probably have to carry out more inspection, and if the requirements of the standard are too stringent he may have to scrap a higher proportion of his output than formerly. In drawing up a standard, therefore, care must be taken not to set unnecessarily severe acceptance limits. It is difficult to satisfy all users of the standard. Those who are using tool steels for onerous duties may wish to specify precise limits of composition and quality which lead to an increase in production cost. The improvement in performance may compensate for an increase in cost to such users but this will not be welcomed where the steel is required for less critical service conditions. These remarks apply to standards in general but they apply with particular force to high-speed steel, for example, where significant reduction in carbide segregation or decarburization can add considerably to the cost of production.

BASIS FOR SELECTION OF STEEL

The best known classification of tool steels is that of the American Iron and Steel Institute which groups the steels according to type. This system is widely used in this country and abroad and has formed the basis for other systems of classification. It was decided that it should also be used as the basis for the British Standard, retaining the American symbols which have now become well known, prefixing them with the letter *B*.

Not all the American compositions have been included and some of those that have been are modified slightly to make them conform to British practice. The intention has been to restrict the number of compositions including only those most widely used in this country and bearing in mind that one advantage hoped for is a reduction in the number of tool steels in use. One of the difficulties has been to decide which steels to include, since little information is published on tool-steel usage in this country. The selection has therefore been based on information obtained from the principal manufacturers and users. It may be found that some additional steels should have been included; if so they will have to be added in subsequent revisions. The introduction of new steels will probably prove easier than the deletion of others which are falling into disuse. Revision of the tool-steel standard will perhaps need to be more frequent than is customary with British Standards because of the changing pattern of tool-steel practice. Even while the standard has been in course of preparation there have been changes in the usage of high-speed steel which perhaps ought to have been taken into account by the drafting committee.

The new standard includes most types of tool steel: high-speed, hot-work, cold-work, shock-resisting, special-purpose, and carbon tool steel. High-speed steels are probably the most important, and since this is a cutting tools conference attention will be concentrated on this class of steel. Table 1 gives a list of the steels selected for the standard and also gives their compositions.

A British Standard has been prepared for tool steels based on the American system of classification, but modified to conform with British practice. Composition limits and hardness values are specified and appropriate heat treatments are recommended. These have been based on current practice. No more rational procedure has been possible owing to the absence of sufficient reliable information on the influence of manufacturing variables on cutting performance. It has been appreciated that if the standard of quality is set too low users who require a high-quality product will be dissatisfied, if it is set too high other users will be paying more than is necessary for their requirements. Two sets of tolerances for dimensions and decarburization have therefore been specified, and for carbide segregation a set of standard photomicrographs has been prepared. It is left to the purchaser and supplier to agree which set of tolerances, and which of the standard structures, shall be used.

669.14.018.252.3:389.6(42)

The author is with the Brown-Firth Research Laboratories, Sheffield

COMPOSITION LIMITS

The American system of classification quotes typical composition only. In the British Standard it was decided to go a step further and specify composition limits so that users ordering to the standard could expect a consistent response to heat treatment. Permissible limits for residual elements have also been specified. Some degree of control is obviously desirable for elements such as phosphorus and tin which if present in excess could cause brittleness. Other elements such as nickel and molybdenum, although not normally regarded as harmful could, if present in unusually large amounts, lead to variability in response to heat treatment. It would be illogical with tungsten high-speed steel, for example, to control the tungsten content to close limits and then permit wide variations in molybdenum content. There are considerable economic advantages when melting tool steel to use a high proportion of scrap, but as a result of scrap contamination elements such as Ni, Co, and Mo are found in steels where they are not wanted. To restrict these elements to very low limits means that virgin alloy must be used in the charge adding to the cost of production. Nickel is an example of an element which can cause a lot of trouble. It is not deliberately added to high-speed steel, or to many other tool steels, but because of its wide use in other steels and alloys it contaminates tool steel. This in moderate amounts causes no significant difference in behaviour, in large quantity it can lead to difficulty in softening and to increased austenite retention during hardening and tempering. A limit of 0.40% has been set, higher than many users would like and lower than most manufacturers would wish or think necessary. In setting composition limits the drafting committee have been hampered by the lack of reliable information on the influence of variation in composition on tool performance, and have to a large extent been guided by precedent.

PROPERTIES

One of the difficulties in preparing a standard for tool steel is that there are no simple tests which can be applied to prove the suitability of the steel for use. Hardness tests can be used of course and the standard specifies maximum values for the steel as supplied to the customer, and minimum values which the steel must be capable of giving after a specified hardening and tempering treatment. The first shows that annealing has been adequate, the second that the steel is capable of attaining an adequate hardness in the condition in which it is intended to be used.

Hardness is not the only quality, however, which a tool needs in order to cut efficiently. Toughness, resistance to repeated stressing and to abrasion are also desirable properties, but there are no generally recognized tests to assess these qualities and not much information on how they are related to cutting performances. Tensile tests which are used to indicate that an adequate standard of quality has been attained in the production of constructional steels, and which give the engineer a parameter he can use in design, are of no value for tool steels since in the heat-treated condition their ductility is too low for a tensile test to give a result which has any meaning. Other tests such as unnotched impact, torsion impact, and bend tests have been proposed to measure toughness but they have not been generally accepted and insufficient is known of their correlation with cutting performance for any of these tests to be regarded as having a useful place in a tool steel standard.

· Hot hardness is of interest but its relationship with room-temperature hardness does not vary much from one steel to another provided the tests are made below the temperature at which the steel has been tempered. Consequently, for the purpose of assessing a steel's suitability for service, hardness tests at elevated temperatures offer no worthwhile advantage over the much simpler test at room temperature.

Cutting performance tests are so difficult to carry out that not only are they out of the question as acceptance tests, but very little is known of the influence on cutting performance of carbide segregation or small variations in composition and heat treatment. If such information had been available it would have provided a rational basis for specifying limits of composition and segregation. Instead the standard has had to be based on current practice which again has been derived from what users consider to be necessary and manufacturers consider to be practicable.

CARBIDE DISTRIBUTION

The question of carbide distribution led to considerable discussion when preparing the standard. There is general agreement that excessive segregation is objectionable and that it is desirable to have some safeguard to prevent the supply of excessively segregated steel, particularly for severe conditions of use. Some users already specify what degree of segregation they are prepared to accept, but there is no general agreement either on the method of defining segregation or on the level which is considered to be acceptable.

The two principal methods of assessing carbide segregation are the measurement of the width of carbide bands, typified by German practice, and by comparison with standard photomicrographs. In well worked steel, where the carbide particles are strung out in the direction of rolling, there is little difficulty and either method is adequate. In larger sections, where the carbide distribution is irregular, the position where the band width is to be measured can be difficult to decide; comparison with standard photomicrographs is often easier with such sections, although it must be admitted that structures are frequently encountered which are so different in pattern from the standard, that there can be scope for disagreement. In the end it was decided that standard photomicrographs should be used.

The problem of specifying the permissible degree of segregation has been evaded by leaving this to be agreed between purchaser and supplier. If a high standard had been set the cost of production would have been increased and for less critical applications users would be paying more than

Table 1 Percentage composition of high-speed tool steels

	C	W	Mo	Cr	V	Co
BM1	0.75–0.85	1.0–2.0	8.0–9.0	3.75–4.5	1.0–1.25	0–0.60
BM2	0.80–0.90	6.0–6.75	4.75–5.5	3.75–4.5	1.75–2.05	0–0.60
BM4	1.25–1.40	5.75–6.5	4.25–5.0	3.75–4.25	3.75–4.25	0–0.60
BM15	1.45–1.60	6.25–7.0	2.75–3.25	4.5–5.0	4.75–5.25	4.5–5.5
BM34	0.85–0.95	1.7–2.2	8.0–9.0	3.75–4.5	1.75–2.05	7.75–8.75
BM42	1.00–1.10	1.0–2.0	9.0–10.0	3.5–4.25	1.0–1.3	7.5–8.5
BT1	0.70–0.80	17.5–18.5	0–0.70	3.75–4.5	1.0–1.25	0–0.60
BT2	0.75–0.85	17.5–18.5	0–0.70	3.75–4.5	1.75–2.05	0–0.60
BT4	0.70–0.80	17.5–18.5	0–1.0	3.75–4.5	1.0–1.25	4.5–5.5
BT5	0.75–0.85	18.5–19.5	0–1.0	3.75–4.5	1.75–2.05	9.0–10.0
BT6	0.75–0.85	20.0–21.0	0–1.0	3.75–4.5	1.25–1.75	11.25–12.25
BT15	1.40–1.60	12.0–13.0	0–1.0	4.25–5.0	4.75–5.25	4.5–5.5
BT20	0.75–0.85	21.0–22.5	0–1.0	4.25–5.0	1.4–1.6	0–0.60
BT21	0.60–0.70	13.5–14.5	0–0.70	3.5–4.25	0.40–0.60	0–0.60
BT42	1.25–1.40	8.5–9.5	2.75–3.5	3.75–4.5	2.75–3.25	9.0–10.0

Table 1   Stahl-Eisen Werkstoffblatt 320–63

Designation New	Old	Werkstoff no.	Chemical composition, % C	Co	Cr	Mo	V
S3–3–2	ABC III	1·3333	0·92–1·02	..	3·8–4·5	2·5–2·8	2·2–2·5
S2–9–1	BM09	1·3346	0·78–0·86	..	3·5–4·2	8·0–9·2	1·0–1·3
S2–9–2	BM09V	1·3348	0·92–1·02	..	3·5–4·2	8·0–9·2	1·8–2·2
S6–5–2	DMo5	1·3343	0·78–0·86	..	3·8–4·5	4·7–5·2	1·7–2·0
S6–5–3	EMo5V3	1·3344	1·15–1·25	..	3·8–4·5	4·7–5·2	3·0–3·5
S6–5–2–5	EMo5Co5	1·3243	0·78–0·86	4·5–5·0	3·8–4·5	4·7–5·2	1·7–2·0
S10–4–3–10	EW9Co10	1·3207	1·15–1·30	10·0–11·0	3·8–4·5	3·5–4·0	3·0–3·5
S12–1–2	D	1·3318	0·83–0·93	..	3·8–4·5	0·7–1·0	2·3–2·6
S12–1–4	EV4	1·3302	1·20–1·30	..	3·8–4·5	0·7–1·0	3·5–4·0
S12–1–4–5	EV4Co	1·3202	1·25–1·40	4·5–4·5	3·8–4·5	0·7–1·0	3·5–4·0
S18–0–1	B18	1·3355	0·70–0·78	..	3·8–4·5	..	1·0–1·2
S18–1–2–5	E18Co5	1·3255	0·75–0·83	4·5–5·0	3·8–4·5	0·5–0·8	1·4–1·7
S18–1–2–10	E18Co10	1·3265	0·72–0·80	9·0–10·0	3·8–4·5	0·5–0·8	1·4–1·7

best performance in machining high-tensile construction materials for aircraft, spacecraft, and missiles. This M42 shows an original hardness of 62–65 HRC after quenching and contains about 25% of retained austenite which can be broken down by quadruple tempering to about 2% with an increase in hardness to 69–70 HRC.

Table 6 shows some super-high-speed steels which have been developed from this basic steel. All these steels show a vanadium to carbon ratio which is lower than that of standard high-speed steels and it is interesting to consider how a similar ratio would affect such a standard high-speed steel, e.g. M2 (S6–5–2). In theory, after suitable heat treatment, such a modification should lead to higher hardness and better resistance to tempering and wear.

The hardening temperature is, in fact, slightly reduced with the upper limit lying at about 1220°C (2228°F). Due to good solution properties this steel contains a comparatively high proportion of retained austenite in the as-quenched condition and this is reflected in its lower hardness. Multiple tempering (of at least × 3) lifts the hardness to higher values than in the normal M2 type (i.e. its secondary hardness is higher). It was finally settled in American reports that hardenability, temper resistance, and wear resistance were improved over normal M2. Besides this, the steel with the higher carbon level should have a smaller grain size and better carbide distribution.

Soon afterwards, reports began to appear in German literature of trials on higher carbon S6–5–2 (M2). Again the lower hardening temperature was mentioned and also the increased difference between original and secondary hardness due to the fact that more than half of the carbon goes into solution resulting in better tempering resistance. In spite of this, the elevated carbon leads to greater amounts of undissolved carbide which increase the wear resistance. In Germany, the grain size was found to be coarser with higher carbon levels.

Naturally, these improvements can only be attained with some loss of toughness due to the coarser grain and the increased carbide content but this need not affect tool performance too much because, in the meantime, there has been a lot of experience with carbide which is far less tough.

So far we have considered published data which did not always agree and so a series of trials were conducted, particularly with steel of the M42 type. There were, of course, many failures which could be attributed to outside causes

Table 2   Gradation of high-speed steels

German standard Stahl-Eisen Werkstoffblatt 320

Edition	For normal performance	For high performance	For highest performance Rough cutting	Fine cutting
1963	S3–3–2 S10–0–1 S2–9–1 S2–9–2	S12–1–2 S6–5–2	S6–5–2–5 S10–4–3–10 (S12–1–4–5) S18–1–2–5 S18–1–2–10	S6–5–3 S12–1–4 S12–1–4–5
1969	S3–3–2	S6–5–2 SC6–5–2	S6–5–2–5 S10–4–3–10 S18–1–2–5	S6–5–3 (S10–4–3–10) S7–4–2–5 S12–1–4–5

Table 3   Graduation according to alloys

Stahl-Eisen Werkstoffblatt 320 Alloy type	Edition 1963	1969
Mo–W alloyed	S3–3–2	S3–3–2
	S2–9–1	..
	S2–9–2	..
	S6–5–2	S6–5–2
	..	SC6–5–2
	S6–5–3	S6–5–3
	S6–5–2–5	S6–5–2–5
	..	S7–4–2–5
	S10–4–3–10	S10–4–3–10
Medium W alloyed	S12–1–2	..
	S12–1–4	..
	S12–1–4–5	S12–1–4–5
High W alloyed	S18–0–1	..
	S18–1–2–5	S18–1–2–5
	S18–1–2–10	..

Table 4   Survey of deliveries

Designation	Delivery, % 1965	1966	1967	1968
S3–3–2	6·1	7·0	6·5	5·8
S2–9–1	0·3	0·01	1·1	1·0
S2–9–2	0·3	0·8	0·9	0·4
S6–5–2	46·8	53·9	45·6	61·1
SC6–5–2	..	0·02	0·05	0·3
S6–5–3	0·3	0·2	0·2	0·6
S6–5–2–5	2·9	3·5	3·5	3·3
S7–4–2–5	..	..	0·01	0·3
S10–4–3–10	1·5	2·0	3·5	3·0
S12–1–2	0·1	0·2	0·1	0·5
S12–1–4	0·5	0·2	0·1	0·2
S12–1–4–5	0·3	0·4	0·8	0·5
S18–0–1	24·5	18·9	24·2	15·0
S18–1–2–5	6·6	4·7	7·1	3·0
S18–1–2–10	2·3	1·7	3·5	2·4
Sonstige	7·5	6·4	2·8	2·6
Σ in t	4215,7	3679,6	3769,0	4227,7

Table 5   Carbon content of high-speed steels

Stahl-Eisen Werkstoffblatt 320

Designation	Edition 1950	1959	1963	1969
S3–3–2	1·00	0·92–1·02	0·92–1·02	0·95–1·03
S2–9–1	..	0·78–0·86	0·78–0·86	..
S2–9–2	..	..	0·92–1·02	..
S6–5–2	0·85	0·78–0·86	0·78–0·86	0·84–0·92
SC6–5–2	..	..	..	0·95–1·05
S6–5–3	..	..	1·15–1·25	1·17–1·27
S6–5–2–5	..	..	0·78–0·86	0·88–0·96
S7–4–2–5	..	..	..	1·05–1·15
S10–4–3–10	..	..	1·15–1·30	1·20–1·35
ABC II	0·80	0·78–0·86	..	..
S12–1–2	0·85	0·82–0·90	0·83–0·93	..
ECO3	0·90	0·77–0·85	..	..
S12–1–4	1·30	..	1·20–1·30	..
S12–1–4–5	1·35	1·25–1·40	1·25–1·40	1·30–1·45
S18–0–1	0·75	0·70–0·78	0·70–0·78	..
S18–1–2–5	0·80	0·75–0·83	0·75–0·83	0·75–0·83
S18–1–2–10	..	0·72–0·80	0·72–0·80	..

Table 6   First generation of super-high-speed steels

AISI	C	Cr	Mo	V	W	Co
M41*	1·10	4·20	3·75	2·00	6·80	5·00
M42†	1·10	4·00	9·50	1·20	1·50	8·00
M43	1·23	3·80	8·70	2·00	1·80	8·20
M44	1·15	4·20	6·25	2·25	5·80	12·00

* S7–4–2–5 † Hypercut

such as inadequate heat treatment or machines with insufficient rigidity but, apart from this, it became apparent that only a few tool types can be operated at hardnesses over about 67 HRC even with steels in the M40 series. It is not surprising, therefore, that only a few applications are covered by this class of steel (primarily M42) where it is used to some extent to replace the 4%V type with its grinding problems. Although M41 has found a place in the new addition as S7–4–2–5 it is only for rather special applications, e.g. taps and roughing face cutters.

Apart from this development which was imported from the USA there was a parallel movement initiated in Germany by the demand for a specific ratio between carbon and vanadium. With 1·60–1·80%V, a carbon level of 1·03–1·08% should be the optimum in order to achieve maximum hardness (920 HV and more). Small grain size gives good toughness which allows the use of higher hardnesses and such steels should be comparable in performance with cobalt-alloyed steels of the same basic composition.

This demand has not been recognized in the new edition because there has been no real success yet in this field. Additionally, there were teething difficulties in the production of these steels in larger sizes. The grain size in forged bar was not fine enough while carbide size and distribution would not meet an ideal specification.

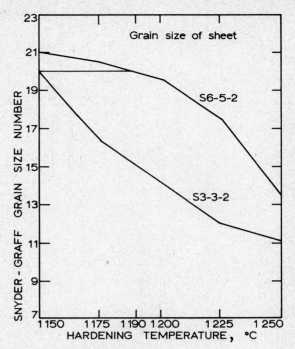

2   Grain size of high-speed steel sheet

In spite of these difficulties, many investigations and trials were started. The change from 0·85%C to the theoretical figure of 1·08%C for M2 seemed to be too big a risk for many people who, therefore, commenced trials with intermediate carbon levels of 0·90–0·95% or 1·00% while only a few started with 1·10%C. All of this happened at a time when conjecture was rife and the steel manufacturers could hardly oppose the wishes of their customers, which were not always reasonable.

The aim of the national standard data sheets is not only to guide the choice of the most suitable steel but also to reduce the number of steels in order to streamline production and stockholding. Thus it was high time for a new edition since development was running in just the opposite direction. In addition, it was time for a revision of carbon limits on standard steels since nobody would have accepted a steel like S6–5–2 (M2) or S6–5–2–5 (M35) with a carbon content lower than 0·80%.

It was clear that everybody was aware of the difficulty in attaining 64 HRC minimum if there was insufficient carbon present for solution in the matrix. In this connexion it is interesting to consider balance of carbon content against carbide formers.

Theoretically, it is possible to calculate the optimum carbon content from the sum of the contents of the carbide-forming elements provided that adjustment is made to allow for their various reactions with carbon. Using the formula $\Sigma SC = (\%W) + (1·9 \times \%Mo) + (6·3 \times \%V)$ a graph showing optimum carbon saturation can be derived as in Fig.1.

This shows most standard types of high-speed steel to lie below the optimum, particularly S6–5–2 (M2) and S6–5–2–5 (M35), thus indicating insufficient carbon to allow the highest possible hardness from their alloy content.

Far above the line lies S3–3–2 which is hardly used anywhere outside Germany. Although this is a low-alloy steel it has a very good performance, particularly as hacksaws or circular saws. It has a tendency to grain growth making it imperative to avoid overheating or oversoaking and thus the heat treatment is rather critical.

From our experience of this steel we would go further and say that all steels with a carbon level of over 1·0% will have a tendency to grain coarsening unless the carbon is matched by a higher vanadium addition. For a long time now we have

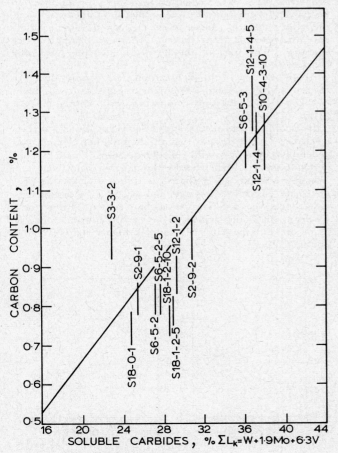

1   Carbon saturation of high-speed steel

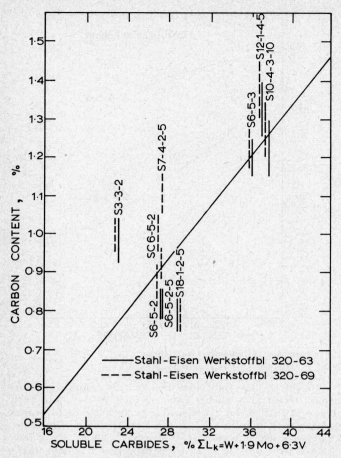

3   Carbon saturation of high-speed steel

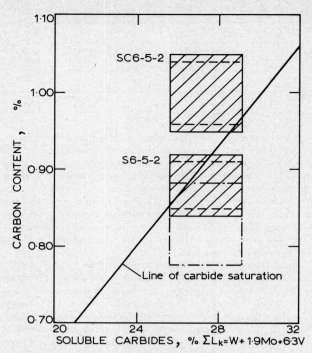

4   Carbon saturation of M2 and HiC M2 alloy range

been checking the grain size of production sheets and the results obtained on S6–5–2 and S3–3–2 are shown in Fig.2.

The difference in grain size is quite evident and a customer who uses both of these steels could well complain of coarse grain in S3–3–2 but there is little that the steel manufacturer can do since this is an inherent feature of this quality.

If the revised carbon limits as stated in the new edition are plotted on the graph of carbon content versus soluble carbide we arrive at Fig.3 which shows some of the steels to be just as far above the line as S3–3–2 which we have just considered.

Difficulties may thus arise from the grain size (which indicates the toughness of the tool) and while this is not very likely with S6–5–2 (M2) or S6–5–2–5 (M35) it could easily be a problem with S7–4–2–5 (M41) and perhaps even with SC6–5–2 (HiC M2). Indeed, it is exactly these two steels where one finds reference to coarse grain in published literature and we therefore felt we should conduct our own investigations into grain size comparisons between S6–5–2 (M2) and SC6–5–2 (HiC M2).

Before showing these results we would like to mention the influence of the permitted variation in carbon and alloy contents. In the new edition a range of 0·08% is allowed on carbon levels up to 1·00%, 0·10% on levels up to 1·20%, and 0·15% on levels over 1·20%. It is fortunate that steelworks do not require the whole of this permitted range since it would otherwise be very difficult to guarantee uniform hardening properties within any one brand. This is shown quite well in Fig.4 which plots all permissible variations in analysis for S6–5–2 and SC6–5–2.

As can be seen, either of these steels may lie within a quite large field and yet the properties of that steel should always be the same, irrespective of whether its composition places it in the bottom right or the top left corner. Hardening and

tempering should produce the same results regardless of whether it has the lowest permitted carbon level with the highest permitted alloy contents or vice-versa.

It is important, therefore, that all steelworks should try to narrow this range as we do and this is indicated on the graph. Normally a customer will find little difference in hardening properties between batches from the same supplier although he may find a difference between rolled and forged steel since the latter tends to a coarser grain. He may also find a difference in materials from different sources of supply since analysis is only one of the factors influencing hardening characteristics.

About 100 tons of the high-carbon type were produced in the second half of 1968. The graphs in Fig.5 show the grain size and the hardness in comparison with normal S6–5–2 (M2).

It should be explained that the S6–5–2 type had a carbon level of 0·82–0·88% and that the high-carbon version was generally 0·95–0·97% with 1·00% maximum. This difference is not so great as that fixed by the new Werkstoffblatt 320. The above results are all based on one size of flat bar produced for circular saw inserts and can thus be assumed to reflect only the difference in carbon content. From these results there would appear to be hardly any difference in grain size but the HiC M2 gives a higher hardness after double tempering.

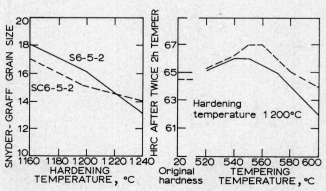

5   Grain size and tempering properties of HiC M2 and M2

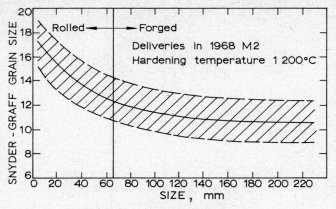

6 Grain size in dependence of size (round bar)

It is not possible to draw conclusions, from the above results, on the effects of the higher carbon in the new edition nor of the effects of forging in larger sizes. Unfortunately, we have not yet had a proper opportunity to test this HiC M2 in larger quantities nor in bigger sizes. There have been limited orders for these high-carbon types, particularly in the larger sizes because the customers' interest has been virtually killed since the German Special Steel Authorities agreed on a 10% increase in price for these steels to compensate for the reduced yield in production.

We have seen that this question of austenite grain size is one of the most important subjects for investigation in high-speed steel quality since it depends not only upon hardening temperature and soaking time but also upon composition, metallurgical treatments such as inoculation, amount of deformation and also, to a large extent, upon the thermal history both before and after hot forming.

Predelivery checks by the manufacturer and incoming material control by the customer should both include determination of grain size as an essential prerequisite to release of the steel. In our experience, however, many users take no account of the influence of material size upon grain size. It is well known that larger sizes of bar (e.g. over 100 mm dia.) must be produced from comparatively large ingots which require several heatings with long soaking times during hot reduction. This can well result in grain size numbers of 10–12 or sometimes even lower, i.e. at the lower limit of the acceptance range, even though the amount of reduction is sufficient.

The results of all the grain size tests which were conducted on S6–5–2 (M2) production during 1968 have been plotted and the resultant graph is shown in Fig.6 which clearly illustrates a direct relationship between grain size and bar size.

The curve and scatter band thus obtained is not so attractive as similar published literature. Since our curve has not been corrected in any way, we believe that graphs published elsewhere have either been taken from a small number of samples or that the results have been 'smoothed' to better values.

A further source of coarse grain, in addition to production method and size, is the annealing treatment, particularly where long soaking times are applied at higher temperatures. A rather special case of grain size trouble is the mixture of large and small grains side by side which can be caused by insufficient austenitization in an isothermal anneal.

In the new edition, the hardening and tempering temperatures will be listed as before but there will also be shown the HRC which can be obtained with each quality after tempering (*see* Table 7).

Allowance is made for the slight increase in carbon level by reducing the hardening temperature for S6–5–2 (M2) at its upper and lower limits by 10°C and for S6–5–2–5 by lowering the upper limit. In spite of this data, however, the tool manufacturer will still need to determine for himself the optimum hardening temperature in conjunction with soaking time.

For the first time, the S6–5–2 steel is proposed for cold forming tools in addition to the usual cutting-tool applications. High-speed steels have already been used for such tools for a considerable time but cold forming (cold extrusion) as well as hot forming is on the increase and opening new fields every day. Additionally, the raw materials which are being shaped by these methods are becoming more highly alloyed and of higher strength.

Maximum attention should be paid to this problem by the steelworks in order to provide the very best material for this application since, until now, it has been thought in many cases that these demands could not be met by steels. In a lot of instances, however, the heat treatment of cold-working tools has not been suitable for the type of loading which is encountered and because of this the new edition proposes controlled under-hardening from 1 180°C (2 156°F).

Our experience leads us to suggest a slightly lower temperature of 1 170°C(2 138°F) followed by tempering at 520°C (968°F) or 580°C (1 076°F) if a hardness of less than 63 HRC is required. It is important not to go too low with the hardening temperature where the tools have to withstand heavy pressure loads. Only a structure which is sufficiently transformed can stand high pressures and it is often necessary to compromise between full transformation and under-hardening in order to obtain the optimum balance between toughness and compressive strength.

A further item taken into consideration for the first time in the new edition of the Stahl-Eisen Werkstoffblatt is the addition of sulphur. Sulphurized high-speed steels were developed many years ago and are being used where good machinability is desirable. Gear-cutting hobs and milling cutters should have a smooth surface after machining on the backing-off lathes. This backing-off, or relieving, is done at low speeds and high-speed steels in the normally annealed condition do not give a fine surface finish. It is, therefore, common practice for the rough machined tool to be pre-hardened to 38–42 HRC or to use a sulphurized version of the high-speed steel. Many toolmakers continue to pre-harden even when using these free-cutting types.

There is no universally accepted sulphur level since

Table 7 Hardening and tempering temperature

Designation	Stahl-Eisen Werkstoffblatt 320 1963 Edition Temperature, °C Hardening	Tempering	1969 Edition Temperature, °C Hardening	Tempering	HRC*
S3–3–2	1 180–1 220	530–550	1 180–1 220	520–550	62–64
S6–5–2	1 200–1 240	540–560	1 190–1 230	530–560	64–66
SC6–5–2	. .	. .	1 180–1 220	530–560	65–67
S6–5–3	1 200–1 240	550–570	1 200–1 240	540–570	64–66
S6–5–2–5	1 210–1 250	560–580	1 210–1 240	540–570	64–66
S7–4–2–5	. .	. .	1 180–1 220	540–570	66–68
S10–4–3–10	1 210–1 250	560–580	1 210–1 250	550–580	65–67
S12–1–4–5	1 210–1 250	560–580	1 210–1 250	550–580	65–67
S18–1–2–5	1 260–1 300	560–580	1 260–1 300	550–580	64–66

* Hardness after tempering

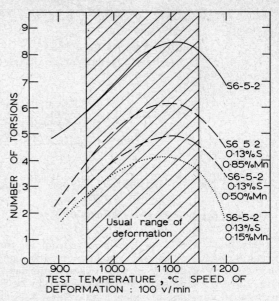

a relieved hob; *b* turning tests; *c* sulphur inclusions at 0·12%S

7   Sulphurized M2 (S6–5–2)

8   Hot-forming properties of high-speed steels M2 and 0·13S and different Mn content

American markets demand a figure of 0·12–0·16%, while in Germany a lower figure of 0·06%S is the norm. This difference is probably accounted for by the fact that in Germany, prehardening is usually carried out in any case, because this not only improves the surface finish obtainable but also acts as a stress-relieving treatment after the rough machining, thus giving a reduction in distortion problems during the later, genuine hardening treatment. The advantages to the toolmaker of sulphur additions are quite obvious from even a simple test as is shown in Fig.7.

These steels are not very popular with the steel producer, however, because it is not always possible to recoup the higher production costs. As an illustration of this we would say that the yield from sulphurized steels is reduced by about 1% for every 0·01% increase in sulphur. These increased manufacturing losses usually occur during the initial stages of hot deformation and also through the need for a greater discard from the bar ends to ensure a sound bar to the user.

Hot torsion tests generally confirm the lower plasticity of sulphurized high-speed steels and the results of such tests are shown in Fig.8. The effect of the manganese content is also demonstrated at the same time with a steady improvement in the properties as the manganese level is raised.

We are pleased that the German Special Steel Committee have proposed an increase of about 10% on the price of these steels since this merely means acceptance of the facts.

Indeed it is surprising to us that only for the higher carbon-alloyed type SC6–5–2 (HiC M2) is the addition of sulphur provided. It may be that, in Germany, there are toolmakers who use this type but we only know of customers using sulphurized S6–5–2 (M2) and S6–5–2–5 (M35).

Higher prices are also fixed for the steels SC6–5–2 and S7–4–2–5 since the hot deformation characteristics of these steels are also worse than with the normal carbon level steels. This is illustrated by the hot torsion figures shown in Fig.9 and when one considers these curves it is no wonder that the output is lowered and the production costs raised with the carbon content.

It was, in any case, high time for a new issue of the Stahl-Eisen Werkstoffblatt 320 which would take into account the general trend in high-speed steel development. It was also advisable, in our opinion, that the steel manufacturers and the Special Steel Committee should ask for higher prices for those types of S6–5–2 (M2) which contain more than 0·92%C and those types of S6–5–2–5 (M35) with more than 0·96%C. It was felt that this would reduce the number of different types of high-carbon high-speed steels and prevent customers

from changing to qualities which bring more difficulties into the steel production without giving any real advantage to the tool performance.

Thus far we have spoken of the developments as reflected by the changes in composition but, in spite of the continuing demands for higher performance there has also been much work done to improve the properties of the general high-speed steels and different steelworks have chosen different ways to achieve this object, but all efforts have, primarily, aimed at achieving the utmost homogeneity.

Special metallurgical methods, as well as higher rates of deformation have been applied with many trials and investigations being made. We would first like to mention two of the most important metallurgical methods. The first one is ingot inoculation using an agent such as titanium nitrides to produce a homogeneous structure in the as-cast ingot, and this is used by many steelworks, the main success being in the elimination or reduction of transcrystallization and formation of large primary crystals. Crystal segregation is kept to a low level and the grain size characteristic is quite different to that of an aluminium deoxidized steel.

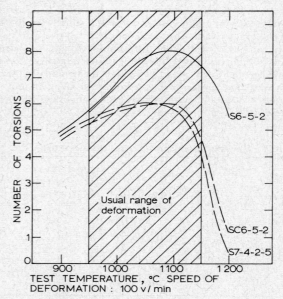

9   Hot-forming properties of high-speed steel

a beginning; b end

10 Upsetting process of original ingot

This treatment must be conducted with care since over-inoculation can give ingot segregation leading to ghost lines. Although the grain is comparatively small the knots of the carbide network can be thick which lead to short but wide carbide lines which can still be seen after high rates of deformation.

The second metallurgical treatment is to remelt the steel by the vacuum consumable electric-arc or by the electroslag processes. We have investigated both of these methods and would say that the only real improvement given is in the degree of cleanness. There is no improvement in carbide distribution unless a higher rate of hot working is applied. Since we have very little trouble with non-metallic inclusions in melting high-speed steel by the electric-arc process, it is not economical to produce by the electroslag process (particularly in large quantities) unless we were to receive a higher price. Even for special purposes however, in our experience, no customer is prepared to pay extra for the better quality since, in most cases, he in turn cannot ask a higher price for offering a better tool.

The same reasons of economy prevent us from applying very large amounts of hot reduction although we have done a lot of work in this field, arriving at a special forging technique which allows the production of a better steel at reasonable costs. The initial ingot, which must be of the highest soundness and metallurgical quality, is upset to provide large amounts of hot reduction, even in bars of larger sizes, without the use of large ingot moulds which can never be a satisfactory starting point.

Some idea of the stresses involved during the upsetting of the original ingot can be gained from Figs.10 and 11 which

show the beginning and end of this process respectively. It requires no great imagination to visualize what would happen to the inside of an unsound ingot during this sort of deformation on a relatively weak, as-cast structure.

This operation could be repeated several times to obtain the very best carbide distribution but this would be accompanied by rising production costs which, as already mentioned, the customer would be unwilling or unable to bear and a compromise is therefore made which gives sufficient hot reduction to guarantee a primary structure which is satisfactory for most applications.

In this way we are able to meet the various specifications issued by different toolmakers from our standard production. A typical specification, designed in this instance by a leading twist-drill manufacturer in Germany, is presented here for interest with approximate English sizes given in parenthesis:

Condition of delivery: black bar, annealed to a tensile strength of 230–280 HB

A. Decarburization:
   13–30 mm dia. ($\frac{1}{2}$in–$\frac{13}{16}$in); 0·4 mm max. (0·015in)
   30–50 mm dia. ($\frac{13}{16}$in–2in); 0·6 mm max. (0·023in)
   50–70 mm dia. (2in–2$\frac{3}{4}$in); 0·9 mm max. (0·035in)
   70–100 mm dia. (2$\frac{3}{4}$in–4in); 1·2 mm max. (0·047in)

B. Carbide distribution:
   tested to Stahl-Eisen Prufblatt 1615
   up to 20 mm dia. ($\frac{3}{4}$in)   grade 1*l*, 1*d*
   over 20–30 mm dia. ($\frac{3}{4}$in–1$\frac{13}{16}$in)   grade 1*l*, 1*d*
                                           2*l*
                                           2*l*, 2*d*
                                           3*l*
   over 30–40 mm dia. (1$\frac{13}{16}$–1$\frac{9}{16}$in)   grade 1–3 *l*+*d*
                                           4*l*
   over 40–60 mm dia. (1$\frac{9}{16}$–2$\frac{3}{8}$in)   grade 1–4 *l*+*d*
                                           5*l*
   over 60–100 mm dia. (2$\frac{3}{8}$–4in)   grade 1–6 *l*+*d*
                                           7*l*

C. Carbide size:
   up to 10 mm dia. ($\frac{3}{8}$in); 0·006 mm max. (0·0024in)
   over 10–20 mm dia. ($\frac{3}{8}$in–$\frac{3}{4}$in); 0·008 mm max. (0·0031in)
   over 20–40 mm dia. ($\frac{3}{4}$in–1$\frac{9}{16}$in); 0·012 mm max. (0·0047in)
   over 40–60 mm dia. (1$\frac{9}{16}$–2$\frac{3}{8}$in); 0·014 mm max. (0·0055in)
   over 60–100 mm (dia. 2$\frac{3}{8}$–4in); 0·020 mm max. (0·0078in)

D. Crystal segregation:
   tested to Stahl-Eisen Prufblatt 1612
   up to 13 mm dia. ($\frac{1}{2}$in) grade 1
   over 13–40 mm dia. ($\frac{1}{2}$in–1$\frac{9}{16}$in) grade 1–2
   over 40–100 mm dia. (1$\frac{9}{16}$–4in) grade 2–3

E. Non-metallic inclusions:
   tested to Stahl-Eisen Prufblatt 1570
   up to 20 mm dia. ($\frac{3}{4}$in) max. grade 1
   over 20–100 mm dia. ($\frac{3}{4}$in–4in) max. grade 2

This specification includes the steels S6–5–2 (M2) and S6–5–2–5 (M35) where this customer asks for carbon on the upper limit (0·87–0·92% on M2 and 0·90–0·95% on M35) and for vanadium at the lower limit (1·7–1·9V).

This specification is not very difficult to meet but tighter specifications usually cause trouble only on the carbide distribution and perhaps on the carbide size. It is no wonder, therefore, that all efforts at improving high-speed steel quality have been aimed at surmounting this obstacle and the use of a sintering technique appears to be a very interesting method of producing the highest possible homogeneity throughout the whole of the material but information on this development is scarce and it is not yet possible to estimate whether such a method would be economical.

Particle-metallurgy high-speed tool steel

E. J. Dulis and T. A. Neumeyer

It is well known that particle metallurgy provides a means of making new alloys and composites of tool steels which cannot be made by conventional melting, casting, and hot-working methods. Major advances in new tool materials will most likely come from applying the P/M process. However, in addition to providing a method for making new alloys, the P/M process provides a means of significantly improving properties of existing tool steels. In high-speed tool steels, a problem that has existed for many years is that of carbide segregation in the conventionally solidified ingot that persists in relatively large-sized endproducts. The segregation of hard, brittle carbide regions in final large tool-steel sections results in undesirable characteristics such as out-of-roundness after heat treatment, regions of lowered impact toughness, non-uniform hardness, variability in grinding characteristics, etc., that have an effect on performance either directly or indirectly. To overcome this problem, production by P/M offers an excellent solution.

To upgrade the quality of conventional high-speed tool-steel mill products, Crucible developed a procedure for making large rounds by particle metallurgy. A pilot process has been in continuous operation for almost two years. The procedure involves gas atomization of a prealloyed molten stream, drying, screening, and blending the powder. Our pilot gas atomization unit (Fig.1) is capable of making 600 lb lots; 800 lb lots of blended powder are put into cans and consolidated. Typical compacted 800 lb billets are about $9\frac{1}{2}$in (24 cm) diameter by 42in (107 cm) long (Fig.2), and these are processed to final mill product sizes on conventional mill facilities.

Although most of our earlier work has been on M2S, other high-speed alloys made include M1, M3, M7, M7S, M41, M43S, and T15. For the studies covered in the investigation described here, high-carbon M2S with a nominal composition as follows was used: $1 \cdot 0C–0 \cdot 30Mn–0 \cdot 30Si–4 \cdot 15Cr–2 \cdot 0V–6 \cdot 5W–5 \cdot 0Mo–0 \cdot 15S$.

AS-CAST STRUCTURES

During normal commercial ingot solidification of high-speed tool steel, appreciable carbide segregation occurs that is associated with slow cooling rates. Typical as-cast microstructure for a conventional ingot of M2S high-speed steel shows heavy carbide networks (Fig. 3a and b). In powder atomization, the very fine particle sizes (less than $0 \cdot 02$in ($0 \cdot 5$ mm) diameter) result in a fast solidification rate so that microsegregation is substantially minimized (Fig. 3c and d). Thus, alloy homogeneity and structural uniformity are achieved in the fine-sized prealloyed particles or microingots of the P/M process.

1 Pilot gas atomization unit at Crucible Materials Research Center

2 As-compacted 800 lb billets of high-C M2S to be further processed on conventional mill facilities

A general description of a particle-metallurgy pilot-scale method of producing 800 lb billets of high-speed tool steel M2S is presented. The elimination of carbide segregation found in conventional products and the improved properties of P/M steel are described. The effects of processing temperature and time on the microstructure of P/M steel are presented. Tool life, out-of-roundness, rate of heat-treatment response, and C-notch impact toughness of P/M versus normal M2S steel are presented. The development of ultrafine grain size in P/M steel by a temper-anneal treatment is described. Intermittent-cut lathe tool life increase of threefold was found in the ultrafine grained steel compared to conventional P/M steel.

669.14.018.252.3:621.762.3/.5

The authors are associated with the Crucible Materials Research Center, Colt Industries, USA

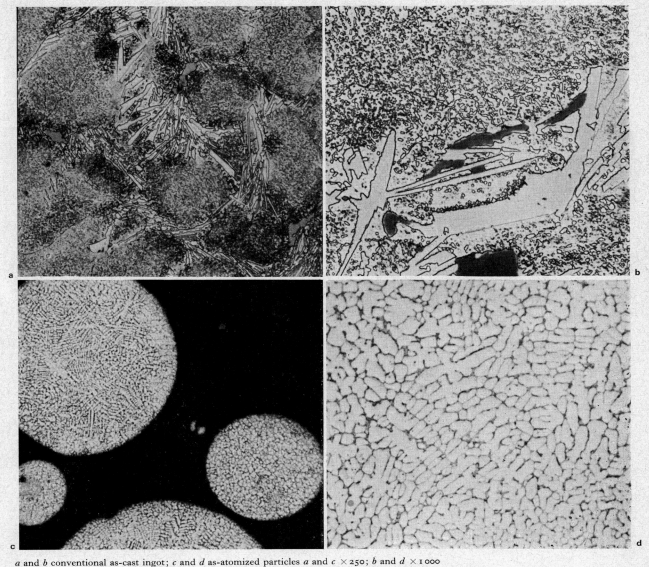

a and b conventional as-cast ingot; c and d as-atomized particles a and c ×250; b and d ×1000

3 As-cast microstructure of conventional ingot and atomized particles of high-C M2S steel

A comparison of the macrostructure of an as-compacted P/M product and an as-cast ingot of similar size (Fig. 4a and b) shows the typical uniform cross-section of the P/M product and segregation even in a small cast ingot. Macrosegregation was still clearly observable in the conventionally cast product after hot working to 5in (12·7 cm) bar size (Fig.4d). The P/M product at 5in (12·7 cm) bar size remains uniform in macrostructure (Fig. 4c).

CARBIDE SIZE AND DISTRIBUTION
Carbide microstructure in conventional cast and wrought products (Fig.5) exhibits typical segregation and large carbide size in a 5in (12·7 cm) diameter bar (Figs. 5a and b) and a more uniform distribution of finer carbides in a ¾in (1·9 cm) bar (Fig.5c and d). The effect of substantial amounts of hot work on final carbide size and distribution in conventional high-speed steel has been well known for a long time. In comparison, the carbides in 5in (12·7 cm) diameter P/M material are small in size and completely uniformly distributed (Fig.5e and f). Interestingly, the effect of hot work on carbide size and distribution in P/M steel is minimal (Fig. 6a and b). The main difference between as-compacted steel (a) and the same steel after hot working 75% (b) is the change in the sulphide morphology. In the as-compacted material the sulphides are spherical (a) whereas in the hot-worked

material the sulphides are elongated (b). The microstructures exhibited are for the steel in the quenched and tempered condition after processing.

Primary carbide diameters were measured on a Quantimet (QTM) image-analysing computer. By suitably etching tempered metallographic samples, excellent contrast was achieved between primary carbides (white) and the tempered matrix (black). No difficulty was encountered in the measurement of carbides in conventional high-speed steels. However, the carbide diameters in P/M products frequently approached the limit of resolution on the QTM ($\sim$1μm). Thus, QTM measurements were supplemented with measurements made on a Tukon filar-eyepiece microscope. Typical primary carbide sizes of conventional and P/M high-carbon M2S bars of 3–5in (7·6–12·7 cm) diameter are given in Table 1.

RESPONSE TO HARDENING TREATMENT
The austenitizing time required for full hardening is reduced in P/M steel so that shorter austenitizing times during heat treatment can be used (Fig.7). This is related to the smaller sized primary carbides. The effect of primary carbide size on improved heat-treatment response is related to the increased interface area between primary carbides and the austenite matrix, so that less time is required for diffusion of alloying elements, from carbides to matrix. Therefore, in a given

a as-compacted P/M 9½in (24 cm) billet; *b* as-cast 10in (25·4 cm) ingot; *c* P/M product, 5in (12·7 cm) diameter bar; *d* conventional product, 5in (12·7 cm) diameter bar

4   Macrostructure of high-C M2S in the as-compacted condition for P/M and the as-cast condition for a similar size ingot; also, both starting conditions after hot working to 5in (12·7 cm) diameter bar

austenitizing time, the P/M steels dissolve a greater percentage of alloy, and hence, a greater amount of secondary carbide precipitation occurs during the subsequent multiple tempering of the quenched steel. This effect is evident in comparing the difference in response of the conventional ¾in (1·9 cm) and 5in (12·7 cm) diameter steels as well as the P/M steel in which primary carbide sizes are significantly different. The difference in carbide sizes is shown in Fig.5.

SIZE CHANGE AFTER HEAT TREATMENT
Because of an inherently superior microstructural uniformity, i.e. uniformly distributed carbides and minimum segregation, P/M products undergo less distortion during heat treatment. Out-of-roundness (OOR) or deviations in cross-section

roundness is minimized. Improved OOR for large-sized tools, such as hobs and milling cutters, is an important factor for the production of unground hobs and for minimizing the amount of final grinding required in ground large tools. A comparison of OOR for conventional and P/M large bars is given in Table 2.

It should be clear that the effect of hot-working amount and mode is important, probably of primary importance, in controlling OOR pattern and magnitude. However, the added benefits of P/M should also be recognized.

TOOL LIFE AND PERFORMANCE CONSISTENCY
Many lathe-tool life tests have been conducted on tools prepared from P/M and from conventional bars. In continuous-

Table 1 Typical primary carbide sizes of conventional and P/M high-carbon M2S bars of 3–5in (7·6–12·7 cm) diameter

Product	Primary carbide diameter, μm	
	Mean	Maximum
P/M	1·5–2	3–5
Conventional	3–5	16–24

Table 2 Comparison of OOR for conventional and P/M large bars

Product	Diametrical OOR in 3–5in (7·6–12·7 cm) bars
P/M	0·000 3–0·000 5in (0·000 7–0·001 3 cm)
Conventional	0·001 0–0·001 6in (0·002 5–0·004 1 cm)

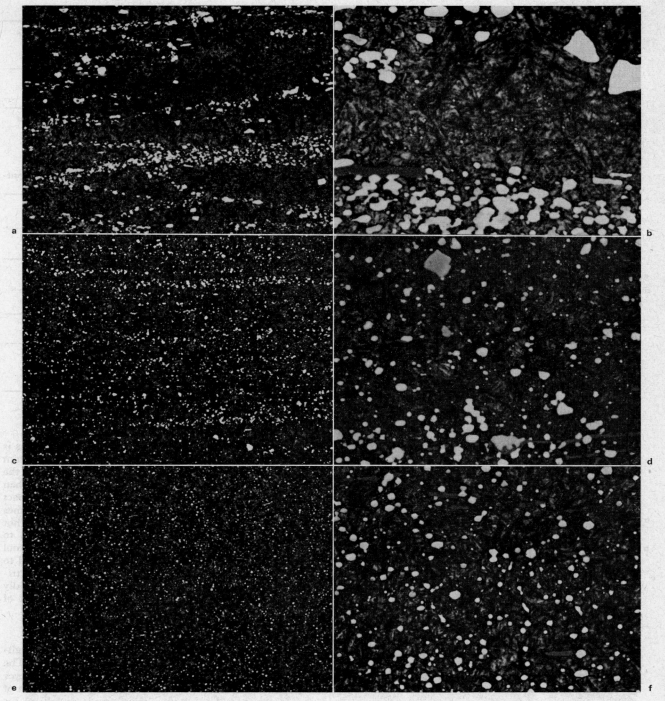

a and *b* conventional 5in (12·7 cm) bar (mid radius), carbide size: mean 4 μm, max. 24 μm; *c* and *d* conventional ¾in (1·9 cm) bar, carbide size: mean 3 μm, max. 16 μm; *e* and *f* P/M 5in (12·7 cm) bar, carbide size: mean 2 μm, max. 4 μm *a*, *c* and *e* ×250; *b*, *d*, and *f* ×1000

5   Carbide size and distribution in conventional high-C M2S bars and in P/M products

cut testing, the standard workpiece was H13 at 300 HB, and the endpoint was a tool wear land of 0·010in. In intermittent-cut testing, the standard H13 (300 HB) workpiece was slotted at 90° intervals to provide four intermittent cuts per revolution. The endpoint in this test was a wear land of 0·015in or complete nose failure, whichever occurred first. For the tests conducted in this investigation, sections were machined from the transverse radial direction of 5in (12·7 cm) diameter bars and prepared as lathe tools at a hardness of 66 HRC. Both tests have been used by our laboratory for measuring quality in P/M investigations of different variables. The intermittent-cut test, because of its severity, was particularly useful in

discriminating differences between materials and/or heat treatments. Results of these tests are summarized in Table 3.

The consistency and level of performance in P/M tools is shown by the summarized results of comparative intermittent-cut tests in Table 4.

IMPACT-TEST RESULTS
C-notch impact tests conducted on transverse radial specimens from 5in (12·7 cm) diameter bars and on specimens from ¾in (1·9 cm) bars show little difference between P/M and conventional high-carbon M2S (Table 5).

**Table 7 Effect of ultrafine grain size on interrupted-cut lathe
tool life**

Sample	Snyder-Graff grain size	Av. no. of impacts to failure
A	30	3310
B	13	1270

For these tests, a test bar of H13 at higher than normal hardness, 340 HB, was used. The significant increase in tool life as measured by this severe test shows the potential for substantial improvement in tool performance when P/M high-speed steel plus temper-annealing treatment are used.

The effect of final hot-working conditions on response to the temper-anneal treatment is such that very fine subgrains may develop rather than true ultrafine grains. Thus, a seemingly fine grain size of Snyder-Graff 40, (Fig.8e), is in reality a Snyder-Graff 13 with subgrains (Fig.8f). Intermittent-cut lathe tool-life test results showed that the subgrains did not exert any appreciable effect on tool life and the steel behaved as a steel that had a grain size of 13. Thus, the control of final hot-working parameters is important to the development of true ultrafine grain sizes.

SUMMARY

The results of the investigations on high-speed tool steel made by P/M reported in this paper are as follows:

1 800 lb P/M billets can be produced and further processed to various mill products on conventional steel-mill facilities.

2 Uniformly distributed fine-sized primary carbides are consistently found in P/M steel.

3 Response time to austenitizing heat treatment is shorter for P/M steel in comparison to conventional steel.

4 OOR dimensional change during hardening heat treatment is about one-third that of conventional steel.

5 Tool life as determined by both continuous- and intermittent-cut tests showed P/M steel to be superior to conventional steel. Intermittent-cut tool life of P/M steel was consistently about $2\frac{1}{2}$ times that of conventional steel.

6 C-notch impact-test results of P/M steel were equal to or slightly better than comparable conventional steel.

7 Minimization of oxygen content in P/M steel is important. If oxygen is present in amounts of 500–1000 ppm, forging following hot compaction is needed to obtain good properties.

8 Ultrafine grain size (Snyder-Graff 30) P/M steel had three times the tool life of conventional P/M steel in intermittent-cut lathe tests.

Cast cobalt alloy cutting metal

M. Riddihough

In the present field of metal cutting, the cast alloys of cobalt, chromium, and tungsten occupy a field between high-speed steel and the carbides where difficult conditions tend to chip the carbide yet greater production is needed than is available from high-speed steel.

Cast-alloy tools were developed during the 1914–18 war, and they were then the only material available to give greater production rates than high-speed steel until the advent of sintered tungsten carbide some 10 years later. The subsequent improvements in design of machine tools and their greater numbers has resulted in a continued demand for cast-alloy tools for specialized operations which include the machining of stainless and high-alloy steels with heavy cuts and feeds, particularly when interrupted cuts, vibration, sand, scale, or rough surfaces are evident.

The typical composition of some cast-alloy tools is shown in Table 1. The alloys are made from commercially pure metals and revert of known analysis from previous melts. Melting is by high frequency or indirect arc in a basic furnace lining, and the metal is cast to the finished tool shape with appropriate grinding allowances. To attain satisfactory cutting ability, a chill mould is necessary, usually graphite, and

TABLE 1   Nominal composition of cast cutting alloys

Alloy no.	C	Co	Cr	W	B	
1	2·5	41	32	17	0·8	1%Ni–0·7%Mo–3%Va
2	2·0	45	32	18	0·2	
3	2·0	46	28	18	0·2	5%Ta
4	2·0	43	34	19	1	

Fig.1a and b illustrates the difference in microstructure obtained when alloy number 4 (Table 1) is cast in graphite and alternatively a sand mould.

The large primary ideomorphic carbides of the sand-cast structure which are relatively brittle do not have time to form during chill casting which gives a refined structure with greater toughness and cutting ability. This structure is not altered by subsequent heat or heat treatment, and the casting technique must be developed to give cast tools free from harmful defects. Production tests include radiography, hardness, transverse strength, shock resistance, and cutting ability, which are carried out on each melt.

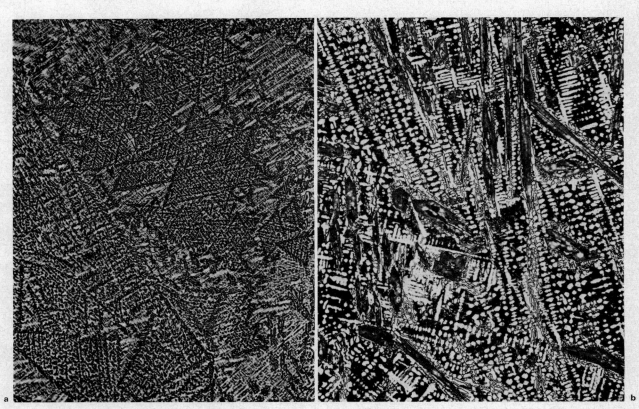

a graphite cast, 825 HV; *b* sand cast, 720 HV

1   Cast Co–Cr–W alloy no.4 (Table 1)   × **100**

Cast cobalt cutting metal occupies the field where difficult conditions cause premature failure of carbide tools. The metal is electrically melted, chill cast to shape, and ground with normal wheels. Graphs are presented to compare the hardness at high-temperature and cutting ability with high-speed steel, and typical machining operations are illustrated.

669.255′26′27–194–14:621.91.025

The author is with Deloro Stellite Ltd

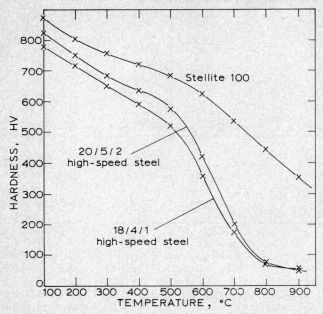

2   Comparative hot hardness of Co–Cr–W alloy no.4 (Stellite 100) and high-speed steel

4   Rough turning S110 austenitic stainless steel thermocouple sheaths with Stellite 100

5   Parting off ⅞in diameter cored bar at 230 rev/min for conveyor spindles

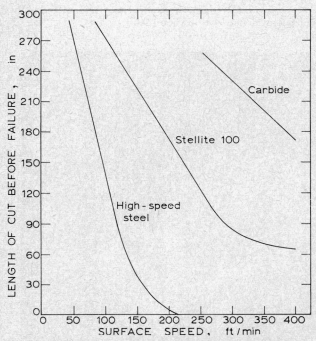

3   Cutting tests on 9in diameter alloy steel billet, 21 HRC, cut 1/16in, feed 3/32in, top rake 0°, approach angle 15°, front cutting edge 6°, corner 3/32in radius, clearance 6°

6   Rough and finish turning crankshafts from the solid in S106 nitriding steel

Figure 2 compares the hot hardness of Stellite 100 with two high-speed steels. The retention of hardness by the Co–Cr alloy above 600°C in comparison with high-speed steel points to its greater ability to withstand heat and thus give greater output. Production control cutting tests in comparison with high-speed steel and carbide under one specific set of cutting conditions are illustrated in Fig.3.

Cast-alloy tools are limited in shape by their method of production and are thus available as tool bits, parting blades, tips, and milling blades. Typical cases of their use in industry are given in Figs.4–6. Triangular section drills (Fig.7) are

8   Shearing hot steel billets with Stellite welded blades

7   Production drilling of hardened steel components to
±0·001in limits with Stellite

used for drilling and enlarging holes in hardened steel (over
51 HRC). Weld deposits of softer grades of the alloy are
used on shear blades (Fig.8), and punches and dies for hot-
and cold-metal cutting. The alloys are ground without
difficulty using normal free-cutting wheels.

Review of current hardmetal technology

E. Lardner

The last time that a comprehensive set of papers on the metallurgy of hardmetals was presented at any conference in England was in 1947. The papers presented at that time and published in *ISI SR38* served as an important landmark in the progress of an industry which in its early years before World War 2 was noted for the obscurity in which it conducted its metallurgical work, and which then experienced a period of intensive development largely behind a war-time security screen.

The 1947 papers presented a picture of an industry which had reached an advanced stage in technology, but which still was most uncertain about much of its scientific foundations. Thus in 1947 the papers by Brownlee and others described the well established use of large vacuum furnaces operating at temperatures over 2000°C. Yet at the same time it was only possible to present a very weak and largely incorrect explanation of the fundamental effect of the addition of titanium carbide to the WC–Co alloys.

The general picture presented by the technology of the industry today would certainly look very familiar to anyone whose experience of the industry had ended in 1947. The basic processes for both the production of powder and the sintered products are essentially the same. The products too would look familiar; typically small parts which form the working surfaces of some variety of tool. Only on closer examination would it be seen that the type of tips being produced and the purposes for which they would be used were often very different.

The changes in technology during the last 23 years would seem to be comparatively small, but major developments have taken place in the sophistication with which these operations are carried out and in the application of the product and in the basic understanding of the manufacture, properties, and applications of hardmetal.

In 1947 the output of hardmetal and its various applications was not known with any certainty, but it was probably of the order of 1 500 tons.

The applications were, excluding the use of armour-piercing shot, confined almost exclusively to metal-cutting tools and to wire- and bar-drawing dies. Other applications were quite exceptional, although rumours of possible use for mining were current in the industry.

Today, and again excluding any armament use, the western world produces about 10 000 tons of sintered hardmetal annually, of which only about 25% is used for cutting tools. During the winter of 1968–69 approximately two thousand million hardmetal tyre studs were produced, alone accounting for about 20% of the hardmetal output. The use of hardmetal for various mining tools now accounts for nearly 50% of the world output, and the balance is used for various forming tools and wear-resistant applications.

It would perhaps seem strange that an industry now only about fifty years old should have changed so little in its products and processes in the last 60% of its life. This is largely because of the unique properties of the simple WC–Co alloys. No other materials can show such high combinations of hardness and strength as these alloys and no way has yet been found of improving upon their combination of properties. Any alloying additions result in a poorer combination of hardness and strength, the advances in this field have all been based upon structural modifications of the basic alloy. When alloys other than the simple binaries are used, strength and/or hardness has been sacrificed in order to obtain some special property such as resistance to corrosion, diffusion wear, or better high-temperature properties. Occasionally small amounts of other constituents are deliberately added, but the main purpose is to assist in processing, such as when better grain-size control is required.

However, there have been major advances in our understanding of the production processes, of the mechanical and physical properties of the hardmetal alloys, and of the wear mechanisms that control their performance. Thus the products, although superficially similar to their predecessors of twenty-three years ago, are much more consistent in properties and give substantially better performance.

The major fields of progress in the hardmetal industry during the last two decades are described. The changes in the processes of the industry would seem modest, but they have resulted in a substantial improvement in quality and consistency of the product. This has been especially significant in the case of the steel-cutting grades where in particular there has been a substantial improvement in strength and toughness. These advances have resulted mostly from a better appreciation of the importance of the control of reaction between compact and furnace atmosphere in both presintering and sintering, and a better understanding of those features of hardmetal that significantly influence their properties. In mechanical operations there has been a considerable increase in sophistication both in presses and press tools and this has made possible the production by direct pressing of components of increased size and complexity of shape. Isostatic pressing has been introduced on a large scale. Not only has it resulted in better quality and dimensional control, but has also largely superseded the hot-pressing method for making larger components. The beginning of a 'secondary hardmetal' industry is mentioned. In cutting-tool applications the outstanding innovation has been the introduction of the 'throwaway' tip, and the influence of its special requirements on hardmetal technology is described. In the same period there have been major advances in knowledge of the technological basis of the industry. The basic phenomena concerned in hardmetal production have been studied intensively and the production processes are well understood. Considerable progress has been made in the study of the complex wear phenomena involved in metal cutting and the role of titanium and tantalum additions is much better appreciated. Considerable attention has been directed to the study of the strength of hardmetals as it is now more clearly realized that this is a crucial feature affecting performance. The influence of grain size and composition variations on strength have been well established but there is still no universally accepted explanation of the effect of these factors or of the surprisingly high strength of WC–Co composites compared with similar systems. Recent developments such as composite and coated tips, and hardmetal based on titanium rather than tungsten carbide, are described.

669.018.25:661.878.621

The author is with Wickman Wimet Ltd, Coventry

PRODUCTION TECHNOLOGY
Production of hardmetal powders
Most hardmetal is still made by producing a pure tungstic oxide or ammonium paratungstate, reducing it in hydrogen to tungsten powder, carburizing by heating with carbon black, and then producing the final powders by lengthy ball milling with the appropriate amount of cobalt powder.

In this old established route minor improvements have been effected to reduce costs and improve quality. The major change is undoubtedly the insistence upon grain size control throughout the production process as the important effect of grain size on the mechanical properties of the resulting hardmetal has been more clearly recognized.

Only one major change in this pattern of production has been introduced, and the distinction for this lies with the Kennametal Company,[1] who in 1963 announced their introduction of macrocarbide.

In this process tungsten carbide is produced directly from ore by aluminium reduction. A charge of ore, calcium carbide, aluminium powder, iron oxides, and carbon, is prepared such that the heat of the reduction reaction will produce a temperature of at least 2475°C. As a result of the highly exothermic reaction a metallic mass consisting of WC crystals in a matrix of iron is produced. The WC is recovered by a chemical leaching operation. The resulting product is a very coarse crystalline WC containing less than 0·2% iron.

This process differs drastically from the conventional methods of producing WC for hardmetal manufacture. In particular, none of the normal methods used for grain size control can be applied, and the carbide produced is vastly more coarse than that normally handled by the industry. To use this product and achieve the fine and consistent grain size required in hardmetal will have needed a completely new concept in grain size control, probably based on communition methods. Fine WC produced in this manner will undoubtedly be highly susceptible to surface oxidation and demand special handling procedures in tip production. The development of this novel extraction method must have brought in its train many technological problems right through the process to the final sintering stage.

Production of sintered components
The fabrication of powder into tips still follows procedures familiar twenty-three years ago and although there has been considerable sophistication introduced into some of the procedures the basic methods are but little changed. In compacting the major innovation has been in the introduction of isostatic pressing. This has greatly reduced mould costs when large components are required, and is especially valuable when small quantities of large components are required. This has resulted in a marked decline in the amount of hot pressing. The other advantage is the high degree of uniformity in density of an isostatically pressed compact. This eliminates all the distortion difficulties resulting from frictional effects in conventional die compaction. Thus, when components are produced from presintered isostatically pressed compacts, considerably improved dimensional control is obtained. In 1947 it was general for larger pieces, especially dies of about 3in diameter and upwards to be made by hot pressing, but now quite massive pieces up to 12in or more in diameter are commonly made by isostatic pressing. This pressing method, of course, would not be sufficient without the availability of furnaces of sufficient size. The typical sintering furnace of twenty-three years ago was a small-diameter continuous-stoking tube furnace, generally no more than 4–5in diameter. The development of batch-type vacuum furnaces has now made it possible to sinter the large pressings which can be obtained conventionally from isostatic pressing.

Pressing
The pressing of tips of simple shape required in such large numbers has been carried out since the earliest days of the industry on automatic tabletting presses. Originally these machines were simple crank presses essentially the same as those used by the pharmaceutical industry. Owing to the simplicity of these presses and the very poor pressing properties of hardmetal powder the shapes that could be produced were very restricted. Until comparatively recently most such tips were generally presintered and subjected to at least one simple machining operation such as the provision of a clearance angle. With increasing labour rates and all other costs the incentive to eliminate machining and to produce every possible component by direct pressing has been very strong.

Progress in this field has resulted from three major approaches. The first is careful consideration of all tip designs with close collaboration between customer and manufacturer to ensure that wherever possible a design is used that can be pressed. The second has resulted from a study of the pressing behaviour of the powder combined with detailed work on tool design. Third, and perhaps most important, has been the introduction of sophisticated press designs which allow complex shapes to be produced. To some extent this has resulted from demands from the whole powder metallurgy industry, but this was not entirely satisfactory as hardmetal pressing has some specialized requirements that are not wholly compatible with conventional powder metallurgy.

For many years the industry was too small to justify the design of special-purpose tools for hardmetal. However, this is no longer true, especially with the demand that arose for new equipment for the industry after the introduction of the indexable insert and its large-scale acceptance. In consequence several presses specially produced, or at least modified for the special requirements of hardmetal, are now available.

Hardmetal powders, unlike the familiar iron- or copper-base powders cannot be pressed to near full density but only to about half of the sintered density. Hence they contract about 20% on all linear dimensions on sintering and the sintered dimensions are in consequence very sensitive to local variations in pressed density. This is especially liable to result from non-uniformity in powder fill and from frictional effects at die walls and core-pin surfaces. Like other metal powders the compact will press 'down' in the direction of pressing very readily but there is very little pressure produced at right angles to the pressing direction, and it is almost impossible to obtain any movement of powder in this direction. Thus ideally, the profile of the powder fill should follow the profile of the completed pressing exactly in the compression ratio.

When this condition is not possible the minimal lateral movement of the powder results in areas of varied densification which, in the extreme, can result in press cracks if one zone reaches the limit of its compressibility before another. If this does not result in cracking it causes distortion as the most highly compressed zone contracts least on sintering. This effect is well known in powder metallurgy, but in many products it is alleviated by the ductility of the material being pressed, but this is not possible in hardmetal powders.

Thus the improvements in presses for hardmetal have been largely aimed at producing more uniform and consistent powder fill, and providing more complex and controllable bottom and top punch and die motions so that this can be made possible. At the same time the balancing of both top and bottom pressures is an obvious requirement if uniform density is to be obtained. A major innovation has been the introduction of hydraulic rams into the bottom and top punch motions which are connected hydraulically so that at the completion of the pressing cycle the pressures are exactly equalized.

None of these improvements would be effective without corresponding improvements in powder feed arrangements and especially the provision to use 'overfill' or 'prefilling' methods. Naturally this is not entirely in the pressmaker's province, good flow properties and a consistent packing density of the powder are clearly the responsibility of the powder manufacturer. With hardmetal this can be an especially troublesome area of operation. Hardmetal powder in its natural state will not flow at all; it must be converted to a coarser granulated product with free-flowing properties. To obtain satisfactory free-flowing granules which also have excellent pressing properties is not easy and it is unlikely that any hardmetal manufacturer is fully satisfied with this part of powder production technology.

Sintering

Sintering is a critical operation in hardmetal production, and the one which has obviously altered more than any other during the past twenty-three years. In 1947 most hardmetal was sintered in fairly small continuous-tube furnaces, either alumina tubes with molybdenum windings or a carbon-tube short-circuit furnace. While furnaces of this type, but generally of more sophisticated construction, still account for a large proportion of the hardmetal produced, a prominent feature of most hardmetal sintering plants is the vacuum furnace. For many years the relative merits of hydrogen v. vacuum sintering have been argued in the industry, and the echoes of these arguments which have filtered through to the user are confusing in the extreme, especially as some suppliers have used changes in sintering methods for sales promotion purposes.

The effect of vacuum sintering and relative advantages and disadvantages differ somewhat between grades with and without titanium carbide. However, the problem is basically the same. We have a product in which the C content must be controlled to a high order of accuracy, typically $\pm 0.04\%$ on an average value of 6%. If the work of Suzuki[2] on the strength of hardmetal is fully confirmed it may well be desirable to improve this control still further. The product is chemically reactive and is readily oxidized, reduced, carburized, or decarburized, and it has a surface area which is expressed in terms of 5×10^4 cm^2/g. Any metallurgist can see that we have a fantastically critical version of the bright annealing of steel problem, which when TiC is present, becomes similar to the problem of bright annealing stainless steel.

With such materials surface effects are of extreme importance and such powders cannot be handled in air without even the thinnest of oxide films on each particle resulting in a significant total weight of oxygen. In fact an oxygen content in the region of $0.1-0.2\%$ is commonplace. Unfortunately this cannot be controlled with any precision, varying with many factors of which the most important are grain size, the presence of excessive 'fines', and atmospheric temperature and humidity. This oxygen is rapidly removed when the compact is heated in hydrogen,[3] and provided the reaction products are removed in reasonable time, as in a simple counterflow hydrogen-atmosphere furnace, this can occur before it can react with and remove carbon from the compact. Obviously in a vacuum furnace this reduction does not occur and a carbon loss equivalent to the oxide content is to be expected. This is not always experienced because in many furnaces used in the hardmetal industry pumping rates available are not sufficient to reduce the atmosphere, which is predominantly carbon monoxide, to a pressure sufficiently low to prevent it having some carburizing and/or reducing effect. Extreme temperature gradients can exist in a rapidly heated vacuum furnace over the temperature range at which these reactions occur, and it is possible to show that one part of a furnace can be carburizing, and another simultaneously decarburizing to WC under these conditions. Thus before it is possible to vacuum sinter WC–Co alloys, quite complex problems had to be solved in order to obtain a high degree of carbon control.

When sintering in hydrogen the oxygen content of powder is unimportant, but other serious problems in carbon control arise. First, pure hydrogen reacts rapidly with WC at temperatures above $850°C$ producing methane and tungsten. This reaction increases rapidly with temperature and it can only be prevented by introducing the equilibrium quantity of methane into the furnace atmosphere in the correct proportion to prevent the decarburization of WC. In practice this is commonly achieved by using graphite boats and packing powders, generally alumina, containing high proportions of carbon. In this way the furnace atmosphere becomes saturated with methane before it comes into contact with the tips. This method, although apparently crude, has the merit that it is largely self-compensating for deterioration in furnace atmosphere due to imperfect construction, or furnace deterioration which allows the ingress of air in varying amounts to the furnace atmosphere.

The one great defect is that the atmosphere is generally too powerfully carburizing and in consequence varying amounts of free graphite are introduced into the sintered tips. Naturally the carbon pick-up is most severe at corners. With many products a local weakness of this nature is of little consequence; with a typical brazed-tip tool this defective material will be removed at least after the first regrind, but as mentioned later it can be a serious drawback with replaceable indexable inserts.

With grades containing TiC similar considerations apply. The surfaces of particles containing TiC oxidize more readily and the oxide once formed cannot be removed by hydrogen at low temperatures but is removed by reaction with carbon in the compact in the temperature range $1100–1250°C$ in vacuum. However, with those grades containing TiC the need for precise carbon control is not so obvious as in the plain WC–Co alloys. With the latter materials a variation outside a range of $\pm 0.04\%$, depending upon Co content, results in the obvious and harmful formation of either the excessively brittle double carbide or free carbon. With the steel-cutting grades commonly used the variation in C content to give the same effect is typically $\pm 0.2\%$ or even more. Thus the carbon loss that occurs in vacuum sintering is generally insufficient to produce the obvious η-phase. However, variations of C content within these extremes can have important effects upon the sintered hardmetal. In addition to the physical and mechanical effects described in Suzuki,[4] this carbon loss, if uncontrolled, can lead to a marked variation in the properties of the sintered product because C content has such a marked effect on the rate of grain growth during sintering. Figures 1 and 2 show the structure of test-pieces from the same batch of powder vacuum sintered at the same temperature. The first (Fig.1) lost no carbon and was at the limit where free carbon would just appear, while the other (Fig.2) lost about 0.5%C and was just above the bottom limit. The corresponding hardnesses were 1470 HV and 1560 HV. Even if the immediate consequences are less obvious it is clear that it is desirable to limit the carbon loss during sintering of the Ti grades to the lowest possible consistent value.

The benefits of vacuum sintering compared with hydrogen sintering are more immediately obvious with the Ti grades. This results mainly from the fact that any oxidation of the titanium-containing phase is irreversible in hydrogen. When oxygen or water vapour enters the hydrogen atmosphere the result is oxidation of the TiC phase, no matter how strongly carburizing the atmosphere might be to plain WC–Co alloys. Additionally titanium nitrides form readily at temperatures above $1200°C$ so air ingress to the sinter atmosphere also introduces nitrogen into the titanium-containing phase. Hence a hydrogen-atmosphere furnace that might be satisfactory for sintering WC–Co alloys could result in excessive loss of carbon from a Ti grade with the additional defects of oxygen and nitrogen contamination of the Ti-containing constituent. Grades sintered under these conditions are generally rather porous and are characterized by relatively

1 **Structure of TiC–TaC–WC–Co alloy with 8·28%C; sintered
1 h at 1450°C**   ×1500

fine-grain TiC solid solution grains which etch very darkly
in alkaline potassium ferricyanide. TiC grades are also very
reactive towards alumina, the most commonly used tip-
packing material, and this results in serious troubles which
are also mentioned when discussing replaceable indexable
inserts. Thus the sintering of Ti grades in hydrogen is far
from satisfactory and the advantages of sintering in vacuum
much more immediately obvious than for the plain WC–Co
alloy.

Vacuum sintering has played a major part in the success
of the indexable insert. A unique feature of these tips is that
their performance is almost entirely dependent upon the
quality of the hardmetal at the corners of the tip. As explained
earlier, corners are the part of any tip most vulnerable to
reactions with the furnace atmosphere, and these vitally
important parts of the tip are the most likely to be made
defective. In hydrogen sintering the carburizing nature of
the furnace atmosphere, resulting in free carbon at the
corners, was generally not the only serious hazard. Normal
practice has generally been to pack tips in alumina or alumina–carbon powder in the graphite sinter boats. This practice results in some pick-up of aluminium in the surface
layers of the tip, with of course a maximum pick-up at the
corners, where as much as 0·5%Al has been measured. With
grades containing TiC, reaction with alumina is even more
extensive and aluminium contents of up to 5% have been
determined in the surface layers of such tips. This amount

2 **Structure of same alloy as Fig.1 with carbon reduced to
7·74%; sintered 1 h at 1450°C**   ×1500

of aluminium has a most serious effect upon the cutting
properties of carbide.[5]

These defects of high free-carbon content and high aluminium content at corners were generally of little consequence with brazed tools as the defective material would
generally be ground away, if not before the first use certainly
at the first regrind.

In fact this need to have the best possible carbide quality
at the corners of the tip makes it impossible to produce first-
class quality indexible inserts without very stringent control
of compact-atmosphere reactions. In practice this means it is
almost impossible to make them of consistent good quality
by any method other than vacuum sintering.

In addition to the major features already discussed there
are several other matters to consider in connexion with
vacuum v. hydrogen sintering. Sintering temperature must
be closely controlled and this is not easy in a simple continuous-tube furnace with its gradient dependent upon work-
load. This has been improved considerably by the use of
multizone furnaces and modern thermocouple materials, but
the vacuum furnace still shows considerable advantage as the
temperature can be so easily measured. Contrary to experience at lower temperatures where vacuum furnaces show
severe temperature gradients, at the high temperatures used
for sintering hardmetal heat transfer by radiation becomes
most efficient, and this combined with the very effective
thermal insulation possible in a vacuum furnace make it
possible to obtain work chambers large enough to contain at
least 200 kg of hardmetal with a temperature variation of less
than 10°C. This effective thermal insulation also results in
very low power consumption, and with most types of furnace
the change from hydrogen continuous to vacuum-batch
type furnaces has resulted in considerable reduction in electrical power requirements for sintering.

Other advantages of vacuum sintering result from the
better elimination of impurities such as Si, Al, Mg, Ca which
are present in an oxidized form in very slight amounts. These
all tend to be reduced and volatized under vacuum-furnace
conditions. Another advantage which is frequently not realized largely results from the continuous v. batch aspect of
the two sintering methods. Hardmetal on cooling from between solidus and liquidus temperatures shows normal
segregation effects. This is not of much consequence with
small tips but with large tips it can be a very marked effect.
A major influence is temperature gradients. These can be
very severe when tips are stoked through a continuous-tube
furnace with a water-cooled exit. Such segregation can be
very marked in lengthy tips stoked longitudinally. When
cooled in a vacuum-batch furnace temperature gradients are
so low that segregation effects are practically negligible.

Vacuum sintering has also been found to produce surfaces
on the tips that are much more easily wetted in brazing.
Almost all carbide grades can be readily brazed by most
common brazing alloys without special surface preparation
if they have been carefully vacuum sintered. This cannot be
done consistently with tips which have been sintered in
hydrogen. In this case a high carbon content at the surface
can be the source of trouble, but again by far the most serious
trouble results from pick-up of aluminium from the packing
material.

No discussion of vacuum sintering hardmetal would be
complete without some reference to high-vacuum sintering.
Much vacuum sintering has been, and still is being done in
furnaces equipped with mechanical pumps giving a cold
vacuum of no better than 0·05–0·1 torr. However, many furnaces are in use equipped with Rootes-type blowers and diffusion pumps of high capacity. With such equipment claims
to be sintering hardmetal at a pressure of 0·001 torr or even
several orders of magnitude better have been made. However,
there is no doubt that the vapour pressure of cobalt is well
above this pressure at sintering temperatures, and if the
pressure in the work chamber was truly of this level then

intolerable trouble from cobalt loss would result. The common construction of work chambers in furnaces for sintering hardmetal leads to severe pressure differences between work chamber and vacuum manifold, and this seems the most probable explanation of these reports. Other users make more rational claims, and one procedure which is in common use is to heat the compacts under the highest attainable vacuum up to about 1200°C and then back fill with argon or helium to a pressure of around 100 torr for completion of the sintering cycle.

Vacuum-furnace design for hardmetal has resulted in the production of some most complex equipment. A major production feature of vacuum furnaces is their slow cooling rate and most of the complexities have been introduced to reduce the sinter cycle time to something approaching that of the continuous-tube furnace. Furnaces have been designed and are in use which have achieved the purpose. It seems a matter of some doubt whether the advantages of shorter cycle time really justify the complexities in both vacuum and high-temperature engineering which are required to achieve this end.

Secondary hardmetal

Most metal industries have a flourishing 'secondary' aspect of their business based upon scrap; in fact some major metallurgical industries are largely scrap based. For many years there was no equivalent branch to the hardmetal industry, and disposal of scrap hardmetal has always been a difficult task. Various methods of decomposing sintered scrap have been proposed, mainly based upon chemical methods of dissolving the Co binder, yielding WC powder in a form suitable for reuse in powder metallurgy. It is uncertain to what extent these processes were used, if indeed they were used at all.

The most commonly used process for recovery of tungsten from hardmetal scrap has been its oxidation in a fused sodium nitrate–nitrite salt bath, a method resulting in recovery of tungsten in soluble form as a mixture of sodium tungstate and sodium carbonate.

Hardmetal manufacturers generally have gone to considerable lengths in process-control procedures in order to avoid making hard scrap as its disposal was so financially disastrous being worth less in weight of contained tungsten than the original ore.

However, encouraged by a rapidly growing market for cheap carbide for such wear-resisting parts as tyre studs, and with the larger amount of scrap hardmetal resulting from the introduction of the 'throwaway tip', two processes have recently been introduced which go far to realize the objective of recovering hardmetal in a readily usable form.

The first of these introduced by the Shwayder Chemical and Metallurgical Corporation[6] is a more practical version of the old 'dissolving out the cobalt' principle but has eliminated both the use of very strong acids and has speeded up the reaction considerably and makes unnecessary the expensive preliminary crushing or disintegration which was generally advocated before treatment with strong acids. It is also claimed to work just as well with grades containing Ti and Ta as with the plain WC–Co alloys.

This new process uses a fairly weak solution of phosphoric acid, probably 10%, at temperatures of the order of 45–50°C in some form of mill in which the continual mechanical agitation is believed to considerably aid the reaction. Carbide produced in this manner obviously consists of the grains originally present in the sintered product, but they are broken down somewhat by the milling associated with the chemical treatment. Like all scrap-metal processes the product is largely dependent upon the careful segregation of the scrap before treatment. This process is operated commercially and the products offered are freely available in grades segregated according to the TiC/TaC content.

These carbides are truly 'single carbide particle' powders

and can be reused by remilling with Co. The product can be of excellent quality as, of course, if carefully carried out, such a recovery process is essentially another 'refining' operation. Although small accidental contamination with Ti is generally observable, no impurities likely to result in porosity on sintering are introduced and the freedom from porosity of this product can be excellent.

Some loss in carbon can occur during the recovery process and coping with this seems to be the biggest difficulty. Control of grain size is also limited and can only be exercised to a limited degree by the rather unsatisfactory expedients of intensive ball milling to reduce particle size or by coarsening by high-temperature sintering. Nevertheless this process can give a very useful product of good quality, but without the manufacturer being able to exercise that control of composition and grain size generally considered necessary for a critical product.

The second recovery process is of an entirely novel nature and was introduced by Metallurgical International a few years ago and is now commonly referred to as the 'Cold-stream process'. In this process the hardmetal scrap is first mechanically crushed to the −8 to −4 mesh size and then chemically treated to remove contamination from the crushing treatment. This crushed hardmetal is then blown in a high-velocity air stream into an evacuated chamber containing a hard target against which the fragments of hardmetal are shattered. The product from this chamber is carried by an air stream to a primary classifier where oversize material is separated out and recirculated through the blast chamber.

The material from this process is of the same composition as the feed material, without any loss of Co, and it can be repressed and sintered. However, although an average particle size of 1.5–2.5 μm (Fisher subsieve sizer) is obtained much of the powder is not single carbide particle material, and multigrain particles up to the order of 15 μm can be found. The pressing and sintering properties of this product are rather different from conventionally produced carbide powders, and the freedom from porosity of the sintered product is not so good. However, on both counts substantial improvements can be made by blending the reclaim product with comparatively small amounts of conventional powder.

Both processes have arisen at a time when the demand for hardmetal of only modest quality has reached a high level, and it is to be expected that the product in both cases will for some time to come be confined to such applications rather than to the critical metal-cutting applications. However, with much of the available material for such recovery processes existing as throwaway tips it is inevitable that the reclaim material will contain an appreciable amount of expensive TaC. The urge to reuse this in steel cutting is great, and this, together with the inevitable upward spiral in the price of tungsten, will provide a powerful urge to improve these reclamation processes still further.

PROPERTIES AND APPLICATIONS

Although the basic processes and the general appearance of factories making hardmetal have not altered very much in the past twenty-three years it has been shown that these processes are being carried out under much more closely controlled conditions. There has been a great increase in the understanding of the processes and of knowledge concerning those features which are important in giving the required properties in carbides for different applications. Even more important has been the increase in knowledge concerning the metallurgical features which influence the performance of hardmetal. This has played a most important part in producing grades more suited for particular applications, and applying carbide in the most profitable manner.

The improvements in hardmetals, especially in cutting tools, since 1947 have been undramatic but nevertheless very real, as can be seen from the great increase in performances which are revealed by comparing current machine-shop

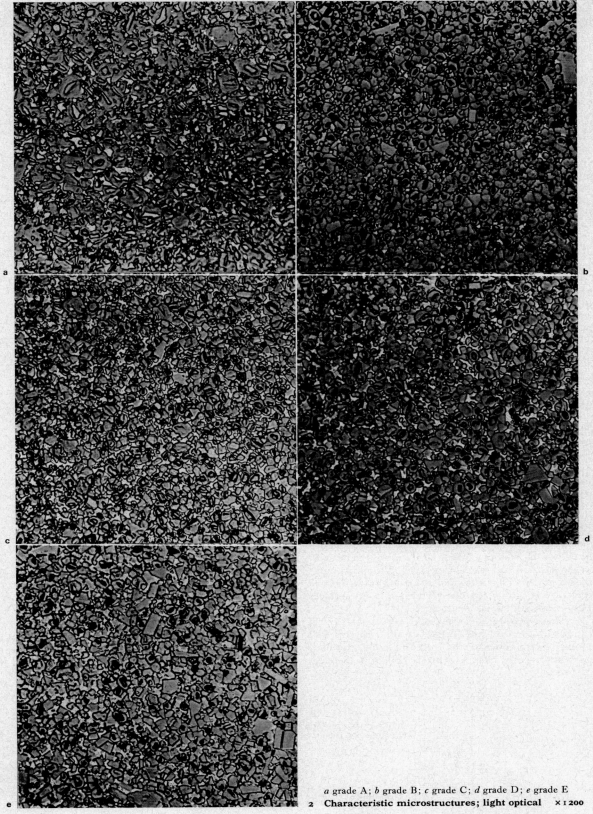

a grade A; *b* grade B; *c* grade C; *d* grade D; *e* grade E

2   Characteristic microstructures; light optical   × 1 200

surface finish and a good chip flow. The main cutting angles were as follows:

cutting rake $-7°$
clearance angle $7°$
front to back rake $-5°$

The workpiece material was an unalloyed carbon steel of the following basic composition in % by weight:

C	Si	Mn	P	S
0·35	0·25	0·6	≤0·030	≤0·030

It was normalized and had a hardness of 150 HB. Steels of this kind are common in workshops, they have good machinability, and wear little cemented carbide. The workpiece dimensions were 600 mm in length, 40 mm wide, and an original height of 130 mm.

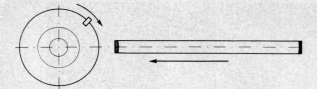

3   Arrangement of mill in relation to workpiece

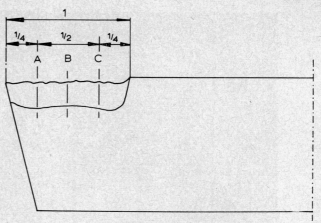

4   Flank wear, showing three zones

The mounting of the workpiece was such that the working plane had dimensions of 600 × 40 mm. The narrow width of the component in relation to the diameter of the face mill, 250 mm, is explained as follows:

(i) disturbing effects from wear phenomena are reduced when the cutting distance goes down

(ii) the number of thermal fatigue cracks increase with decreased workpiece width. This has been shown by Opitz and Lehwald.

The following cutting data were applied:

cutting speed	212 m/min
feed	0·208 mm/tooth
table feed	56 mm/min
cutting depth	2·0 mm

These cutting data lie within those recommended by cemented-carbide producers for the material in question. During the test the axis of the mill was set on the centre line of the workpiece (Fig.3). To avoid deterioration phenomena which can occur in the milling of planes perpendicular to the feed direction, the following precautions were taken. Before commencing the machining with the test grades, the entrance end of the workpiece was milled to the steady state, profile determined by the mill and cutting data. In addition, no test edges were allowed to pass the exit part of the workpiece for similar reasons.

The test sequence of the inserts was chosen at random. After certain intervals permutation was made. Spreading of test results by variations within the workpiece was therefore reduced to a minimum. To secure a level for every individual type of test, it was repeated 4 times with a new insert each time.

Each insert tested was studied with regard to the occurrence of fatigue cracks and wear after test intervals of fixed lengths. The cutting tests were stopped when all the grades to be compared within a group had reached the level of the final number of fatigue cracks. The inspection and counting of thermal fatigue cracks was carried out directly on the insert surfaces tested without any additional treatment.

The investigations were carried out with light optical microscopes with low and high magnification ability as well as a scanning electron microscope. In the light optical studies the magnification varied from × 25 to × 720 and in the elec-

tron optical studies from × 70 to × 3 000. The two kinds of microscopes have different kinds of depth sharpness characteristics, and thus are complementary to each other in an excellent way.

The flank and crater wear of the inserts tested were also measured. Crater wear was determined by means of a tester moved across the chip face at right angles to the cutting edge. Flank wear was measured in accordance with the method suggested by the Swedish Association of the Metalworking Industries in its recommendation on a machinability test.[7] Figure 4 shows the flank wear, viewed at right angles to the side cutting edge, divided into three zones. Their mean width of the worn area in the central zone measured from the original level of the cutting edge has been adopted as the criterion of flank wear.

RESULTS

At a constant binder metal content and grain size of 1·9 μm the number of thermal fatigue cracks increases with increased titanium carbide content (Fig.5). The influence of titanium carbide content is so predominant that it overwhelms the effect of the grain size variations within the range of this study. A small influence of flank and general chip face wear can be traced. The crater wear was so limited that it could not be measured. The flank wear of the grades is given in Fig.5. Increased hardness follows increased titanium carbide content. The transverse rupture strength (TRS) decreases with increased titanium carbide content (Fig.5).

At a high titanium carbide content, grade A, a wide variety of cracks occur. Short and relatively wide ones are illustrated in Fig.6a and b, and narrow ones in Fig.6c, d, and e. Long narrow cracks can be seen in Fig.7a and b. There are many short cracks and relatively few long ones. In grade E with a

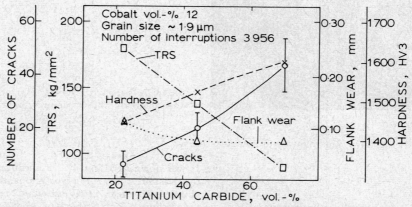

5   Relation of thermal fatigue cracks and of wear, to increase of titanium carbide content at constant binder metal content and grain size

a and *b* short wide cracks; *c*, *d*, and *e* narrow cracks; *a* × 70;
b × 500; *c* × 150; *d* × 500; *e* × 1 500; electron optical

6   Cracks in grade A, high TiC content

low titanium carbide content a few long cracks occurred. They were slightly wider and fewer than the corresponding ones of grade A.

In addition to different kinds of wear, oxidation also disturbs the formation and growth of thermal fatigue cracks. The oxidation increases with decreasing titanium carbide content. Oxide layers have often been found to have more cracks than the carbide material it covers. Many cracks in the oxide layers are so thin that they only can be detected at very high magnifications. Figure 7*c–f* forms a series illustrating this. The fact that some cracks are only localized to the oxide layer means that there is some uncertainty concerning a systematic error in the direction of too high a number of thermal fatigue cracks. The thickness of the oxide layers has

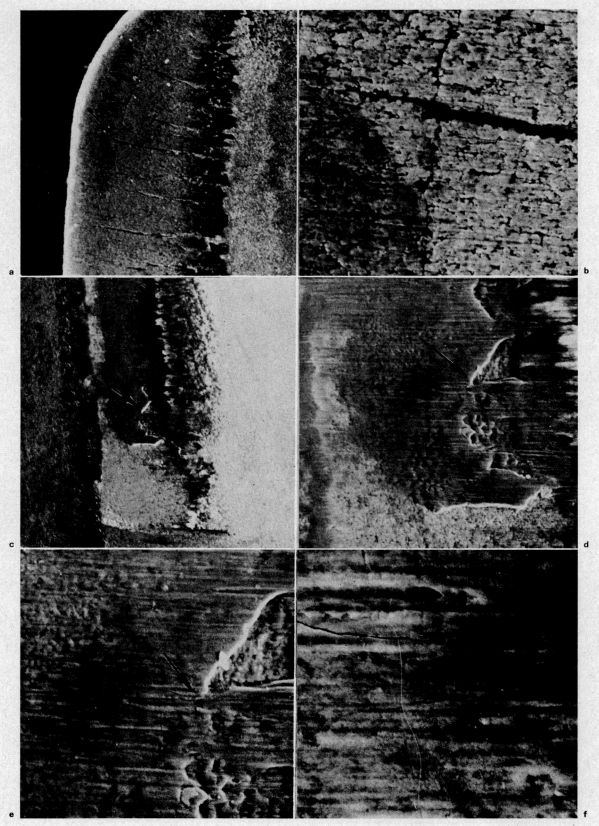

a and *b* long narrow cracks; *c, d, e,* and *f* cracks in oxide layer; *a* ×70; *b* ×500; *c* ×70; *d* ×200; *e* ×500; *f* ×1 500; electron optical

7   Cracks in grade A and in oxide film

been estimated to be up to 2–3 μm. On some inserts tested oxide fragments have flaked and exposed the carbide material (Fig.8). Such flaking is often connected with cracks.

At a constant grain size of 1·9 μm and a constant ratio between the two hard constituents added the number of fatigue cracks increases with increased content of binder phase (Fig.9). The three grades B, C, and D making up this test group had not only an increasing content of binder phase but also a decreasing content of γ-phase from 78–68% by volume and likewise a decrease of the volume ratio between the

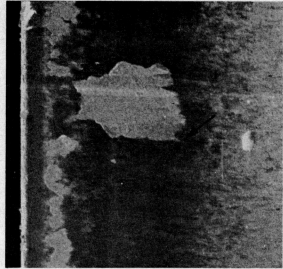

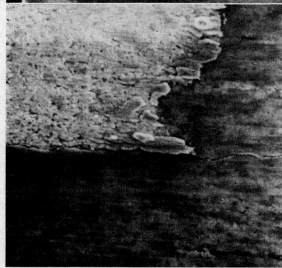

a × 100; b × 600; electron optical

8   Flaked oxide films, on grade E, highest WC content, exposing carbide material

γ-phase and α-phase from 5·5:1 to 4·5:1. This means that the content of tungsten in γ-phase diminishes a little with increasing binder metal content. Concerning the measured lattice parameters (a) of the γ-phase they lie within the expected range, where unfortunately the tungsten content only has a slight influence on the γ-phase lattice parameter.[12]

Also for this series, the grain size ranges studied have only a secondary influence on the sensitivity to thermal fatigue relative to the composition variations.

Variations in binder metal content have a strong influence on the formation and growth of thermal fatigue cracks in the same way as variations in titanium carbide content.

In the grade of the lowest binder metal content, grade B, the cracks are few, long, and very thin, and in character comparable to those of grade E with the lowest titanium carbide content (Fig.10).

With a high binder metal content, grade D, not only a few long cracks but also many short ones occur. Both kinds of cracks give an impression of having grown quite a lot in width through deterioration mechanisms in the fracture surfaces (Fig.11). In this series the disturbing influence of oxidation is obvious. The sensitivity for oxidation increases with increased binder metal content.

In accordance with what has been found in earlier investigations the growth boundaries of the thermal fatigue cracks coincided with the isotherms of the heat flow (Fig.12). It has been established that the initiation of the majority of the cracks is located at that part of the contact area between the chip and the tip where the chip leaves the tip face.

The grain size of all the derivatives of the principal grades studied have varied within the range 1–3 μm.

With a high titanium carbide content, grade A, the sensitivity of thermal fatigue cracking passes a maximum at a grain size of about 2 μm (Fig.13). The range of the measuring values is indicated in the figure. With a decreasing titanium carbide content the grain-size dependence of the sensitivity to thermal fatigue cracking decreases. For the two principal grades C and E the sensitivity to thermal fatigue cracking increases with increasing grain size, but less so for grade E (Fig.17).

The influence of the Co content on the sensitivity to thermal fatigue cracking conforms with the influence of the titanium carbide content, grade B, C, and D (Figs. 14, 15, and 16). The grade with the highest binder metal content, grade D, shows a maximum sensitivity to thermal fatigue cracking at a grain size of about 2 μm.

CONCLUSION

As has been mentioned, grade C has a composition and a microstructure typical for grades which are marketed for the application group P30. The grade which has the most pronounced difference from commercial alloys is grade D. Grade A is also an unusual kind of a cemented-carbide cutting grade.

The results indicate that the influence of the composition and the microstructure has a primary importance concerning sensitivity to thermal fatigue cracking.

The results illustrated in Figs. 5 and 9 indicate there is no

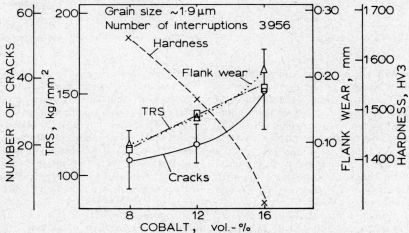

9   Relation of thermal fatigue cracks, and of wear, to increase of binder phase at constant ratio of carbides and constant grain size

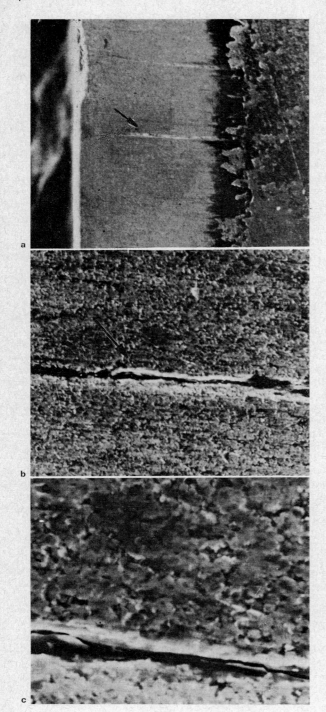

$a \times 70$; $b \times 500$; $c \times 1500$; electron optical

10 Cracks in grade B, lowest binder content

$a \times 70$; $b \times 500$; $c \times 1500$; electron optical

11 Cracks in grade D, highest binder content

direct connexion between transverse rupture strength and sensitivity to thermal fatigue cracking. In fact the most fatigue-sensitive grades, A and D, show a maximum difference in transverse rupture strength. For this reason it is also impossible that a direct relationship can exist between sensitivity to thermal fatigue cracking and mechanical fatigue strength at ambient temperature.

Concerning the hardness at room temperature, it cannot have a direct influence on the thermal fatigue strength, which is clearly illustrated in Figs. 5 and 9. In this case also, A and D show a maximum difference.

The thermal expansion coefficient increases with increasing titanium carbide content[8] and even more with increasing binder metal content.[9] The heat conductivity decreases with increasing TiC[10] content as well as increasing Co content.[11] The two grades most thermal-fatigue sensitive, A and D, have within their comparison groups unquestionably the highest thermal expansion coefficients and lowest heat conductivities, which means that the necessary conditions for the highest thermal fatigue stresses are satisfied. These two grades also have, owing to their composition and the methods under which they are made, a γ-phase characterized by a decreasing content of tungsten. This is especially valid for the grades with a constant binder metal content and an increasing titanium carbide content. An increased tungsten content in the γ-phase will have an improving effect on the binding strength between the γ-phase and the β-phase.

As outlined earlier, it has been established that thermal

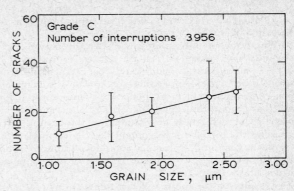

15 **Influence of grain size on thermal fatigue cracking in grade C**

12 **Coincidence of thermal fatigue crack with heat flow isotherm, in grade B; electron optical** ×150

fatigue cracks mainly propagate along grain and phase boundaries. This may depend on differences between the thermal expansion coefficients of different phases, not only between themselves but also in different crystal directions. The latter fact is especially valid for the α-phase.

With a bigger grain size the differences in thermal expansion increase between the opposite sides of the grain and phase boundaries, which will mean increased microstresses

and thus increased risks for the initiation and propagation of cracks.

Oxidation and wear have an obvious influence on the mechanisms leading to thermal fatigue cracking. The special effect of these phenomena increase with decreasing titanium carbide content and even more with increasing Co content. Under conditions which are valid for a cutting edge in use, the oxidation and wear resistance decrease for the three separate phases in the following order: γ, α, β. Grades D and E consequently have thermal fatigue cracks, which to a high extent have been changed by these phenomena. This is most predominant in grade D.

Owing to the fact that this grade has the highest β-phase content it also has the highest capacity for absorbing energy since it has the highest volume available for plastic deformation. The more energy that can be absorbed by plastic deformation, the more the propagation of cracks can be slowed down. The fact still exists that grade D has a high sensitivity of crack initiation. A slow crack growth combined with wear

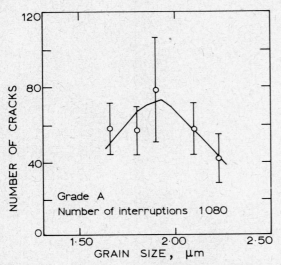

13 **Influence of grain size on thermal fatigue cracking in grade A**

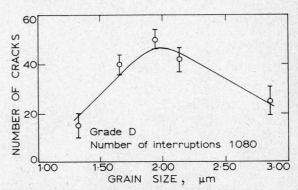

16 **Influence of grain size on thermal fatigue cracking in grade D**

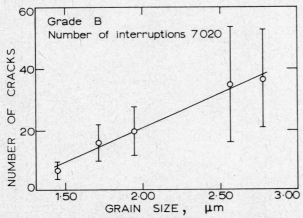

14 **Influence of grain size on thermal fatigue cracking in grade B**

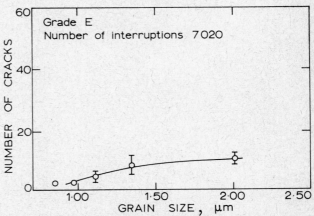

17 **Influence of grain size on thermal fatigue cracking in grade E**

and oxidation have in certain cases concerning grade D caused obliteration of cracks.

Milling tests have been performed with the aim of keeping the conditions constant for all the grades. However, it has not been taken into consideration that cemented-carbide grades of different compositions at constant cutting conditions have different heat generation and cutting force conditions around the edges. Increasing the titanium carbide content reduces the heat generation and the cutting forces. Increasing the binder-phase content has the opposite effect. This means that grade A has had more favourable conditions than grades C and E, while grade D has had more unfavourable conditions than C and B.

The results may be summarized as follows:

1 Increasing titanium carbide content leads to increasing sensitivity to thermal fatigue at constant Co content.

2 Increasing Co content leads to increasing sensitivity to thermal fatigue at constant titanium carbide–tungsten carbide ratio.

3 Increasing titanium carbide content as well as cobalt increases the grain size dependence of the thermal fatigue.

4 For two grades it has been observed that the sensitivity to thermal fatigue has a maximum at a certain grain size. For the other grades the thermal fatigue sensitivity increases with the grain size.

5 Wear and oxidation combined with the composition of the cemented carbide influence the conditions under which the cutting edges operate. This means that the sensitivity to thermal fatigue of cemented-carbide tips is not a purely thermal phenomenon.

6 The relations established show that further work in this field must be comprised of cemented-carbide grades covering a larger composition and grain size range than what has been done in this investigation. The conclusions made here have only a limited validity.

REFERENCES

1 ANONYMOUS: *Metalwork. Prod.*, 1960, Sept. 28, 77–78
2 H. OPITZ AND W. LEHWALD: 'Untersuchungen über den Einsatz von Hartmetallen beim Fräsen'; 1963, Köln und Opladen, Westdeutscher Verlag
3 Z. JEFFRIES *et al.*: *Trans. AIME*, 1916, **57**, 596–607
4 R. L. FULLMAN: *J. Met.*, 1953, March, 447–452
5 ISO Technical Committee 119, Powder metallurgical material and products, Document 119 N 17
6 Book of ASTM Standards, Part 2, 1531–1535; 1954, Philadelphia, ASTM
7 N. Å. HÖRLIN: Machinability test, 1964, PM Jan. 2, Stockholm, Swedish Association of Metalworking Industries
8 R. KIEFFER AND F. BENESOVSKY: 'Hartmetalle', 197; 1965, Wien, Springer-Verlag
9 R. KIEFFER AND F. BENESOVSKY: *ibid.*, 179
10 R. KIEFFER AND F. BENESOVSKY: *ibid.*, 196
11 R. KIEFFER AND F. BENESOVSKY: *ibid.*, 178
12 A. G. METCALFE: *J. Inst. Met.*, 1947, **73**, 591–607

Titanium carbide cutting tools: development and performance

J. E. Mayer, D. Moskowitz, and M. Humenik

The high hardness and melting point of pure titanium carbide, both of which exceed tungsten carbide, have been an impetus for its use as a cutting tool almost since its first synthesis by Moissan.[1] The first patents covering its use as a tool material were obtained by Schwarzkopf and Hirschl in Austria in 1931[2] and in the USA in 1933.[3]

A composition containing $42 \cdot 5$TiC–$42 \cdot 5$Mo$_2$C–14Ni–1Cr, based on a solid solution of Mo$_2$C in TiC, was produced under this patent by the Metallwerk Plansee in Austria and marketed in Europe under the trade name Titanit, as well as in the USA and England under the name Cutanit. The failure of this as well as other early TiC-based materials to gain acceptance, at a time when the steel-machining grades of WC were successfully entering the market, was due mainly to their low strength.

Improved grades of tungsten carbide were subsequently developed during the decade of the 1930s,[4] but little further development of titanium carbide took place until after World War 2. It was then that cemented TiC 'cermet' materials were studied, mainly under US Air Force and Office of Naval Research sponsorship, for possible high-temperature structural applications such as jet engine components. It was hoped that these cermets would combine the high-temperature strength and toughness of each constituent into one useful composite. Many of the requirements for a successful high-temperature material, such as adequate oxidation and thermal shock resistance as well as stress-rupture strength, were in fact well on their way to solution by the mid-1950s. In the final analysis, however, despite much development work, the inferior impact resistance of cermets as compared to nickel- and cobalt-based superalloys precluded their use in these applications. Furthermore, their strength was still no greater than 50–60% of that of WC–Co materials.

At about the same period, research into the wetting properties of liquid metals on refractory substrates such as carbides was being conducted by a group at the Ford Motor Company Scientific Laboratory. This work culminated in the publication of a series of papers[5-7] which outlined the importance of wettability in controlling the microstructure of systems such as WC–Co. Since the strength of cemented carbides is determined primarily by their microstructure, this opened the way for the development of TiC–Ni–Mo materials of comparable strength to the WC–Co counterparts. An intensive development program was undertaken, and by 1959 a finishing grade of cemented TiC had been successfully developed and evaluated.[8,9]

A USA patent covering a range of TiC–Ni–Mo materials was granted in 1961[10] and reissued in 1965.[11] Since then, a number of companies have been manufacturing finishing grades of TiC cutting tools, licensed under the above patent. More recently, a titanium carbide grade for the general purpose machining of steel has become commercially available.

MICROSTRUCTURE OF CEMENTED CARBIDES

It has been shown that the fracture behaviour of cemented carbides is related to the size of the carbide particles.[7,12,13] In both WC and TiC based materials, the crack passes mainly through the carbide grains when their size exceeds about 3 μm but proceeds around them for finer grain sizes. Lower strength is obtained in materials characterized by the former case, i.e. a transgranular mode of fracture.

The carbide grain size of WC–Co materials is determined mainly by the particle size of the milled powder, since little grain growth occurs during sintering. A major problem encountered when bonding TiC with Ni, Co, or Fe, on the other hand, is the excessive grain growth that takes place during sintering. The group at Ford studying wettability established that grain growth in this type of system is influenced by the degree to which the solid carbide particles are wet by the liquid metal. Where complete wetting takes place, such as in Co–WC, little carbide grain growth is noted. On the other hand, when wetting is incomplete, as in Ni–TiC, significant grain growth of the carbide phase takes place by a process involving coalescence of the carbide particles followed by rapid liquid phase sintering at the junctions of the coalesced particles. It was also shown that the wetting of TiC by liquid Ni could be enhanced by molybdenum additions to the Ni, resulting in a refinement of carbide grain size, as illustrated in Figs. 1 and 2. Since the fracture behaviour of cemented carbides, and therefore their strength, is related to the microstructure, control of the microstructure in cemented titanium carbides has resulted in the development of higher strength materials.

PHYSICAL AND MECHANICAL PROPERTIES
Strength
Figure 3 shows the transverse bend strength as a function of vol.-% binder for both WC–Co and titanium carbide-based compositions. In the latter, the Mo:TiC ratio was fixed at 1:7, and the Mo additions were made as Mo$_2$C, so essentially the variation in binder content is obtained by variable nickel additions. Data for the WC–Co system represent the range of values reported by Schwarzkopf and Kieffer,[4] while the points for the titanium carbide compositions represent average values for compositions having a carbide particle size of about 1 μm and 2–3 μm.

From these results, it can be seen that not only do we find a similar strength–binder relationship for the two systems, but the increase in optimum strength at higher binder contents for the titanium carbide compositions having a smaller

The basic concepts important to the control of microstructure and mechanical properties of new cemented titanium carbide materials are briefly reviewed. More specific attention is directed to three grades of titanium carbide developed for metalcutting applications. Their performance is demonstrated by results from laboratory machining tests and production machining operations. The applicability of titanium carbide grades is shown in rough, semifinish, and finish cutting of various workpiece materials. Increased resistance to flank wear and crater wear, plus improved surface finish, are shown to yield markedly greater tool life relative to commercial tungsten carbide tool grades.

621.91.025.7:669.018.95:661.882.621:539.538

The authors are with Ford Motor Co., Dearborn, Michigan

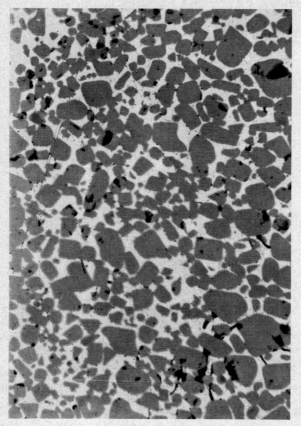

1 Microstructure of a 40Ni–60TiC composition, sintered 1 h at 2500°F in vacuum; alkaline ferricyanide etch   × 1 000

2 Microstructure of a 40(62·5Ni–37·5Mo)–60TiC composition, sintered 1 h at 2500°F in vacuum; alkaline ferricyanide etch
× 1 000

carbide grain size is consistent with the results of Gurland and Bardzil[14] for WC–Co materials. Although only the average strength values are shown for the titanium carbide compositions, it might be pointed out that for the 2–3 μm compositions, maximum individual values of strength approaching 400 000 lb/in² were obtained, while for the 1 μm compositions, maximum individual values exceeded 400 000 lb/in².

Hardness

The hardness as a function of the binder content is shown in Fig.4 for the titanium carbide compositions having a 2–3 μm carbide grain size. Data for WC–Co compositions are also given in Fig.4 and represent the range of values reported by Schwarzkopf and Kieffer. Here again, a similar hardness–binder relationship is obtained for the two systems. Generally, the hardness of the titanium carbide based materials appears to be somewhat greater than the average of the WC–Co at equivalent binder contents. However, this is not too surprising in view of the higher hardness of titanium carbide (3 200 kg/mm²) compared to tungsten carbide (2 400 kg/mm²).[15]

Density

The densities of the titanium carbide base dcompositions are shown in Fig.5 as a function of the binder content. Although the values for WC–Co are not shown, the densities of the titanium carbide compositions over the range of 3–47 vol.-% binder are $\frac{1}{2}$ to $\frac{2}{3}$ lower than WC–Co compositions at equivalent binder contents.

Abrasion resistance

The wear resistance of TiC cutting tools has been measured by obtaining the volume of material lost after 1 300 revolutions at 100 rev/min by rubbing 30 grit aluminium oxide against the specimen using a 1020 steel wheel at a constant load of 50 lb.* Figure 6 shows the abrasion-resistance factor of TiC grades compared to that of WC steel-machining grades. The data imply that TiC compositions may not be particularly suited for applications requiring great abrasion resistance, e.g. wear parts and dies such as are used for pressing ceramic powders. Although the resistance to abrasion, under the conditions described above, of TiC grades currently available is somewhat inferior to that of WC materials,

* CCPA Standard P-112

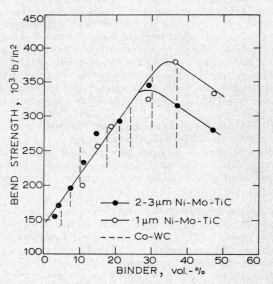

3 **Transverse rupture strength–vol.-% binder of TiC–Ni–Mo and WC–Co materials**

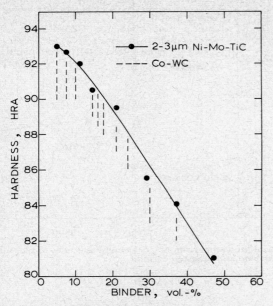

4 Relationship of HRA hardness and vol.-% binder of TiC–Ni–Mo and WC–Co materials

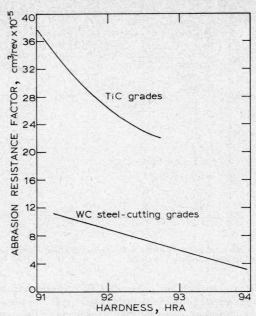

6 Abrasion-resistance factors of TiC base and WC base steel-cutting tool grades

newer TiC compositions are in the process of development† which are comparable in wear resistance to WC steel-cutting grades.

CUTTING-TOOL PERFORMANCE
Cutting-tool grades

At the present time, three grades of titanium carbide, designated 4J, 5H, and 7G, are being applied in production-machining operations. All three compositions are manufactured by the standard techniques of cold pressing and vacuum sintering as practiced in the carbide industry. The properties of these grades and the areas of application according to the general USA and ISO standard classifications for carbide grades are given in Table 1.

It might be noted that the P10 classification overlaps the 4J and 5H grades. In P10 applications the harder 4J titanium carbide should be considered first; however, in the event of chipping, then the tougher 5H grade would be more appropriate. A similar overlap is found in the P20 classification for the 5H and 7G grades of titanium carbide.

Tool-life tests

The superior flank wear resistance of the three titanium carbide grades compared to tungsten carbide based tools when machining steel has been demonstrated in laboratory turning tests. Figure 7 shows a tool-life comparison of steel-finishing grades of titanium carbide (4J) and tungsten carbide (C–8) over a range of cutting speeds. These tools have comparable hardness and strength levels. The tool-life endpoints are based on the specified flank wear land. The tool geometry designation is given in the American system, the order being: back rake, side rake, end relief, side relief, end and side cutting edge angles, and nose radius. For the cutting speed range of 800–1 200 sft/min shown, a 5–7-fold improvement in tool life is obtained for the 4J titanium carbide grade. This result is about the same with and without the coolant. Figures 8 and 9 show the performance of the steel-semi-finishing grade of titanium carbide (5H) and of the steel-roughing grade of titanium carbide (7G) respectively. Each grade is compared to a tungsten carbide grade having comparable hardness and strength. Both of these titanium carbide grades have superior tool life at all cutting speeds shown, relative to the tungsten carbide grade with which each is compared. Note that 6-fold improvements are obtained for the 5H and 7G titanium carbide grades at 750 and 550 sft/min respectively.

In addition to the superior flank wear resistance when machining steel, cemented titanium carbide tools have superior cratering resistance compared to cemented tungsten carbide tools. This is readily apparent in Fig. 10, which shows macrographs of tungsten carbide and titanium carbide tools after cutting 180 and 819 pieces respectively in a production steel-machining operation. The 4J titanium carbide grade was employed in this test. To date, it has been established in

† to be published

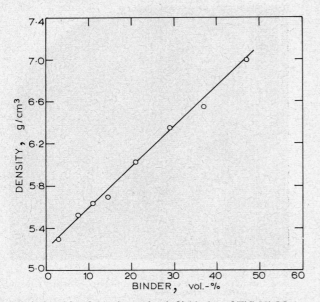

5 Relationship of density and vol.-% binder of TiC–Ni–Mo materials

Table 1 Properties of TiC cutting-tool grades

Tool designation Ford	USA	ISO	Hardness, HRA	TR strength, lb/in² ×10³	Density, g/cm³	Young's modulus, lb/in² ×10⁶
4J	C7, C8	P01, P10	92·8	200	5·55	63·9
5H	C6, C7	P10, P20	92·0	235	5·63	62·5
7G	C5, C6	P20, P30	91·0	275	5·80	59·9

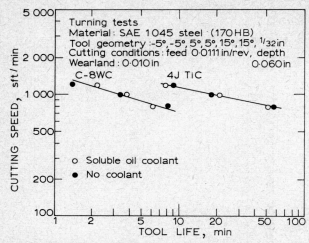

7 **Tool-life comparison of steel-finishing grades of titanium carbide and tungsten carbide**

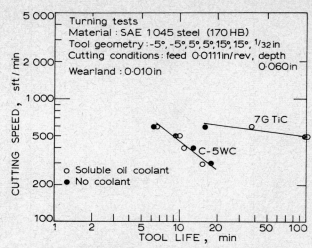

9 **Tool-life comparison of steel-roughing grades of titanium carbide and tungsten carbide**

a variety of machining operations on different steels that titanium carbide tools show little crater wear; therefore, tool life is determined essentially by flank wear. This is not too surprising, since improved cratering resistance of WC–Co tools containing titanium carbide additions is well known and therefore further improvement in cratering resistance might be expected in cutting tools wherein the carbide phase is essentially titanium carbide.

Although primarily developed for steel-machining operations, cemented titanium carbide tools also have been found to possess superior flank wear resistance when machining cast iron compared to cemented tungsten carbide tools. Figure 11 shows flank wear–cutting time results for the 4J titanium carbide grade and a C–4 tungsten carbide grade in laboratory turning tests on SAE111 cast iron. A 5-fold improvement in tool life is obtained for the 4J titanium carbide grade when compared at a 0·015in wear land.

PRODUCTION-MACHINING RESULTS
With the knowledge of the superior tool performance from laboratory machinability tests, the application of titanium carbide tooling in manufacturing operations at the Ford Motor Co., USA, has been in progress for some time. Specific applications will now be cited which demonstrate benefits of titanium carbide tools in production-machining operations.

Improved tool-life applications
The following examples show typical tool-life improvements

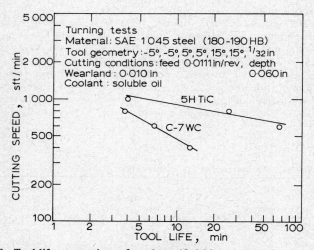

8 **Tool-life comparison of steel-semifinishing grades of titanium carbide and tungsten carbide**

a C–7 WC 180 pieces; *b* TiC 819 pieces
Finish turn OD 620 sft/min, 5140 steel forging, 187–217 HB, feed 0·011 in/rev, depth 0·047in

10 **Relative cratering resistance of TiC base tool and steel-cutting grade of WC–Co**

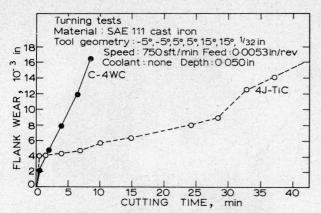

11 Tool-wear comparison of 4J titanium carbide and C–4 tungsten carbide grades when machining cast iron

obtained with the 4J titanium carbide grade in production operations. In these applications, the tool life is determined primarily by the tool wear resistance.

A turning and facing operation on a steel part which has cutting speeds up to 1649 sft/min is illustrated in Fig.12. All cutting conditions, steel type, and part name are given in this figure and those which follow. Five cutting tools perform this operation as the part is rotated in a lathe. Formerly this part was machined with C–7 and C–8 tungsten carbide tools. A change to the 4J titanium carbide grade increased

tool life dramatically by a factor of 4.5:1. This gain in tool life reduced the cost of this operation by 26%. In addition, the productivity (pieces/h) was increased by 21% due to reduction in the number of tool changes per interval of time.

Titanium carbide tools also perform extremely well in steel-grooving operations where the groove form and location are critical. One such operation is illustrated in Fig.13. The 4J titanium carbide grade has replaced the C–7 tungsten carbide grade in this application and provides a 5-fold improvement in tool life.

A rough-turning operation on a cast-iron part is illustrated in Fig.14. The 4J titanium carbide grade has outperformed C–2 tungsten carbide tools in this application by a factor of 5:1 in life. It might be noted that this operation is performed at a relatively low cutting speed of 200 sft/min.

Improved surface-finish applications

Titanium carbide tools have also been found to provide superior surface finishes in production-machining operations on cast gray, nodular, and malleable irons. The improvement is attributed to the absence of any significant build-up on the cutting edge of titanium carbide tools. The following applications demonstrate the benefits obtained with the 4J titanium carbide grade in production cast-iron operations.

Figure 15 shows the improved surface finish obtained in a turning operation on a SAE111 cast-iron part for titanium carbide compared to tungsten carbide tools. In this test identical cutting conditions and tool geometry were employed

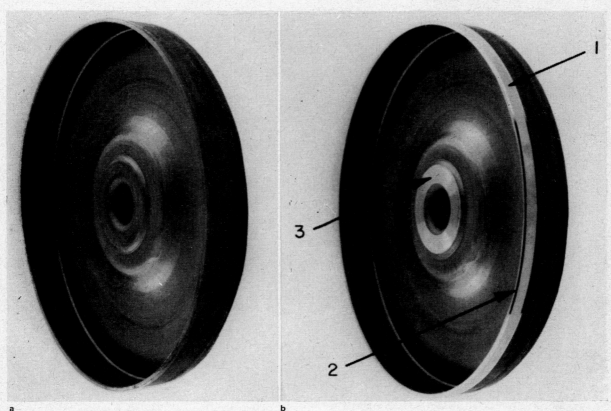

a   b

a before operation; *b* after operation

Surfaces machined	Tools	Speed, rev/min	sft/min	Feed, in/rev	Depth, in	Length, in
1	Finish turning	540	1649	0·0084	0·04	0·35
2	Rough and finish facing	540	1649	0·0084	0·09, 0·02	0·15
3	Rough and finish facing	540	425–595	0·0084	0·04, 0·02	0·60

12 Turning and facing operation on transmission converter cover assembly, SAE1022 steel

a before operation; b after operation
Speed 394 sft/min (3·9in dia.); feed 0·002in/rev; depth 0·060in; width 0·065in

13   Grooving operation on transmission clutch race, SAE5060 steel, 189 HB

After operation a; speed 200 sft/min (7·1in dia.); feed 0·013in/rev; depth 0·11in; length 1·3in

14   Turning operation on transmission drum, SAE111 cast iron

for both tool materials. This figure plots the number of successive parts machined with each tool on the abscissa. The ordinate represents the average surface-finish reading obtained on each machined part. In this application the 4J titanium carbide grade provides and holds about a 39% lower

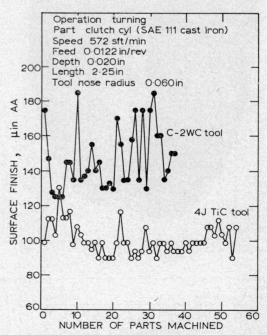

15   Effect of titanium carbide and tungsten carbide tools on the surface finish when turning a cast-iron part

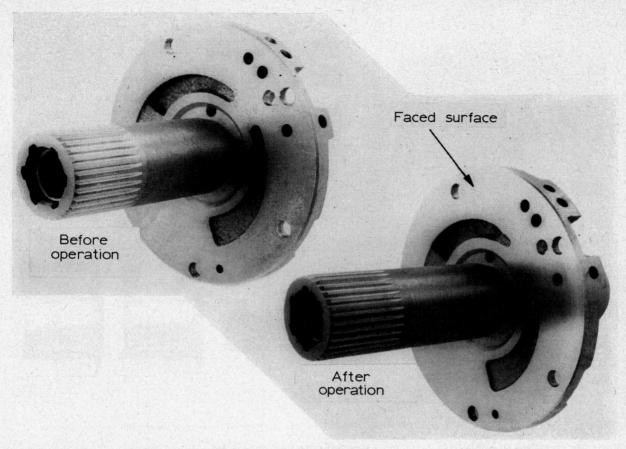

Speed 251–841 sft/min (565 rev/min); feed 0·005in/rev; depth 0·015in; length 2·0in

16 Finish-facing operation on transmission stator support, SAE111 cast iron

surface-finish reading than the C–2 tungsten carbide tool. It also provides less finish variation from part to part, the range being about 50% smaller. The finish specification for this surface is 80–150 μin AA, therefore Fig.15 indicates that the 4J titanium carbide grade offers an improvement in tool life of more than 2-fold, since the tool life in this operation is based primarily on finish.

A very impressive application of cemented titanium carbide which provides superior surface finish is the finish-facing operation on a SAE111 cast-iron part illustrated in Fig.16. Note that the surface machined has many cut interruptions. Figure 17 shows the surface finish produced on successive parts by the 4J titanium carbide grade and by the tungsten carbide tool previously employed on this job. The titanium carbide tool provides an initial μin finish on the first part which is one-half that obtained with the tungsten carbide tool and results in a 6:1 tool-life improvement based on the maximum allowed 100 μin AA limit. Flatness requirements of

0·0002in convex and 0·0008in concave, both maxima, are also met. The finish and flatness being produced by the titanium carbide tools on this operation permitted a subsequent finish-grinding operation to be eliminated, thus providing substantial savings in addition to the savings obtained from the longer tool life.

Applications of 5H and 7G titanium carbide grades

The following applications demonstrate the strength and superior wear resistance of the 5H and 7G titanium carbide grades in production steel-machining operations.

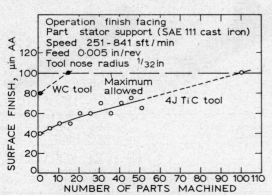

17 Effect of titanium carbide and tungsten carbide tools on the surface finish when facing a cast-iron part

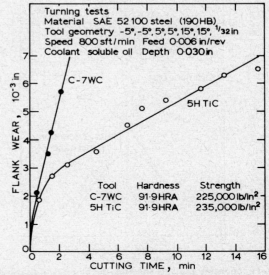

18 Tool-wear comparison of 5H titanium carbide and C–7 tungsten carbide grades when turning SAE52100 steel

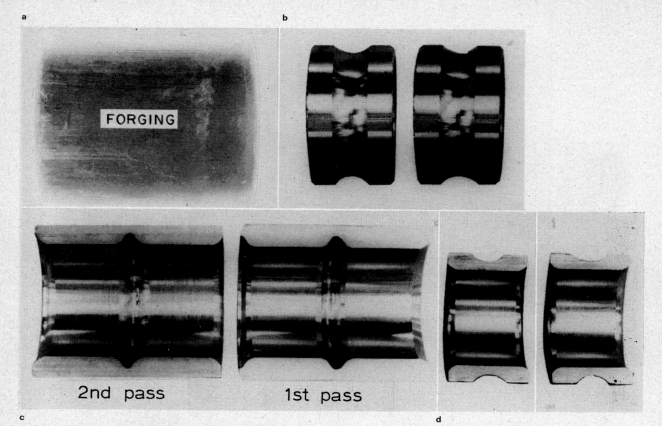

a before operations, length 2·48in, i.d. 1·41in, o.d. 2·12in; b after operations; c 1st operation, bore i.d., turn radii, face one end; d 2nd operation, turn o.d., ball track and chamfers, cut in two, face ends

19   Machining operations on ball-bearing forged inner ring, SAE52100 steel, 165–200 HB

The application of single-point throwaway-insert titanium carbide tools to the production of ball-bearing inner rings on newly developed multispindle, contour boring, and turning machines paid off in lower machining costs and higher productivity. The 5H titanium carbide grade is employed for the major portion of this process. The superior wear resistance provided by this grade compared to a C–7 tungsten carbide tool when machining SAE52100 steel, the bearing ring material, is demonstrated by the laboratory turning test results shown in Fig.18. A 5-fold tool-life improvement is obtained when compared at a 0·005in wear land. The two machining operations performed in the production process are illustrated in Fig.19. The starting workpiece is a hot-forged ring. Each forging is subsequently machined into two bearing rings.

The first operation in Fig.19 is performed in two passes. Two 5H titanium carbide tools are employed in this operation, one for the first pass (roughing) and the other for the second pass (finishing). Each tool machines the entire inner surface. In the roughing pass, the tool machines into the forging wall to generate the contours shown. Also in the first operation, a third tool is employed to face one end of the forging. The 4J titanium carbide grade is used for this cut which does not require the higher strength of the 5H grade.

The second operation in Fig.19 is carried out in one pass. One 5H titanium carbide tool is employed in this operation

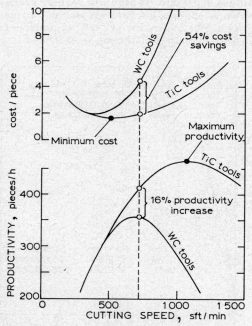

20   Tool economics analysis showing the benefits of titanium carbide compared to tungsten carbide tools for 2nd machining operation on ball-bearing forged inner ring

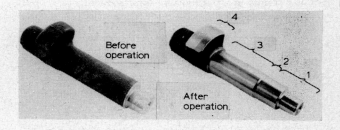

Surfaces machined	Speed, rev/min	sft/min	Feed, in/rev	Depth, in	Length, in
1	1410	310–465	0·02	0·09–0·12	2·1
2	1410	465–495	0·0083	0·06–0·09	0·2
3	1410	495	0·02	0·06	2·8
4	1410	1115–1185	0·0083	0·14	1·0

21   Rough-turning operation on steering gear sector shaft, SAE6118H steel forging, 156–207 HB

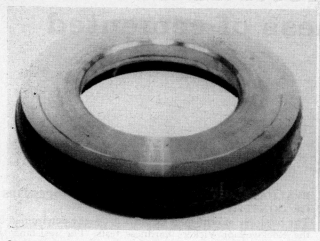 

a   b

a part way through operation; *b* after operation; speed 510–840 sft/min, feed 0·015in/rev, depth 0·02in, length 1·7in (8·6in o.d., 5·2in i.d.)

22   Facing operation on ring gear, SAE8615 steel forging, 143–187 HB

to turn the entire outer surface. The ball track is roughed out and then finished by this same tool during the pass. Also in this second operation, three 4J titanium carbide tools are employed: two for end facing and one for cutting the attached bearing rings apart. The cut-off tool does not require higher strength because the part is chucked on its entire bore throughout the cut.

The results of a tool-economics analysis performed on the second machining operation in Fig.19 to determine the optimum cutting speed is shown in Fig.20. This figure plots production cost per piece and productivity as a function of the cutting speed. The cost per piece was computed by summing the machining time cost, idle time cost, tool change time cost, and tool cost. The productivity was computed by considering the machining time, idle time, and tool change time. Since the minimum-cost cutting speed differs from the maximum-productivity cutting speed, a compromise cutting speed of 720 sft/min was selected for this operation using titanium carbide tools. At this cutting speed, Fig.20 shows that titanium carbide tools provide a 54% cost saving and a 16% productivity increase compared to tungsten carbide tools.

The motivation behind developing this new process at the Ford Motor Company, USA, for producing ball-bearing forged inner rings, was research findings which demonstrated improved bearing life compared to conventional ball bearings which are made from seamless tubing. The new machining process for forged bearing rings described above, using all titanium carbide tooling, is considered a first in the bearing industry.

The application of the 7G titanium carbide grade to two different production steel-machining operations will be discussed next. An application which simultaneously imposes both a severe cut interruption and a high cutting speed is the rough-turning operation illustrated in Fig.21. A single tool performs the entire operation in one pass; however, the sector portion has the most difficult cutting conditions. In this application, the 7G titanium carbide grade achieved a tool life of up to 1·7 times that of the C–5 tungsten carbide tools employed on this job. Another application of the 7G grade is the facing operation illustrated in Fig.22. A tool insert which has a ground-in chip breaker is required on this job. Therefore a roughing carbide tool grade is necessary to provide sufficient strength to the reduced included angle of the cutting edge to prevent breakage. In this application the 7G titanium carbide grade provides a 2-fold tool-life improvement compared to the C–5 tungsten carbide tool previously employed on this job.

CONCLUSIONS

From extensive production-finish, as well as semifinish, machining data on steel, it has been established that an average improvement in tool life of about 320% is obtained with the 4J titanium carbide grade. This material is now specified for all finish and semifinish machining of steel in Ford, USA, where previously four different steel-cutting grades of tungsten carbide (C–80, C–8, C–70, C–7) were used. In the area of rough machining on steel, the tougher grades of titanium carbide (5H, 7G) have been evaluated on numerous production operations and improvements in tool life ranging from 50–400% have been obtained. Based on the results to date, it is anticipated that cemented titanium carbides will become the predominant cutting-tool materials for steel-machining operations.

In addition to steel-machining operations, cemented titanium carbide tools have shown successful performance on a variety of cast gray, nodular, and malleable irons, particularly where superior surface finishes are required. More extensive evaluation will be required on the high-carbon iron alloys to determine the full potential of titanium carbide tools in these applications.

ACKNOWLEDGMENTS

The authors wish to acknowledge the valuable assistance of Mr S. Cowell and the numerous tool engineers within Ford Motor Co. who have contributed data on the performance of titanium carbide tooling in production-machining operations.

REFERENCES

1 H. MOISSAN: *Compt. Rend.*, 1895, **120**, 290; 1897, **125**, 839
2 Deutsche Edelstahlwerke AG, Austrian patent no. 160,172 1931
3 P. SCHWARZKOPF AND I. HIRSCHL: US patent no. 1,925,910, 1933
4 P. SCHWARZKOPF AND R. KIEFFER: 'Cemented carbides', 8; 1960, New York, Macmillan
5 M. HUMENIK AND N. M. PARIKH: *J. A. Ceram. Soc.*, 1956, **39**, 60
6 M. HUMENIK AND N. M. PARIKH: *ibid.*, 1957, **40**, 315
7 N. M. PARIKH: *ibid.*, 335
8 *Production*, 1959, **43**, 5, 99
9 *Iron Age*, 1959, **183**, 12, 101
10 M. HUMENIK AND D. MOSKOWITZ: US patent no. 2,967,349
11 M. HUMENIK AND D. MOSKOWITZ: US patent reissue no. 25,815
12 J. R. LOW: *Trans. AIME*, 1956, **206**, 982
13 G. S. KREIMER AND N. A. ALEKSEYEVA: *Fiz. Met. Metallov.*, 1962, **13**, 14, 609
14 J. GURLAND AND P. BARDZIL: *Trans. AIME*, 1955, **203**, 311
15 P. SCHWARZKOPF AND R. KIEFFER: 'Refractory hard metals', 86, 160; 1953, New York, Macmillan

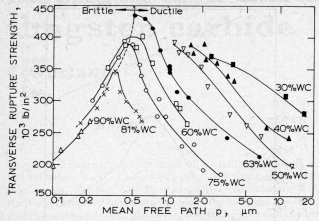

3 Transverse rupture strength v. mean free path (12)

Table 1   Comparison of strength of cemented carbides[14]

Carbide	Elastic modulus, $\times 10^6$ lb/in^2	Transverse rupture strength, $\times 10^3$ lb/in^2 10 vol.-%Co	37 vol.-%Co
WC	104*	241	346
TaC	91*	163	269
TiC	72*	162	254
NbC	49	132	205
VC	37	83	183
ZrC	20	37	113

* From reference 20

which takes into account the contiguity C, a measure of the degree of contact between WC particles.

As indicated in Fig.3, the curves show a transition between brittle and ductile behaviour. To the right of the strength maximum, the 'ductile' side of the diagram, failure takes place between the limit load of the composite based upon the yield strength of the cobalt and the limit load corresponding to the ultimate strength of the cobalt.[3] The fracture strength of the composite depends on the critical stress for crack initiation in the carbide particles or elsewhere in the microstructure. The fracture mechanism is ductile in the sense that failure occurs by an accumulation of damage in the structure under increasing loads, for instance by cracking of carbide particles, as observed by Nishimatsu and Gurland,[6] until the aggregate fails when a critical number of particles are broken and the remaining material can no longer support the load.

The fracture mechanism to the left of the strength curve is brittle in the sense that failure occurs by catastrophic propagation of fracture from a critical flaw. It may be seen from Fig.1 that the fracture strength in tension lies below the proof stress compression in the relevant composition range. The constraint factor is so high that plastic deformation does not take place to any appreciably extent, with the consequence that local stress concentrations are not relaxed by plastic flow and instead contribute to fracture initiation and propagation. It may be argued that an increase of the mean free path raises the fracture strength because it decreases the constraint factor and the peak stresses in the aggregate.

In brittle materials in general the fracture strength is determined by the critical conditions for crack propagation and the Griffith–Orowan criterion provides a quantitative correlation between fracture strength and microstructural parameters of brittle alloys:

$$\sigma = k\left[\frac{Ep}{l_C}\right]^{\frac{1}{2}} \quad\quad\quad\quad\quad (3)$$

where σ is a critical fracture stress, k is a constant, E is the modulus of elasticity, p is the total work of fracture, and l_C is the length of the initiating crack or other discontinuity. The application of the Griffith or Griffith–Orowan criteria to cemented carbides is most extensively developed by Kreimer and his co-workers,[2] and is also the basis of fracture theories proposed by Exner and Fischmeister,[11] Gurland,[12] and Nabarro and Luyckx.[13] In order to justify the relevance of this fracture criterion to cemented carbides, the separate influence of the physical parameters E, p, and l_C will now be examined.

The elastic modulus E enters into equation (3) through its contribution to the strain energy, and it appears reasonable that this represents the modulus of the alloy and is it so used by Kreimer. The major contribution to the modulus of an alloy of tool grade composition, i.e. rich in WC, is the value of the modulus of the carbide constituent; if all other terms are constant, the fracture strength should be proportional to that modulus. The results of one study[14] support this relation to some extent, as shown in Table 1. The table compares the strengths of various carbide–cobalt alloys at two different Co contents and shows that the strengths increase in the same order as the values of the moduli of the carbides.

The major part of the work of fracture, p, is undoubtedly contributed by the work of plastic deformation of the cobalt

direct contribution of the carbide becomes dominant or cracking occurs, at which point other factors would be determining.

Admittedly, the continuum treatment is based on a highly idealized and simplified model; it has, however, the advantage that the postulated mechanism of strengthening can be checked on models. Comparison of the calculated yield strength with the measured proof stress of Fig.1 shows fair agreement of strength ratios in the range from 20–50%Co, although the level of strength is underestimated. Part of the discrepancy may simply be due to the wrong value chosen for the yield strength of cobalt, but part may also be attributed to the size effect for the yield strength of cobalt in layers so thin ($\sim$0.1 μm) that bulk properties can no longer be used as reference values. According to Drucker[3] this size effect becomes important when the dimensions of the cobalt layers are so small that the number of dislocations in each layer becomes a relevant variable. Continuum theory and dislocation theory overlap at this level of structure.

The microscopic treatments of flow strength which have been proposed, summarized in reference 2, also reference 10, are generally based on dispersion hardening theories. It has not yet been shown directly by observation that the specific dislocation mechanisms of dispersion hardening actually take place in the cemented carbides; and these theories involve either correction terms or constants, the numerical values of which have not yet been confirmed by theory or independent measurements. While the various approaches agree that the flow stress is determined by some measure of the thickness of the cobalt layers, it is clear that more work is required in order to establish unequivocally and quantitatively the relation between flow stress and structural parameters.

Fracture strength

The fracture strength is best discussed by referring to Fig. 3, which presents the transverse rupture strength as function of the nominal mean free path, i.e. the average distance in the cobalt between carbide particles:

$$p = l\frac{f}{1-f} \quad\quad\quad\quad\quad (1)$$

where p is the mean free path, l is the mean linear intercept particle size of WC, and f is the volume fraction of cobalt. This representation has the advantage of combining both composition and particle size into one parameter. The nominal mean free path, as used by Gurland and Bardzil,[7] is based on an idealized structure in which all particles of WC are regarded as non-contiguous. In a similar representation, Exner and Fischmeister[11] introduce a 'true' mean free path:

$$p_p = \frac{l}{1-C}\frac{f}{1-f} \quad\quad\quad\quad\quad (2)$$

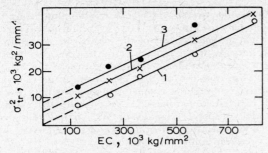

Mean carbide particle size (1) 1·64 μm; (2) 3·3 μm; (3) 4·95 μm (after reference 2, Fig. 74)

4 Plot of square of transverse rupture strength (σ_{TR}) v. product of modulus of elasticity (E) and cobalt content (C)

in the crack path. Figure 4 reproduces the data of Kreimer et al.[2] as a plot of fracture strength (σ) against the product of Co content in weight percent (C) and the elastic modulus (E). The observed linear relation permits Kreimer et al. to propose the following equation for the strength of brittle cemented carbides:

$$\sigma^2 = AEC + K \quad\ldots\ldots\ldots\ldots\ldots\ldots\ldots\ldots\ldots\ldots\ldots\ldots(4)$$

where A and K are constants. The equation is derived from equation (3) by postulating that the plastic work of fracture is proportional to the area fraction of the fracture surface in the cobalt phase, which in turn is proportional to the weight fraction, more accurately, volume fraction, of cobalt in the alloys. It is interesting to note that the slope constant A is independent of particle size and that the particle size appears only in the intercept value K. According to Kreimer,[2] the characteristics of the constants mean that the plastic work in the cobalt is not affected by the particle size of WC at a given composition and that the value of K is related to the contribution to the fracture stress of the tungsten carbide itself. Indeed, it is observed that K increases with particle size, as would be expected since the fracture path preferentially traverses the larger WC particles.

Little is known about the physical significance of the parameter l_C in equation (3), which, in Griffith's model, is the critical length of the crack nucleus. In Kreimer's derivation of equation (4), l_C is contained in the constant A together with the work of fracture per unit of Co content. It is experimentally difficult to separate the two variables and it is often necessary to assume the magnitude of one of them. Gurland,[12] for instance, plotted the maximum strength of four compositions, 63, 69, 75, and 81 vol.-%WC, from the data of Fig.3 against the reciprocal of the square root of the particle diameter of WC. As shown in Fig. 5, the relation is approximately linear, within the limited range of the data. From the

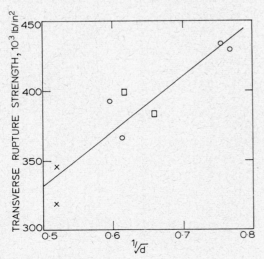

5 Transition strength

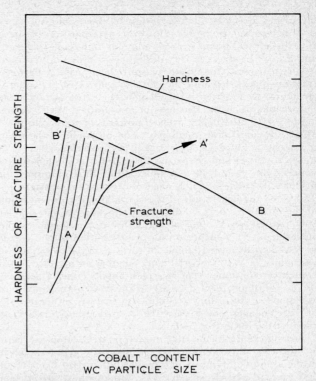

6 Schematic representation of variation of strength and hardness with composition and particle size

Griffith–Orowan equation and the slope of the line of Fig.5, a value of 9000 ergs/cm² is calculated for the work of fracture. Whether or not this is a reasonable value depends on the validity of the assumption that l_C is proportional to the particle size of WC. In Gurland's work, this assumption was based on the observed cracking of large WC particles in these alloys[6] and the postulate that the maximum strength is attained at the transition from the 'ductile' mode of fracture to cleavage. As in most brittle fracture theories, the value of l_C cannot be verified unless the work of fracture is independently established, and vice versa.

Here again, it must be concluded that further work in relating the strength to structural parameters is required before the Griffith–Orowan criterion can be applied quantitatively to cemented carbides. In the meantime, the equation of Kreimer and co-workers (4) serves as a useful approximation describing the observed behaviour.

DISCUSSION OF THE CONTROL OF MICROSTRUCTURE

The preceding discussion suggests certain guidelines for alloy improvement. The argument might be based on Fig.6, a schematic summary of Figs.1–3, and on the desire to improve on the properties of the alloys on the solid line AB. Extrapolating line A to the right (A′) leads into a region of

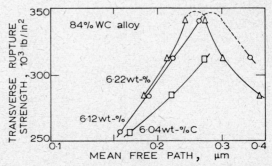

7 Strength as function of carbon content and mean free path (after 5 and 9)

increasing strength but decreasing hardness and therefore does not appear profitable. However, extrapolation of curve B to the left (B′) points towards alloys of both higher strength and greater hardness, a very desirable direction. This would imply, of course, the ability to shift the transition between fracture modes towards the left by control of the microstructure. As a matter of speculation, the extrapolation to the limit could reach the hardness of fine-grained WC (>2000 kg/mm² HV) at strength levels of upwards of 700000 lb/in².[15] While this ideal limit might not soon be attained, the more immediate alloy development will perhaps take place with microstructures and properties located in the shaded region between line A and the extrapolated line B′. Examining, for instance, the data of Fig.3 one notices that the strength increases on the brittle side of the graph and at constant value of the mean free path, with greater Co contents, i.e. smaller WC particle sizes. One of the main requirements is the achievement of a highly uniform distribution of the cobalt in the microstructure, which, in turn, requires reducing the contiguity of WC. Butler[9] points out that an increase of Co content at constant mean free path entails either a decrease of the contiguity or a decrease of the WC particle size, see equation (2). According to continuum analysis, either or both of these changes tend to raise the fracture strength at a fixed value of the mean free path.

Among the problems encountered in attempting to evaluate theories and experimental results is the general neglect in the scientific literature of variables other than Co content and average particle size of WC. Among these are carbon content, impurity content, porosity, particle size distribution, and thermal history. A review of parameters influencing the mechanical properties may be found elsewhere.[16]

As one example of the importance of one of these parameters, Fig. 7 shows the marked effect on strength of variation of carbon content, even within limits generally considered safe. The data is from the work of Suzuki[5] as plotted by Butler,[9] and the shape of the curves is consistent with the plastic constraint and Griffith–Orowan theories. Another example of the large effect of small compositional changes is provided by the work of Montgomery[17] on the effect of various surface coatings on the wear behaviour of cemented carbides. He found that thin coatings (∼100 μin) of gold, tin, silver, manganese, freshly deposited titanium, lead, tin, bronze, and solder on a cemented carbide surface allow a WC–Co slide to operate at least 10000 cycles at conditions which cause surface fatigue of the same slides on uncoated carbide in only 100 cycles. The author suggests that diffusion of the coating material may diminish the tensile stresses normally present in the cobalt and thereby increase the fatigue strength.

Finally, reference should be made to the interesting study of the polar anisotropy of WC by French and Thomas.[18] Following up the discovery of slip and hardness anisotropy[19] in WC crystals, these authors found that WC crystals have different properties along different senses of the crystallographic direction (10$\bar{1}$0). Parallel prism faces (10$\bar{1}$0) and (0$\bar{1}$10) differ in etching characteristics, wettability by polar liquids, and microhardness. It is possible that preferred orientation of the crystallites of WC in the alloys might produce an anisotropy of properties of the aggregate which has not yet been studied or exploited.

It is obvious from the preceding discussion that not enough is known about the design of cemented carbide tool materials in spite of an appreciable increase in knowledge about their properties during recent years. The industry shows signs of entering an era of renewed activity in grade development and materials design and it is to be hoped that some of the new ideas will come from the kind of materials research discussed in this survey.

REFERENCES

1 E. M. TRENT: Metallurgical Reviews, 1968, **13**, 127, 129–143
2 G. S. KREIMER: 'Strength of hard alloys'; 1968, New York, Consultants Bureau (English translation)
3 D. C. DRUCKER: in 'High-strength materials', ed. V. F. Zackary, 795–833; 1965, New York, J. Wiley and Sons, also *J. Mat.*, 1966, **1**, 873
4 R. KIEFFER AND P. SCHWARZKOPF: 'Hartstoffe und Hartmetalle', 146; 1953, Vienna, Springer-Verlag
5 H. SUZUKI: *Trans. Japan Inst. Met.*, 1966, **7**, 112–117
6 C. NISHIMATSU AND J. GURLAND: *Trans. ASM*, 1960, **52**, 469–484
7 J. GURLAND AND P. BARDZIL: *Trans. AIME*, 1955, **203**, 311–315
8 T. FAKATSU AND T. SASAHARA: *J. Japan Soc. Powder Met.*, 1963, **10**, 30–37
9 T. W. BUTLER: PhD thesis, 1969, Brown University, Rhode Island
10 H. DAI *et al.*: *Trans. Met. Soc. AIME*, 1969, **245**, 1457–1470
11 H. E. EXNER AND H. F. FISCHMEISTER: *Arch. Eisenh.*, 1966, **37**, 499–510
12 J. GURLAND: *Trans. Met. Soc. AIME*, 1963, **227**, 1146–1150
13 F. R. N. NABARRO AND S. BARTOLUCCI LUYCKX: Proc. Int. Conf. Strength of metals and alloys, *Trans. Japan. Inst. Met.* (Suppl.) 1968, **9**, 610–615
14 J. GURLAND: *Trans. AIME:* 1957, **209**, 512–513
15 D. GUCER AND J. GURLAND: *Jerkon. Ann.*, 1963, **147**, 111–115
16 H. E. EXNER AND J. GURLAND: *Powder Met.*, 1970, **13**, 13–31
17 R. S. MONTGOMERY: *Trans. ASLE*, submitted for publication
18 D. N. FRENCH AND D. A. THOMAS: 'Int. symposium on anisotropy of single crystal refractory compounds', ed. F. W. Vohldiek and J. Mersal, 1968, New York, Plenum Press
19 T. TAKAHASHI AND E. J. FREISE: *Phil. Mag.* 1965, **12**, 1–8
20 A. KELLY: 'Strong solids', 199; 1966, Oxford, Clarendon Press

New developments in the field of cemented-carbide and ceramic cutting tools

R. Kieffer, N. Reiter, and D. Fister

In the development of cutting alloys with increasing cutting speed, tools of carbon steel, alloy steel, high-speed steel, and stellites were followed by tools of cemented carbides and oxides (see Fig.1).[1]

Table 1 shows three groups of cutting materials: sintered WC base alloys, sintered WC free alloys, and ceramic cutting tools. The first group demonstrates the addition of transition metal carbides like TiC, TaC, NbC, and HfC, usually as carbide solid solutions, to the basic WC–Co composition; the second group shows the complete substitution of WC by other carbides like TiC, Mo_2C, VC, Cr_3C_2, or even nitrides like TiN; the third group demonstrates the progressive increase in toughness obtained with Al_2O_3 based tools and the approach to sub-micron Al_2O_3 and cement-like carbide–oxide tools.

SINTERED WC BASE ALLOYS
The continental cemented carbides, mostly divided into K, M, and P type tools, have the approximate composition of:

(i) K type: 0–5%TiC+TaC (NbC), rest WC–Co
(ii) M type: 6–10%TiC+TaC (NbC), rest WC–Co
(iii) P type: 10–40%TiC+TaC (NbC), rest WC–Co

HfC containing alloys
The first investigations with additions of HfC went back to 1959. Kieffer, Benesovsky, and Messmer[2] proved that HfC can replace either TiC or TaC in WC–TiC–TaC alloys (see Table 2), HfC showing at least the good properties of TaC.

The growing Zr industry and the automatic separation of Zr from Hf brought down the price of HfC and HfC containing solid solutions with NbC, TaC, and WC to the TaC level.

New investigations were concerned with binary, ternary, and quaternary HfC solid solutions and related cemented carbide combinations. Miscibility gaps in the system TiC–HfC, TiC–ZrC, VC–NbC and VC–TaC were discovered[3,4] and the influence of the TiC–HfC gap was investigated in the systems WC–TiC–HfC (Fig.2), WC–TiC–NbC–HfC, and WC–TiC–TaC–HfC.

Figure 3 gives details about the miscibility gap TiC–HfC and Fig. 4 shows the phase distribution in the quaternary system WC–TiC–NbC–HfC and the appearance of 3 phase regions.

The composition and the properties of NbC(TaC)–HfC containing WC–Co alloys are evaluated with standard WC–TiC–TaC–Co alloys in Table 3. NbC(TaC)–HfC additions (preferably in the form of solid solutions) are equivalent or superior to pure TaC. Table 4 shows the results of continuously replacing TaC in WC–TiC–TaC alloys by HfC. An addition of the solid solution of TaC–HfC (1:1) proved to be definitely better than HfC or TaC rich alloys.[5]

Cutting tests were run with alloys according to Table 5 where similar volumes of NbC–HfC and TaC–HfC solid solutions were added to alloys 1 and 2 as well as 4 and 5. As standards WC–TiC–TaC alloys of similar composition were used. The summary of about 80 cutting tests (speed as well as time versus cratering and flank wear) allowed the classification of the types with increasing wear resistance from WC–TiC–TaC–Co to WC–TiC–NbC–HfC–Co to WC–TiC–TaC–HfC–Co. The HfC containing alloys showed approximately 10–20% better wear resistance, giving about the same amount of improvement as obtained 20 years ago with WC–TiC–TaC alloys compared with TaC free WC–TiC alloys.

Carbide-coated and nitride-coated throwaway tips
For several years, the carbide producers tried to make sintered-carbide compound materials with a tough WC–Co interior and a sintered-on layer of harder TiC(WC)–Co alloys.[6,7] A great deal of effort, especially in the UK, was made but without achieving a definite breakthrough.[8]

During the last few months, throwaway tips have appeared on the market showing a similar, thin TiC deposit to that obtained by Münster et al.[9-15] in the treatment of steel parts with $TiCl_4$, CH_4, etc. in collaboration with the Metallgesellschaft Frankfurt/Main. G. Schumacher[16] described the production of 'Widia Extra' tips with 5–10 μm thick TiC layers produced by chemical vapour deposition according to Metallgesellschaft patents.[17-21] Other carbide producers are doing the same or applying other coating techniques (see the Coromant S_4C gamma-coated tips).

Our investigations in collaboration with the Titanit Fabrik Krefeld have shown that TiN coatings are superior to TiC coatings, especially in that the TiN coatings have a much lower tendency to crater. The yellow-golden TiN layers (thickness 5–20 μm) are produced at 900–1 200°C by the reaction of $TiCl_4$ with H_2 and N_2 or with NH_3. Red-golden layers of carbonitride are formed by adding CH_4 to the reaction gases. Figure 5 shows the photomicrograph ($\times$ 500) of a TiN coated tip.

We agree with G. Schuhmacher that the coating of cemented carbides will give longer life and less cratering to throwaway tips and may decrease welding-on effects on the cutting edge. The coatings may even be useful for wear-resistant parts.

The 'submicron idea' in the field of cutting tools
Since the first appearance of sintered WC–Co alloys, the carbide producers have been aware of the great importance of

The paper reviews the hard carbides used in sintered and cemented tool materials, the so-called hardmetals, classified as those containing and those without tungsten carbide. Throwaway tips, carbide or nitride coated, and the importance of grain size are briefly considered, and more recent carbide systems are referred to. A brief mention is made of ceramic (oxide) types, stressing their sensitivity to shock.

669.018.95: 661.883.262.1

The authors are with the Technische Hochschule, Vienna

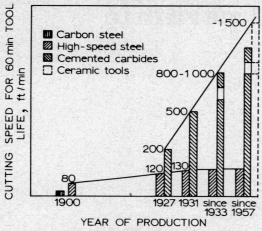

1   **Increase of cutting speeds in turning steel since 1900**

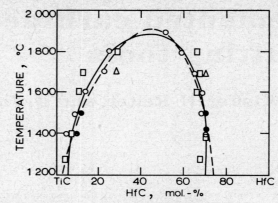

2   **Pseudo-ternary system TiC–HfC–WC at 1600°C**

the grain size of the primary WC powder and of the size of the WC phase in the finished products.[1] For tough grades or percussion drill bits, WC powders with 2–8 μm were used; for tools for special hard-cast iron, WC powders of about 1 μm and less were applied.

Some years ago, Ciba, Basel, introduced 'submicron carbides' (WC, TaC, TiC, etc.) to the market with a grain size of 0·01–0·1 μm and even finer.[22] The Co cemented submicron WC types showed higher hardness values when grain growth was inhibited by the addition of V carbides, VC being the most efficient inhibitor.[23] Later and independently, Du Pont of Nemours offered through the Baxtron Comp. hot-pressed submicron WC–Co grades that showed excellent cutting performances on hard cast iron. It is our strong belief that the submicron idea is well grounded and will be applied one day to solid solutions of carbides, e.g. WC–TiC, WC–TaC, WC–TiC–TaC, and WC–NbC(TaC)–HfC.

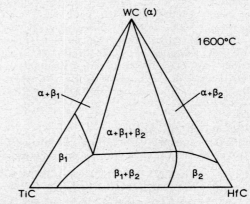

3   **Miscibility gap in the TiC–HfC system**

Table 1   Development of sintered WC base alloys, WC free alloys, and ceramic cutting tools

WC base alloys Composition	Year	WC free alloys Composition	Year	Ceramic cutting tools Composition	TRS, kg/mm²	Year
WC–Co	1923–25	TiC–Mo₂C–Ni, Cr, Mo	1929–31	Al₂O₃ for dies and tools	15–25	1912–13
WC–TiC–Co	1929–31	(solid solutions)		Sinterkorund	15–25	
WC–TaC(VC, NbC)–Co	1930–31	TaC–Ni	1930–31	Al₂O₃	20–35	1930–31
WC–TiC–TaC(NbC)–Co	1932	TiC–TaC–Co	1931	Sinterrubin		
WC–TiC–HfC–Co	1959	TiC–Cr, Mo, W, Ni, Co	1931	Al₂O₃+Cr₂O₃	30–40	1937–38
WC–TiC–TaC(NbC)–HfC–Co	1968–69	TiC–VC–Ni, Fe	1938	Al₂O₃ hot pressed	50–70	1944–45
WC–TiC–NbC(TaC)–HfC–Co	1968–69	TiC–NbC–Ni, Co	1944	Al₂O₃+0·5–1%MgF₂	40–50	1948–51
		TiC–VC–NbC–Mo₂C–Ni	1949	Al₂O₃–Mo₂ C–(Mo)	35–45	1951–59
		TiC(Mo₂C, TaC)–Ni, Co–Cr	1950	Al₂O₃+Ti, TiC, TiC/WC	40–55	1955–58
TiC and TiN coatings on WC alloys	1955–70	(WZ alloys)				
Submicron WC and solid solutions cemented with Co (Ciba and Du pont/Nemours resp. Baxtron)	1967–70	Cr₃C₂–Ni, Cu, Cr	1951–60	Al₂O₃ submicron	70–90	1968–70
		TiC with steel binder (heat treatable)	1952–61	Al₂O₃+TiC, Ni hot pressed	80–100	1968–70
		2TiC–1 TiB₂	1957			
		TiC–Mo₂C(mixtures) –Ni, Mo	1965–70			
		(Ti, Mo)C₁₋ₓ–Ni, Mo, Cr	1968–70			
		TiC–TiN (solid solutions)–Ni	1969–70			

Table 2   Composition and properties of HfC containing cemented carbides

Composition, wt-% WC	TiC	HfC	Co	Specific gravity, g/cm³	HV 60, kp/mm²	TRS,* kp/mm²	Magnetic saturation, $4\pi\sigma$
87	..	5	8	14·50	1 450	150	150
83	..	10	7	14·40	1 520	150	135
79·5	..	12.5	8	14·14	1 450	150	149
89	..	15	6	14·35	1 560	130	110
68	..	25	7	13·83	1 500	140	128
84·5	4·5	5	6	13·60	1 750	180	115
69	16	8	7	13·52	1 670	140	115

*±10kp/mm²

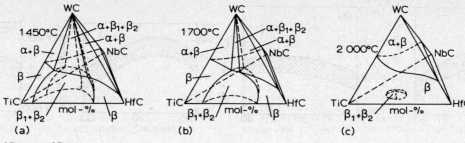

a 1450°C; *b* 1700°C; *c* 2000°C
4   Pseudo-quaternary TiC–HfC–NbC–WC system

Table 3   Standard WC–TiC–TaC–Co alloys, TaC being replaced by HfC–NbC or TaC–HfC solid solutions

| Composition, wt-% | | | | | | TRS,† | HV3, |
WC	TiC	TaC	HfC	NbC	Co	kp/mm²	kp/mm²
75*	7	8	..	..	10	185	1500
75	7	..	4	4	10	190	1600
75	7	2	2	4	10	190	1550
75	7	..	2	6	10	180	1550
69*	16	8	..	..	7	120	1700
69	16	..	4	4	7	145	1750
69	16	2	2	4	7	125	1700
69	16	..	2	6	7	120	1750
45*	40	8	..	..	7	100	1900
45	45	..	4	4	7	105	2050
45	40	2	4	4	7	105	2000
45	40	..	2	6	7	95	1950

*Standards P30, P10, P03
† ±10 kp/mm²

WC FREE ALLOYS

The first and still industrially applied WC free alloys were developed by Schwarzkopf, Hirschl, and Kieffer[1] 1929–31 on the basis of TiC/Mo₂C solid solutions with 10–15%Ni, Ni–Cr, Ni–Mo, and Ni–Mo₂C binders (*see* Table 1). More than a dozen cemented-carbide producers accepted this grade for the fine cutting of steel at high speed. The transverse rupture strength was increased during the last 40 years from 60–100 kg/mm² to 110–140 kg/mm²; some hot-pressed samples even showed values of about 150–160 kg/mm.[2]

TaC–Ni and Cr₃C₂–Ni alloys had no great success, but TiC–VC, TiC–TaC, and TiC–NbC based alloys with various binders were accepted for a time in the carbide industry. Sintered or infiltrated heat-resistant TiC alloys (WZ type) with high binder content (up to 60%Ni–Cr resp. Co–Cr alloys) and heat-treatable TiC alloys with 50–70% steel binder (ferro TiC) found growing interest in the industry.

Alloys on the basis of Ti–Mo–C–Ni

Above we gave a short historical review of the old TiC–Mo₂C–Ni alloys, which contained all the Ti and most of the

Table 4   WC–TiC–TaC–HfC–Co alloys, compared with WC–TiC–TaC–Co and WC–TiC–HfC–Co alloys

| Composition, wt-% | | | | | TRS, | HV3, |
WC	TiC	TaC	HfC	Co	kp/mm²	kp/mm²
77	10	5	..	8	140–160*	1810
77	10	2·5	2·5	8	150–170*	1760
77	10	..	5	8	140–150*	1770
72	15	5	..	8	135–155*	1900
72	15	2·5	2·5	8	180–195*	1810
72	15	..	5	8	135–155*	1800
67	20	5	..	8	130–145*	1960
67	20	2·5	2·5	8	140–155*	1910
67	20	..	5	8	120–135*	1920
73·5	15	5	..	6·5±0·5	120–140†	1970
73·5	15	2·5	2·5	6·5±0·5	140–160†	1770
73·5	15	..	5	6·5±0·5	115–135†	1810
78·5	10	2·5	2·5	6·5±0·5	130–145†	1830
73·5	15	2·5	2·5	6·5±0·5	140–160†	1770
68·5	20	2·5	2·5	6·5±0·5	110–125†	1960

*normal sintered †hot pressed

Mo in the form of a 4:1 to 1:1 TiC–Mo₂C resp. Ti–Mo–C solid solution.

In the last few years, American alloys have been introduced to Europe, which, according to corresponding patents, do not contain Ti–Mo–C solid solutions but consist of a pure TiC phase and a Ni–Mo–C binder alloy.[20] X-ray investigations proved, however, that Ti–Mo–C solid solutions are present when fine TiC is heated to 1400–1500°C in the presence of Mo+C and a liquid Ni or Ni–Mo–C auxiliary metal phase.[25]

Because of the relatively poor reproducibility of Ti–Mo–C–Ni alloys with high TiC contents, we tried to study again the effect of the relationship TiC:Mo₂C and the unknown but very important percentage of bound carbon in the Ti–Mo–C phase.

Some of the results of this research work are plotted in the diagrams in Fig.6*a–d*. The hardness and transverse rupture strength of alloys with vacuum-refined solid solutions and 14%Ni binder were varied from 3:2 TiC:Mo₂C to 9:1 TiC:Mo₂C. Because there are favourable compositions between 3:1 and 9:1, vacuum-refined solid solutions 4:1 were used for the second series of this study. By carefully preparing understoichiometric solid solutions of Ti–Mo–C (mixtures of TiC, Mo₂C, and Mo+carbon were vacuum solution treated at 1700°C), and by cementing them with 14%Ni we obtained optimum hardness and transverse rupture strength values for alloys having 90–96% carbon saturation of the carbide phase. When varying the Ni content from 10–18%, the best cutting performance on steel was achieved with 12–14% binder. On applying the hot-pressing technique to some non-stoichiometric samples, transverse rupture strength single values of 160–170 kg/mm² were measured.

X-ray investigations to determine the equilibrium distribution of Mo between the carbide and the auxiliary metal phase have not yet been made.

TiN–Ni and TiN–TiC–Ni (Mo) alloys

In the course of investigating the stability of transition metal carbides under high nitrogen pressure (30–800 atm) and temperatures between 1500 and 2500°C,[26] we tried to cement some nitrides and carbonitrides with Ni alloys. Some results are summarized in Table 6 and show extremely good

5   TiN coated P15 tip   ×500

Behaviour of ceramic cutting-tool materials

A. G. King

Ceramic cutting tools are useful in the machining of metals because they retain their advantageous properties at the high pressures and moderate temperatures which occur at the tool/workpiece interface. This discussion examines the ceramic cutting tool as a 'material' and from a 'mechanistic' point of view in regard to its serviceability in the machining of metals.

FAILURE MODE

The life of a ceramic cutting edge is determined by fracture rather than by wear processes in the machining of steel. The contrast in behaviour is shown in Fig.1, comparing high-speed steel with carbides and ceramics. On a wear basis, the ceramic is clearly superior to steel and carbide. On a total tool-life basis, the ceramic and carbides are equal at about 1 000 ft/min, with the carbide cutters showing better performance at most practical cutting speeds. From the view of developing better materials, it is clear that the need is not for ceramic materials with greater wear resistance, but for materials with less susceptibility to failure by a fracture mode. We are, therefore, interested in the stresses on the tool edge which produce these fractures, and those processes which weaken the material, allowing the fractures to propagate at lower stress levels.

Stresses on the cutting edge

The classical approach to the cutting process views the tool and workpiece in contact with a chip being formed and with force vectors drawn for the various components which are relevant to the analysis. The classical picture does not take into account that the process is a dynamic one, or that the forces are not homogeneously distributed over the tool contact area. While the real case is much more difficult to analyse, it is possible to obtain estimates of these forces, and of their distribution, from experimental evidence. Most of this evidence depends upon rationalization, and therefore has an accompanying degree of uncertainty.

Figure 2 shows a typical least square fitted curve, plotting the width of the wear land as a function of time. Significant to our discussion, the curve does not extrapolate to the origin, but intercepts the width axis at a finite value. Obviously there is a discontinuity in the curve, and the suggestion of a different wear process which was active for the initial tool/workpiece contact. This initial contact will be between a sharp tool edge and the revolving flat surface of the steel. Since ceramics are rigid materials with high yield strengths, these stresses are concentrated within a very small volume of material. If this stress exceeds the compressive strength of the tool, the small volume of material will crumble until an area large enough to sustain the load is created. At this point in the cutting process, the pressure on the wear land is about 400 000 lb/in² for ceramic cutting tools.

The feed force can be divided into two principle components: the force on the wear land and the friction force on the crater. Now that it is possible to estimate the wear-land components, a correction can be made on coefficient of friction measurements during cutting. Following the reasoning of Kobayashi and Thomsen,[1] this correction resulted in a calculated coefficient of friction of 0·22 as compared to 0·51 without the correction being applied. Separate sliding measurements of alumina on steel at comparable (but somewhat lower) pressures and temperatures yielded results from 0·17–0·24, in good agreement with the corrected figure derived from the corrected metal-cutting experiments. This tends to lend some confidence in the edge-crushing hypotheses, and the rough estimate of cutting pressures.

As cutting proceeds, the width of the wear land increases, distributing the force over a greater area. The overall force in the feed direction also increases with wear, so it is not difficult to believe that the pressures on the wear land are sustained at some large fraction of those initial stresses which we have estimated. And, from the other side of the interface where the metal is shearing, we would expect stresses equivalent to the yield stress of the metal being cut. This again is 200 000 or 300 000 lb/in².

The next three figures are those of wear-land surfaces which were produced by machining steel at 1 200–1 500 ft/min. The conditions at these surfaces were about 200 000–300 000 lb/in² at a temperature of about 1 450°F.

Figure 3 is of a tool wear surface showing areas where grains have been removed, and the flow of the metal curves around features of location topography. This can also be seen in Fig.4, which is a projection on the surface, built up of material which was transported during wear. Figure 5 shows marked evidence of curved wear grooves, nearly at right angles to the sliding direction. These three figures all show evidence of curvature in the wear patterns. The consequences of this observation are not really impressive until the scale of the photographs is considered. In the example of the curvature shown in Fig.5, the metal is abruptly taking a right angle turn over a sharp corner with a 0·001in radius. Since we know the sliding speed and the curvature, it is a simple matter to compute that this metal is being subjected to an acceleration of 4×10^8 cm/s². This is ridiculous! This leads to another surprising conclusion: the primary shear between the wear land and the workpiece is occurring in the metal itself!

Typically the crater and the wear land are separated by an area at the top of the tool edge which appears to be unaffected by the cutting process. Figure 6 shows typical crater wear surfaces which were produced by machining A6 tool steel at increasing speeds. If we reasonably assume that the incidence of wear has some relation to stress distribution, at least at low cutting speeds, there exists a steep pressure gradient across the top face of the tool. Figure 7 is a polished section

The behaviour of a ceramic tool during machining is determined by the stress distribution on the cutting edge, and fatigue properties of the tool material. Stresses are of the order of 200 000 lb/in² compressive, with stress gradients as high as 25×10^6 lb/in²/in during machining. An appreciable component of this stress is cyclic and total tool life correlates reasonably well with fatigue properties of the ceramic measured by independent methods.

621.91.025.7:666.764.32:539.538

The author is with the Zircoa Chemetal Division of the Diamond Shamrock Chemical Co.

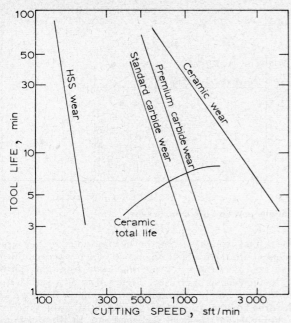

1 Comparison between wear and total tool life for ceramic, carbide, and high-speed steel tools

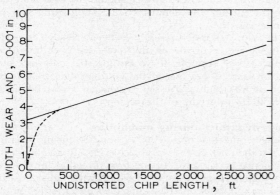

2 Least square fitted wear curve, machining 4340 steel at 1 000 ft/min

3 Electron photomicrograph of a ceramic tool wear surface
×3 000

4 Electron photomicrograph of a ceramic tool wear surface showing projection   ×3 000

5 Stereo electron photomicrographs of a ceramic tool wear surface showing abrupt change of direction   ×3 000

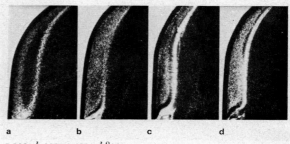

a   b   c   d

a 100; *b* 200; *c* 400; *d* 800

6 Crater wear surfaces after machining type A6 tool steel   ×57

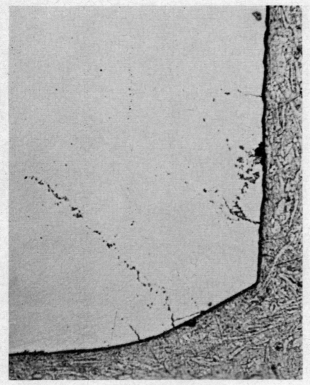

7   Crack pattern in a tool after use, polished section   ×200

of a tool which was cut at right angles to the wear land and crater surfaces. The tool had a negative land ground on to the cutting edge at a 60° angle to the top surface prior to machining. This 0·006in wide land has been rounded slightly on the flank wear surface. The crack pattern that is produced is typical and reproducible for these particular machining conditions. There are a series of small parallel cracks normal to the wear-land surface and a large crack at 45° to the crater

8   Fatigue damage in a tool after use, polarized light view of a polished section   ×200

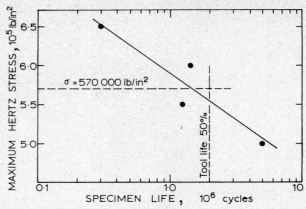

9   Fatigue and tool-life comparison

surface. The distance from the cutting edge to this crack correlates with the feed per revolution which was used during machining, with the crack moving away from the cutting edge as the feed is increased. Figure 8 is of a similar tool cross-section, but photographed with polarized light. The undisturbed material is dark grey in the illustration. Wherever the alumina tool has been damaged due to the machining process, the tool appears lighter coloured. The two areas of damage are on the crater surface and on the wear land, down from the cutting edge about 0·004in. The cutting edge itself is undamaged.

The last two figures present direct evidence that steep stress gradients are occurring on the tool edge over microscopic distances. For example, if we assume that only half of the total pressure is bearing on the cutting edge itself as compared to the highly stressed regions, a stress gradient of 25 m. lb/in²/in occurs in this region.

Stress dynamics during machining

Whenever the tool is quickly withdrawn during machining and the shoulder on the workpiece is closely examined, there is evidence of vibration in the feed direction imprinted on the workpiece surface as small ridges with radial axes from the centre of rotation. Knowing the cutting speed and the wave length of these vibratory marks, the frequency of these vibrations can be determined directly. In the case of machining A6 tool steel at RC52 at 1 200 ft/min, the frequency was 24 000 Hz. The tool cutting edge survived 2×10^6 cycles, on the average, before failure.

Figure 9 presents fatigue data by Parker et al.[2] using hot-pressed alumina in a 5 ball fatigue-test apparatus. Super-imposed upon their data are the results of the machining test, indicating quantitative agreement that fatigue is a reasonable mechanism leading to fracture failure. Moore and Kibbey[3] studied the effects of fatigue using specially designed

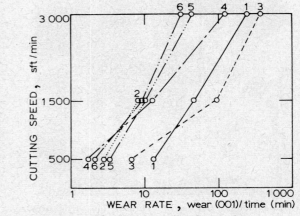

10   Wear-rate data from various vibration modes

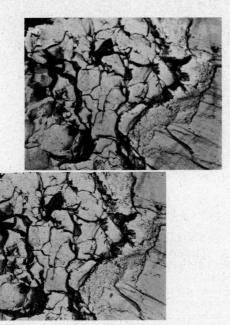

11 Stereo electron photomicrograph of fatigued tool surface after use × 3000

12 Typical ceramic tool fracture × 57

tool holders with variable stiffness and self-induced vibrations. Their data appears in Fig.10. The six curves are for different conditions of vibration direction, amplitude, and frequency. The data spans an order of magnitude in wear behaviour. The curves showing the most rapid wear are for vibrations in the feed direction (curves 1 and 4) and for vibrations of the highest frequency (curves 1, 2, and 3). This would be expected, of course, as vibrations in the tangential direction will have relatively little effect on cutting forces, and the energy contained in a vibration rapidly increases with frequency.

Figure 11 is an electron photomicrograph of a tool surface which spalled after machining. The grain structure of the tool can be seen as each grain is delineated by a black border. These borders are thin filaments of the plastic replicating material which penetrated into open grain-boundary cracking. Propagation of these cracks leads to eventual tool failure.

SUMMARY

The failure mode of ceramics cutting steel is through a fatigue process which nucleates cracks leading to a catastrophic failure, such as can be seen in Fig.12.

Stresses on the cutting edge have been estimated as high as 400000 lb/in² shortly after the initial contact between the tool and the workpiece. An appreciable fraction of this stress is probably still maintained during the life of the tool. Wear patterns, reproducible crack patterns, and fatigue damage indicates that very steep stress gradients occur over very short distances. These may be as high as 25×10^6 lb/in²/in. The fatigue failure has a dynamic fatigue component with the tool being more sensitive to vibrations of high frequency in the feed direction.

This tool behaviour suggests that improved materials will be those with better resistance to fatigue damage and utilizing designs of the cutting edge which minimize stress gradients over microscopically short distances.

REFERENCES

1 S. KOBAYASHI AND E. G. THOMSEN: *Eng. Ind.*, 1960, **82**, 324–332
2 R. J. PARKER *et al.*: NASA Tech. Note TND-2274, 1964
3 H. D. MOORE AND D. R. KIBBEY: Ohio State University, Eng. Expt. Sta. Rep. 169, 1962

Influence of the binder-phase composition of cemented carbides on strength and wear resistance in turning

E. Pärnama and H. Jonsson

During the 1960s, considerable advances were achieved in cemented-carbide research. Following the basic concepts of Gurland,[1] Fischmeister[2] and Exner[3] have contributed to a more quantitative understanding of the relationships between the mechanical properties and the structural parameters of cemented carbides by the application of electron microscopical linear analysis. Dawihl et al.[4] and Kreimer et al.[5] have dealt with the question of carbide skeleton formation and the mechanisms of fracture. Drucker[6] adopted continuum mechanisms to approach the strength problem of cemented carbides. To judge from the recently published works of Luyckx[7] and Doi et al.,[8] fracture mechanism is still a subject that gives rise to controversial results. This is also the case with the Palmqvist[9] toughness method. The interpretation of the experimental results of Palmqvist have been discussed by Dawihl et al.[10] and Exner.[11] The problem of the 'true' surface has thus been taken up to treatment, and also the question of the work hardening of the binder phase during mechanical grinding and polishing. X-ray strain analysis of cemented carbides has been performed by Gurland,[12] Bernard,[13] and Öhman et al.[14] Öhman et al. first applied the two-exposure method and have thereby succeeded in giving an explanation to many results which have been controversial earlier. Their results have been recently confirmed by French.[15] Ramqvist[16] has made an important contribution to the understanding of the wetting properties of the carbides. This subject has been treated earlier by Kingery et al.,[17] Parikh et al.,[18] and Eremenko et al.[19] The influence of the carbide perfection, dislocation densities, and other carbide properties on the cemented carbides have been investigated by Kreimer et al.[20] and Kovalskii.[21] Contributions to these studies have also been made by Anderson,[22] Suzuki et al.[23] and Ramqvist[24] (the last-mentioned worker dealt only with the carbide properties). Lehtinen[25] first succeeded in applying transmission electron microscopy to the study of cemented carbides.

Important work on the influence of the binder-phase composition on the properties of the cemented-carbide alloys have been accomplished by Suzuki,[26] Kubota,[26] and other Japanese workers. Recent contributions to the binder-phase problem have been published by Giamei et al.[27] and Jonsson et al.[28]

FACTORS THAT INFLUENCE STRENGTH PROPERTIES AND WEAR RESISTANCE OF CEMENTED CARBIDES

The strength of cemented carbides is dependent on the composition and the structure of the alloys.

The compositional factors are the amount of binder phase, the composition of the binder phase, both chemical and structural, and the composition of the carbide phases.

The structural factors are the mean carbide grain size ($d_{carbide}$), the mean cobalt path length (l_{Co}), the carbide grain size distribution, the contiguity and the substructural properties of the carbides, i.e. the perfection of the carbides.

All these factors also influence the wear-resistance properties of the cemented carbides. The degree of influence of the different factors depends on wear mechanisms and working conditions.

Among the above-mentioned factors, the influence of the carbide perfection, the contiguity, and the binder-phase composition belong to the factors investigated least.

Figure 1 shows the effect of improvement of the carbide properties (carbide perfection) on the transverse rupture strength of some coarse-grained WC–Co cemented carbide at constant cobalt content and grain sizes. As the fatigue strength of WC–Co alloys is about $0.5 \times \sigma_B$, where σ_B is the transverse rupture strength (TRS), we also obtain an increase in the fatigue strength when the TRS increases.

Due to the different wetting properties of the different carbides and due to the cobalt content we may obtain different degrees of contiguity, K. The degree of contiguity for two-phase cemented-carbide alloys is defined as follows:

$$K = \frac{N_{carbide} - N_{Co}}{N_{carbide}} \dots\dots\dots\dots\dots\dots\dots\dots\dots\dots\dots (1)$$

where $N_{carbide}$ and N_{Co} are the numbers of carbide and cobalt intercepts obtained by electron microscopy.

In a research project, conducted at Chalmers Institute of Technology in Gothenburg and sponsored by Fagersta Steel Ltd in Fagersta, the relationships between mechanical properties and structural parameters of two-phase WC–TiC–Co alloys were investigated.

It was found that there existed a relationship between the

The binder-phase composition in cemented carbides can be determined by magnetic saturation measurements. The accuracy of the determinations depend on the accuracy of the cobalt-content determinations and the calibration curve. The thermodynamic boundary for carbon porosity has been determined to be 1–3 wt-%W dissolved in the Co phase, the η-phase boundary to be 10–13 wt-%W. These boundaries do not necessarily coincide with the appearing of microscopically detectable carbon porosity or η-phase, as this also depends on kinetical factors. The transverse rupture strength as a rule decreases with increasing dissolution degree of W in the binder phase. The presence of η-phase or carbon microporosity does not automatically lead to a decrease in the TRS values as these values also depend on the grain sizes of the η or carbon precipitates. The wear resistance in turning generally increases with increasing degree of dissolution of W in the binder phase up to 13 wt-%W if microscopically detectable η-phase is not present. It also seems to be possible to compensate the low wear resistance in coarse-grained alloys by increasing the dissolution of W in the binder phase.

621.941.025.7:669.018.95:661.878.621:539.538

The authors are at Seco Research Centre, Fagersta Bruks AB, Fagersta, Sweden

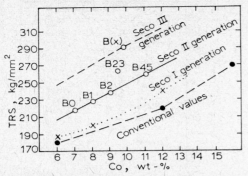

1 Effect of carbide perfection on the transverse rupture strength of some coarse-grained WC–Co cemented carbides

Palmqvist crack-formation energy and the degree of contiguity, K, determined by electron microscopical linear analysis. With increasing K value the crack-formation energy decreased (Fig.2). It was found that contrary to the crack propagation in WC–Co alloys, which according to earlier investigations occurs mainly in the binder phase when the mean grain size does not exceed 4 μm, crack propagation occurs mainly in the carbide phase in two-phase WC–TiC–Co alloys. The higher degree of contiguity and the fact that the crack propagation in these alloys occurs in the carbide phase may explain the lower toughness of these alloys compared to WC–Co alloys with identical cobalt contents and grain sizes.

Suzuki and Kubota showed that the binder-phase composition is mainly a function of the carbon content, and that the strength of cemented carbides at a constant carbide grain size was mainly dependent on the carbon content. They also determined, by selective etching and X-ray diffraction techniques, that the solubility of W in the binder phase varied from 2–10 wt-%W. With increasing carbon content the tungsten content of the binder phase decreased.

EXPERIMENTAL

Test series were designated A, B, and C. (Table 1). Series A consisted of WC–TiC–TaC/NbC–Co alloys with a cobalt content of 9 wt-%. The carbide grain size, $d_{carbide}$, was nearly constant and equal to 1.31 ± 0.06 μm. Series B had the same composition as series A but the mean carbide grain size varied from 1.62–2.28 μm. Series C consisted of WC–Co alloys with 5.5 wt-%Co. The mean carbide grain size varied from 0.60–0.84 μm.

The TRS determinations were made according to the Swedish test procedure, which means that the length of the samples was at least 20 mm, the end area of the samples was $(6.50 \pm 0.25) \times (5.25 \pm 0.25)$ mm², and the punch as well as the two top supports used were bars of cemented carbide with a length of 10 mm and a diameter of 6 mm. The distance or span between the supports was 15 ± 0.5 mm. The samples were ground on a diamond wheel. Values reported here are mean values of 12 determinations.

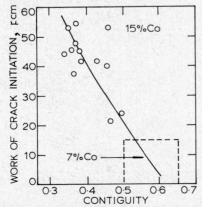

2 The Palmqvist crack-formation energy as a function of contiguity

Table 1 Test series

Series no.	System	Co, wt-%	H_c oe	$\bar{d}_{carbide}$ μm	W, wt-% diss. in Co	TRS kg/ mm²	Tool life T, min
A	WC–TiC–TaC/NbC	9.0			4.2	189	11.9
					5.0	169	8.2
					5.6	164	8.5
					5.9	175	9.8
			136 ±3	1.31 ±0.06	9.7	144	12.5
					11.4	178	15.2
					13.0*	117	10.6‡
					19.0*	90	6.9‡
					25.0*	51	13.2‡
B	WC–TiC–TaC/NbC	9.0	101	2.28	6.4	110	2.6
			110	1.98	7.7	125	.. 7.0
			120	1.69	11.0	149	15.8‡
			114	1.86	9.1	144	17.8
			123	1.62	10.3	109	21.3
C	WC	5.5	237	0.60	13.7	119	16.6
			216	0.72	0.7† (<2.0)	232	9.5 / 11.6
			205	0.78	5.5	200	11.6 / 15.0
			189	0.84	4.6	211	5.6 / 8.0

* microscopically detectable η-phase; † microscopically detectable carbon porosity ASTM C2; ‡ premature failure due to fracture

Wear-test values are given as tool life T in minutes in turning to a wear criterium of mean wear land $VB_{mean} = 0.4$ mm. Series A and B were tested in a steel with a carbon content of 0.60% and HB of 220. Depth of the cut was 2.5 mm, feed 0.4 mm/rev, and cutting speed was 150 m/min.

Series C was tested in cast-iron rods with 450 HB at a depth of cut of 1.0 mm and a feed of 0.2 mm/rev. The cutting speed in this case was 15 m/min.

RESULTS
Use of magnetic measurements to determine structure and binder-phase composition of cemented carbides
The binder-phase composition thus influences the strength properties of cemented carbides. Investigations about the solubility of W in the binder phase, however, have been connected with great difficulties, as the only methods that have been used are selective etching combined with X-ray determination of the lattice parameters of the cobalt phase and the chemical analysis of the selectively dissolved binder phase.

Coercive force measurements
Measurements of magnetic properties and especially of the coercive force (H_c) have for long been used in determining certain structural properties of the cemented carbides, for instance the grain size. Between the experimentally determined values of the mean cobalt intercept lengths (l_{Co})

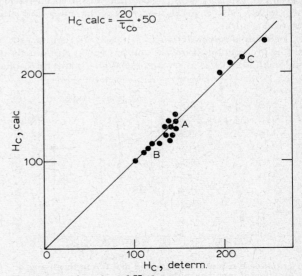

3 H_c calc. as a function of H_c determ.

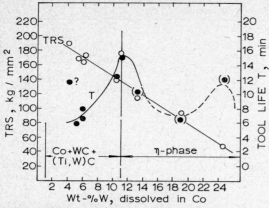

4   Transverse rupture strength and tool life in turning as a function of binder-phase composition

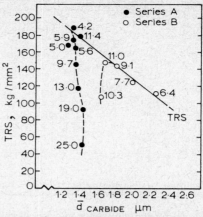

6   Transverse rupture strength as a function of mean carbide grain size; figures give W content of the binder phase

(electron microscopical linear analysis) and experimentally measured H_c values, we have found the following empirical relationship:

$$H_c - 50 = \frac{20}{l_{Co}} \qquad \ldots\ldots\ldots\ldots\ldots\ldots\ldots\ldots\ldots\ldots\ldots\ldots\ldots (2)$$

to be valid for a great number of WC–Co and WC–TiC–TaC–NbC–Co alloys.

To test the validity of the relationship (2) for the three test series we calculated the coercive force H_c with the help of electron microscopically obtained l_{Co} values. As may be seen from Fig.3 a relatively good correlation was obtained between H_c (calc) and H_c (measured) for series A, B, and C.

Magnetic saturation measurements

At the SECO Research Centre in Fagersta we have applied magnetic saturation (MS) measurements to determine the amount of W dissolved in the binder phase.

Contrary to the coercive force, the MS values are dependent only on the cobalt content and on the binder-phase composition (dissolution of W in the binder phase). By calibrating the MS values by using well defined binder-phase compositions, we can by measuring the MS value determine the dissolution degree of W in the binder phase as the MS value decreases linearly with increasing W dissolution in the two- or three-phase regions of the WC–Co and WC–TiC–TaC–NbC–Co systems respectively. Furthermore, from phase-diagram studies and experiments we can conclude that thermodynamically a W dissolution around 2 wt-% could give carbon porosity, and a dissolution of over 11 wt-% should give Co_3W_3C-phase. These phase boundaries do not necessarily coincide with the microscopically detected carbon porosity or η-phase, as microscopically visible occurrence of these phases also depends on the cooling rates and other kinetical factors.

The method is non-destructive, it does not require a special sample preparation, it is fast, and has a good reproducibility. The method is valid for two-phase WC–Co and three-phase

WC–TiC–TaC–NbC–Co alloys, where the dissolution of TiC, TaC, NbC in binder phase is small and may be neglected. Other systems have so far not been investigated. For example, no reliable data exist on the dissolution of W and Ti in two-phase WC–TiC–Co alloys.

Influence of the binder-phase composition on the strength and wear properties of cemented carbides

To test the reliability of the above-mentioned MS method of determining the content of W in the binder phase, and also to study the dissolution effect on the strength and wear resistance of cemented carbides, three series of cemented-carbide alloys with different dissolution degrees were prepared. The results are given in Figs. 4, 5, 6, and 7.

Coercive force, mean carbide grain size, wt-%W dissolved in the binder phase, transverse rupture strength, and wear resistance in tool life T minutes are listed in Table 1.

DISCUSSION OF RESULTS
Determination of the binder-phase composition by MS measurements

The W determination by MS measurements is influenced by the accuracy of the cobalt-content determinations and by the accuracy of the calibration curve. If the Co content of the sample is determined with the accuracy of ± 0.1 wt-%Co the accuracy of the MS measurements is estimated to be better than ± 1 wt-%W, which is of the same order of magnitude or better than the accuracy of earlier applied methods. Carbon porosity may occur at W contents 1–3 wt-%, and η-phase may occur at W contents 10–12 wt-% and higher values.

In series A, η-phase was detected microscopically in specimens with 13, 19, and 25 wt-% dissolved in the binder phase (Table 1). (W contents higher than 12% are as a rule only indicating the presence of η-phase. For instance, an alloy

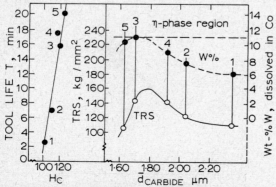

5   Transverse rupture strength and tool life in turning as a function of mean carbide grain size

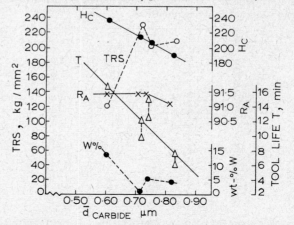

7   Coercive force, transverse rupture strength, tool life in turning, and hardness as a function of mean carbide grain size

with '25 wt-%W dissolved' contains more η-phase than an alloy with '19 wt-%W'. The composition of the Co-phase is probably the same in the two alloys.) In all these specimens we got a premature fracture in turning.

In series B no microscopically detectable η-phase was observed in the sample with 11 wt-%W, but in turning we obtained a premature fracture even with this specimen in spite of its relatively high TRS value.

In series C the specimen with 13·7 wt-%W had no microscopically detectable η-phase but a very low TRS value. The specimen with 0·7 wt-%W dissolution (< 2 wt-%W) had a microscopically detectable carbon porosity of ASTM C2.

The established boundaries for carbon porosity and η-phase formation with 2 wt-%W and 11 wt-%W respectively have been confirmed. In some cases we have obtained indications that the microscopically detectable η-phase formation does not necessarily coincide with the thermodynamic phase boundaries as this also is influenced by kinetic factors. The MS method, furthermore, cannot separate different degrees of carbon porosity (MS values do not vary significantly for W values under 2 wt-%W), whereas it is possible to indicate the amount of the η-phase formation in terms of the W dissolution. With increasing W content exceeding 11 wt-% (η-phase boundary) we will obtain an increase in the amount of the η-phase.

Transverse rupture strength

Transverse rupture strength generally decreases with increasing dissolution of W in the binder phase (Figs. 4, 5, 6, and 7). This confirms the earlier results of Suzuki and Kubota. However, there are some significant exceptions from the rule. In series A the sample with 11·4 wt-%W has a high TRS value, which was also the case with the specimens 3 and 4 of the series B (Table 1) with 11·0 wt-%W and 9·1 wt-%W respectively. On the other hand, sample 3 in the series B had a premature fracture in turning which was not the case with specimen 4 and the above mentioned sample in the A series. Also, in series C the sample with carbon porosity had a high TRS value. These exceptions from the general rule indicate that in case of high dissolution of W in the binder phase or carbon porosity the strength properties also depend on the mode of dispersion, i.e. the distribution and the size of the carbon or Co_3W_3C precipitates respectively.

In case of very fine-grained η-precipitates indicated by MS measurements, the TRS value must seemingly not be low. Carbon porosity ASTM C2 does not decrease the TRS value either. On the other hand we have also obtained indications, such as in series B samples 3 and 5, that a TRS value is not always an appropriate measure of toughness in turning.

From Figs. 6 and 7 we also find that besides the W dissolution, the carbide grain size influences the TRS values as may be expected. For WC–Co alloys the TRS values seem to increase with increasing carbide grain size and decreasing W dissolution. For WC–TiC–TaC–NbC–Co alloys the TRS values on the other hand decrease with increasing carbide grain size (Fig.6). This indicates also that in three-phase alloys a crack propagates mainly through the carbide phases. As mentioned earlier we know that crack propagation in two-phase WC–TiC–Co alloys occurs mainly in the carbide phase. The transverse rupture strength σ_B was in this system found to satisfy the relationship

$$\sigma_B^2 = A \cdot \frac{4}{d_{carbide}} + B \quad \ldots\ldots\ldots\ldots\ldots\ldots\ldots (3)$$

where A and B depend on the Co content and the carbide composition respectively. The relationship (3) is basically related to the proposal put forward by Gurland for the strength of cemented-carbide alloys in the case when crack propagation occurs in the carbide phase, namely:

$$\sigma_B = \left(\frac{E\,\gamma}{\pi} \cdot S_C\right)^{\frac{1}{2}} \quad \ldots\ldots\ldots\ldots\ldots\ldots\ldots\ldots (4)$$

where E is Young's modulus, γ is the surface energy of the carbide phase, and S_C is the specific grain surface of the carbide phase.

The question is now which parameters determine the crack-propagation mechanism. It is evident that it is not the degree of contiguity alone that is the critical parameter, as both in two-phase WC–Co and in two-phase WC–TiC–Co alloys the degree of contiguity decreases with increasing carbide grain size. Instead, it may be assumed as probable that it is the ratio between the carbide grain size and the degree of contiguity that determines whether the crack propagation occurs in the binder phase or in the carbide phase. For a certain system we may then assume that the crack-propagation mechanism is different for different values of the $d_{carbide}$:K ratio.

Wear resistance in turning

For the same carbide grain size, wear resistance increases with increasing dissolution of W in the binder phase (Figs. 4, 5, and 7). On the other hand, when the W content in the binder phase exceeds 11 wt-%W we frequently obtain premature fracture in turning, at least when microscopically detectable η-phase is present. Results in Fig. 4 and 5 still indicate that in many cases highest possible wear resistance may be obtained with high dissolution of W in the binder phase (9–11 wt-%W) when no η-phase is present and sufficiently high TRS values may thus be maintained. Yet this is possible only with a very careful process control.

Anyhow, it seems to be possible to compensate to a certain degree the lower wear resistance of more coarse-grained alloys with high dissolution of W in the binder phase.

ACKNOWLEDGMENTS

The authors thank Dr S. Palmqvist for permission to publish this paper and for valuable discussions. Our thanks also to Mr K. G. Stjernberg who has done work on the two-phase WC–TiC–Co alloys, referred to in this paper.

REFERENCES

1 J. GURLAND AND P. BARDZIL: *J. Met.*, 1955, **7**, 311–315; J. GURLAND: *J. Met.*, 1957, **9**, 512–513; J. GURLAND: *Trans. AIME*, 1963, **227**, 1146–1150
2 H. E. EXNER AND H. F. FISCHMEISTER: *Arch. Eisenh.*, 1966, **37**, 499
3 H. E. EXNER: 'Struktur und Eigenschaften der Hartlegierung WC–10Co', 101; 1964, Jernkontoret U8 no.64–20
4 W. DAWIHL AND B. FRISCH: *Arch. Eisenh.*, 1962, **33**, 1, 61–66; W. DAWIHL AND G. ALTMEYER: *Z. Metallk.*, 1963, **54**, 11, 645–650; W. DAWIHL AND B. FRISCH: *Cobalt*, 1964, **22**, 22–30
5 G. S. KREIMER AND N. A. ALEXEJEVA: *Phys. Met. Metallography USSR*, 1962, **13**, 4, 117–121
6 D. C. DRUCKER: 'High-strength materials', ed. V. F. Zackey, ch.21; 1965, New York, Wiley
7 S. B. LUYCKX: *Acta Met.*, 1968, **16**, 4, 535–544
8 H. DOI et al.: *Trans. JIM*, 1968, **9**, 616–622
9 S. PALMQVIST: *Arch. Eisenh.*, 1962, **33**, 629–634
10 W. DAWIHL AND G. ALTMEYER: *Z. Metallk.*, 1964, **55**, 5, 231–237
11 H. E. EXNER: *Trans. AIME*, 1969, **245**, 4, 677–683
12 J. GURLAND: *Trans. ASM*, 1958, **50**, 1063–1071
13 R. BERNARD: *Jernkon. Ann.*, 1963, **147**, 1, 22–36
14 S. ÖHMAN et al.: *Jernkon. Ann.*, 1967, **151**, 2, 126–159
15 D. N. FRENCH: *J. A. Ceram. Soc.*, 1969, **52**, 5, 267–275
16 L. RAMQVIST: *Internat. J. Powder Met.*, 1965, **1**, 4, 2–21
17 W. D. KINGERY AND M. HUMENIK: *J. Phys. Chem.*, 1953, **57**, 359–363
18 N. M. PARIKH AND M. HUMENIK: *J. A. Ceram. Soc.*, 1957, **40**, 315–320
19 V. N. EREMENKO AND YU. V. NAIDICH: *Russ. J. Inorg. Chem.*, 1959, **4**, 9, 931
20 G. S. KREIMER et al.: 'Hard-alloy cermets for drilling rocks', (Hard metals production technology and research in the USSR), ed. S. I. Bashkirov, 35–63; 1964, Oxford, Pergamon Press
21 A. E. KOVALSKII: *ibid.*, 263–281
22 P. B. ANDERSON: *Planseeber.*, 1967, **15**, 3, 180–186
23 H. SUZUKI et al.: *J. Japan Inst. Met.*, 1969, **33**, 504–509
24 L. RAMQVIST: *Jernkon. Ann.*, 1969, **153**, 1–21
25 B. LEHTINEN: *J. Sci. Instr.*, 1968, ser. 2, 1, 673–674
26 H. SUZUKI AND H. KUBOTA: *Planseeber.*, 1966, **14**, 96–109
27 A. GIAMEI et al.: *Cobalt*, 1968, **39**, 88–96; 1968, **40**, 140–155
28 H. JONSSON AND B. ARONSSON: *J. Inst. Met.*, 1969, **97**, 281–288

Discussion: Tool materials

High-speed steels

Chairman: W. H. Bailey
(BSC, Special Steels Division)

Mr H. C. Child, the rapporteur, presented the following seven papers: 'High-speed steel technology: the manufacturers' viewpoint', by F. A. Kirk, H. C. Child, E. M. Lowe, and T. J. Wilkins; 'Physical metallurgy of high-speed steels', by Dr T. Mukherjee; 'Controlled forging of tool steels', by A. Tomlinson and D. E. Beard; 'A British Standard for tool steels', by P. Jubb; 'Latest developments in high-speed steels in Europe', by A. Füssl; 'Particle-metallurgy high-speed tool steel', by E. J. Dulis and T. A. Neumeyer; 'Cast cobalt alloy cutting metal', by M. Riddihough.

Mr H. C. Child (Jessop-Saville Ltd)
It is not my intention to be provocative and critical of the individual papers, but rather to review them collectively. Perhaps any provocation on my part will be attributed to bias, because I have been asked to present the view of the supplier and, as a supplier myself, I am perhaps naturally inclined, therefore, to err on the side of bias that way. There are seven papers in this session on high-speed steel and I would like to try to separate the main topics that have been covered. First, there are the customer's requirement with regard to specifications, quality, and performance of the tool materials. Second, there is the area of process metallurgy, particularly important from the point of view of the effect this has on the tool quality and on the cost of the product. As a third main area, we have the future requirements of the industry which we ought to consider at a conference like this to see what indications there are where developments are required.

If we can start off, first of all, by considering what these papers have to say on the customer's requirements, it may be surprising to some of you that high-speed steel has lagged so far behind, say, constructional steel with respect to standardization and specification, and that we in this country are only now coming round to agreeing national specifications. Three of the papers before us deal in some depth with the problem of specifications.

Everybody has agreed that we do need to be more selective and have fewer grades, but it is proving quite difficult in practice to restrict these grades.

Kirk et al. in their paper make the point that high-speed steels fall into four categories by application. There is the 14% tungsten steel for light duty, there are the normal duty high-speed steels of the M2 and T1 types, there are the fast-cutting materials of the T4 and T6 type, and the hard-cutting materials of the M42 type. It would be desirable really to have one or at most two high-speed steels for each of these categories by application, but unfortunately, there does not appear at the moment to be a basis for being as selective as this, and Kirk et al. suggest that the way things are going at the moment, we are likely to see three or four grades required for each application except the light duty one, where the 14% tungsten steel is the only one in the running.

It is not surprising therefore to find that the British Standard is proposing to cover 15 grades. I think we must bear in mind that if we have 15 grades, we are not going to get the maximum out of standardization with respect to ease of stocking and optimizing delivery and reducing costs.

Several of the authors make the point that to some extent standardization is impeded by lack of information on the effect of composition on performance. The same situation seems to exist on the Continent. Füssl describes the revision of the Stahl-Eisen-Werk-Stoffblatt 320, and in this, ten grades of high-speed steel are specified even in the revised version. On the Continent, however, there is an official spur to concentrating on molybdenum grades because the T1 specification is being dropped. Füssl comments that in view of the sales of this quality it is somewhat surprising, but as most of you would agree, with the present cost and the ever-present problem of the fluctuating prices of tungsten, there does seem to be a good case for trying to drop the T1 grade.

Jubb discusses in some depth the new British Standard for tool steels. The discussions that have taken place over this proposed standard have revealed surprising ignorance in several areas. One is the precise effect of residuals on service performance and another is the precise effect of structure, including carbide distribution and grain size, on performance. With this background of lack of firm information, there has obviously been some difficulty in agreeing to specification requirements. Indeed, quite a few people believe that the British Standard need only have covered selected compositions and a realistic control of chemistry of both the alloying elements and the residuals, and it is worth bearing in mind that even this approach of closer control of residuals must inevitably put the cost of the steel up, because this involves raw-material selection and scrap selection on the part of the supplier.

The other problem of standardization is, of course, that the customer himself must be prepared to accept a standard quality and there are innumerable examples in the high-speed steel field of customers requiring minor variations of carbon and sulphur within the one quality, such as M2, and if we are going to have a national standard, we must appeal to the manufacturer to accept the same discipline as the supplier.

Turning now to the property side, Kirk et al. discuss the relevance of laboratory properties to service performance and conclude that these really serve only as a very rough guide to cutting performance. The laboratory properties are not really too relevant except perhaps the cold hardness. In view of this, there has been a realistic approach in both the British Standard and the Werkstoffblatt 320 in that they call for only cold hardness as a mechanical property.

Much more controversial in the specifications is the problem of whether carbide distribution should be specified. Inherent in the argument of specifying carbide distribution is the ability to measure this property effectively to get an answer that is fully representative of the material and not just representative of the few samples that are examined by the supplier.

In their paper, Kirk et al. show, as far as carbide band width is concerned on VDE carbide grading, that the variations within the product of an ingot and indeed one sample from the product is such that extensive sampling is indicated as being essential to get a representative picture of the quality. Equally important, of course, is what carbide distribution does to the cutting properties. Here again, there is a surprising lack of published information as to just precisely what carbide distribution does do. Mukherjee concludes in his paper that the uniform distribution of primary carbide is desirable from a mechanical property point of view, and few of us would question this, but as we have said already, there is no clear relationship between mechanical properties and cutting characteristics.

All manufacturers and users are, of course, well aware of the problems that severe carbide segregation cause in uniformity of heat-treatment response, in distortion, and, in the worst cases, in tool cracking or burning. Since, however, this degree of carbide segregation is avoided in normal production, we have to ask ourselves to what extent the user is right in demanding a closer control of the carbide specification. Kirk *et al.* conclude from their limited evidence on drills that carbide distribution variations are not particularly significant on cutting properties, but contrarily, Dulis and Neumeyer show that in the extreme case of intermittent cutting, the much improved toughness that they get from their very fine uniform carbide distribution of the powder metal product does give a significant improvement in cutting performance.

The balanced view is surely that gross carbide segregation must be avoided, but that only where the carbide distribution is really known to affect service performance or the manufacture of the tool should a customer specify something above the normal standard that the industry can supply, because attempts to supply above the standard that is attainable from the present improved standards of ingot design and forging reduction involve one in either material selection or changing to more sophisticated processes which cost more.

The British Standard will, according to Jubb, allow for this situation by providing photographic standards which the user can call up as he thinks necessary.

Turning to grain size, Füssl emphasizes its importance, particularly in view of the trend towards grades containing above 1% carbon. Unless the composition of such grades is balanced by a suitable ratio of the carbide formers tungsten, molybdenum, and particularly vanadium, there is a tendency to grain growth. There is at the moment no intention of specifying grain size in the British Standard and this is perhaps quite wise, but Füssl's paper does emphasize that if there should be a growing use of high-carbon high-speed steels in this country, at the very least we ought to try to ensure, by specifying suitable analytical ranges, that there is a balance between carbon and carbide formers to obviate this grain growth problem.

Füssl comments that in Europe, sulphurized grades containing 0·06% sulphur are preferred rather than the 0·12–0·16% common in the USA and the UK. This is because, in Germany, unground gear-cutting hobs are prehardened before machining. This acts as a stress-relieving treatment and minimizes distortion. It has the important additional benefit of requiring less sulphur, 0·06%, to give the desired surfaced finish, the higher sulphur being necessary for machining in the soft annealed condition. Füssl indicates a 1% loss of metal yield for every 0·01% sulphur added, so there is a strong economic incentive to try and make do with low sulphur contents. In this connexion it is interesting to know that in Germany a 10% premium is charged for sulphur grades because of this reduced yield and a similar premium is charged for high-carbon grades. I understand that this is quite a deterrent to the use of high-carbon grades, but the practice of a premium for the grades that are more difficult to forge is one that might well be adopted in this country.

Several of the papers, for example that by Mukherjee, emphasize the importance of the as-cast structure. The eutectic pools you get in an as-cast ingot may be broken up by repeated upsetting and drawing out, as described by Füssl, but the general relationship between the extent of the original eutectic and the cellularity of large bars or carbide banding in small bars persists.

It is pertinent to consider therefore the properties of as-cast high-speed steels. Unfortunately, none of the papers before us deal with this topic. It is known, however, that considerable use is made of as-cast high-speed steel, e.g. for milling cutters. In general, superior performance is claimed in comparison with the wrought equivalent. I will return to this point later in a general consideration of the importance of structure.

The paper by Riddihough deals with the application of cast Co base tools, and here again, the importance of carbide size and grain size is emphasized, and the production of these tools is confined to chill casting in graphite moulds.

I draw your attention to the claims that these cast alloys are in demand particularly for interrupted cuts and where vibration, sand, scale, or rough surfaces are in evidence. Cast structures seem to have more merit than many of us credit them with, and some doubt must be felt about the general need for fine carbide and fine grain size in wrought products.

Kirk *et al.* discuss the production of high-speed steel by conventional methods and emphasize how, by suitable selection of ingot size and geometry and by adequate forging, generally acceptable carbide banding or in larger bars the degree of cellular carbide, can be controlled. Several of the authors consider the potential of electroslag remelting for improving the cutting efficiency of high-speed steel. Kirk concludes that the improvement, if any, is only marginal. The benefit lies in the avoidance of premature failures due to large non-metallic inclusions or carbide segregates. Füssl indicates that a similar view is held on the Continent, and states that electroslag remelting and vacuum arc remelting do not improve carbide distribution to any significant extent but only improve the cleanness.

Apart from the improved breakdown of cellular carbide by upsetting and redrawing, it is questionable whether the forging process affects carbide significantly other than through forging reductions, that is, the time–temperature aspect of forging appears to be of secondary importance.

Tomlinson and Beard of BISRA have, however, considered the significance of the repeated heating cycles normally used when forging high-speed steel. Repeated heating causes substantial loss due to scale, decarburization, and reduction in the productivity of the forge. The more difficult the alloy is to forge and the narrower the forging temperature range, the more significant this effect is. The penalty for inadequate control is, of course, cracked material during forging, so Tomlinson and Beard have considered all the methods of supplying heat to the work during forging and have concluded that only radiant heat by the use of infra-red heating elements is really practicable.

They have experimented with various grades of high-speed steel and shown that a 4:1 forging reduction in a simple working operation is possible with reduced cracking, oxide scaling, and decarburization. They claim that the scaling and decarburization are about half the corresponding industrial levels.

An interesting development mentioned in their paper was the ability to forge an M2 grade isothermally at 800°C, in the ferrite region. By this somewhat novel procedure, the decarburization was virtually eliminated and scaling was reduced to about 0·2%.

Tomlinson and Beard suggest that infra-red heating is of potential interest, due to the possibility of controlled thermomechanical treatments which should give improved properties. Unfortunately, the work has not yet reached the stage of verifying the possibility, but discussion on this theme would be desirable.

One must commend the authors on an interesting approach to improving the processing of high-speed steel and indeed, other metals. The process will succeed provided the economic benefits outweigh the quite severe practical problems and costs of heating at the press.

One of the most interesting papers is that by Dulis and

Neumeyer on the powder-metal route. Several of the authors comment that this route offers the greatest potential not only for improved carbide size and distribution but for the development of compositions containing more carbon and carbide formers that can be easily incorporated in a forged alloy. Unfortunately, Dulis and Neumeyer tell us very little of their production route other than that it starts with atomized powder, and that an 800 lb billet of 9·5in diameter by 42in long is processed on 'conventional mill facilities'. This is an intriguing situation and one would hope perhaps to hear a little more about the process route used.

The process can undoubtedly fulfil its promise with respect to carbide size and distribution. The high-carbon M2 steel studies had a mean carbide size of about 0·0001in and a maximum carbide size of about 0·0002in even in a 5in diameter bar. This is about one fifth of the carbide size that would be found in conventional 5in diameter high-speed steel.

The Snyder-Graff intercept grain size is about 13 for the conventionally treated powder-metal product, but the combination of the fine carbide size and a 16 h temper anneal at 732°C can give an intercept grain size of 30, which is exceedingly fine.

By this route, the authors have therefore produced the sort of high-speed steel with fine grain and fine carbide that the tool manufacturers have been trying to specify for years. It is interesting therefore to consider what properties go along with this near-perfect structure.

As one would expect, the hardening response is speeded up. Out of roundness is improved by a factor of three, but the authors emphasize that the mode of hot working, i.e. the use of a round slug broken down by undisclosed but conventional techniques, has a lot to do with this improved out of roundness.

Tool life and consistency are improved. Continuous cutting tools show a 30% improvement in life, and intermittent cutting tools 160% improvement. It should be emphasized that this improvement was found in a comparison of lathe tools machined from the transverse radial direction of 5in diameter bar. This is a somewhat unusual source of lathe tools but was presumably necessary to evaluate the bars which are more likely to show beneficial effects from production by the powder-metal route. I think that most of us would feel somewhat disappointed that the cutting efficiency of such a near-perfect structure is not more significantly improved.

Dulis and Neumeyer emphasize the importance of using low oxygen content powder. High oxygen content powders give poor results unless substantial hot working is carried out. I think that we shall see a number of papers in the coming years on this technique, and one would make an appeal at this stage, therefore, to see oxygen contents reported. This comment of Dulis and Neumeyer emphasizes perhaps the difficulty of direct-sintering such powders to give such effective properties as can be achieved by using the route of forging or otherwise reducing the initial powder compact used.

The advent of this powder-metallurgical material enables us to compare the properties, albeit on somewhat slender evidence at this stage, of high-speed steels having a very wide range of carbide size and intercept grain size. For example, cast high-speed steel will have a carbide size of about 5–10 thou, wrought high-speed steel in the range of 2–8 thou, and powder metal 1–2 thou. Similarly, the intercept grain size goes from 2–3 for cast, through 13–20 for wrought, and as high as 13–30 for powder metal. This is quite a range of carbide size and intercept grain size, and broadly speaking, the results of continuous cutting tests, if one looks widely at the literature, show very little significant difference over all this range of structural features. In intermittent cutting, there does seem to be a trend, however, that the very fine carbide, powder-metal material is marginally better.

Somewhat provocatively, perhaps, I would suggest that any improvement in cutting performance over the whole

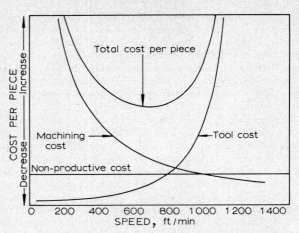

I   **Total cost per piece shown as a product of machining cost, tool cost, and non-productive cost**

range of these structures covered does little more than increase the reliability of a tool under standard cutting conditions and does not allow for any increased speed and feeds which would permit significantly improved outputs to be achieved.

It is appropriate at this stage to consider what future demands will be put on the suppliers of high-speed steels. There will be continuing pressure to improve the general quality of the steel with respect to uniformity of structure and hence uniformity of response to heat treatment allied with the avoidance of cracking and distortion. The pursuit of the ideal structure will give an added bonus of toughness which may or may not help prolong the service life of the tool.

For most applications, these quality improvements will rely on minor improvements to conventional production routes but even so will inevitably result in a marginal increase in cost.

Where the application warrants even higher standards with respect to structure, the powder-metallurgy route offers an attractive possibility, but one hazards a guess at a fairly substantial premium. Perhaps Mr Dulis will enlighten us on this point.

The growth of automation, such as NC machining, will place a premium on consistency of service behaviour, and the pursuit of improved uniformity of structure will be beneficial to this end. Here electroslag remelting may also help by giving a more consistent product at about the same general standard as the best conventionally produced material.

The user will undoubtedly be seeking higher productivity. One must emphasize that machining time is the main factor here. Labour and plant costs and also overheads are at fixed hourly rates and as seen in Fig.I, it pays to accept a shorter tool life to reduce the total cost per piece to a minimum. It is worth emphasizing that, if you do not choose the longest tool life you can get, you optimize this with respect to the overall production cost.

The importance of the slope of the tool life v. cutting speed curve on the economics of metal cutting is well established.

The ideal tool material would not be affected at all by high-speed high-temperature cutting, but would have the same wear characteristics at all cutting speeds. This would give a 45° slope as shown in the 'ideal' curve in Fig.II. Unfortunately, you will see that high-speed steel has a very shallow slope. This means that an improvement of, say, 100% in life on a tool cutting test means very little in terms of the cutting speed that the tool material will stand. It also means that by reducing very marginally the cutting speeds, if you use a high-speed steel, you can achieve a much longer life, or alternatively a greater degree of reliability against premature tool failure.

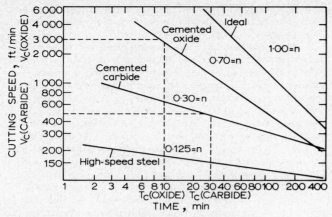

II **Wear characteristics of different tool materials**

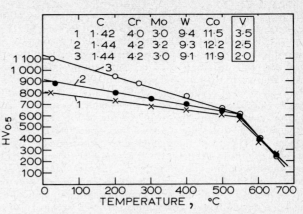

	C	Cr	Mo	W	Co	V
1	1·42	4·0	3·0	9·4	11·5	3·5
2	1·44	4·2	3·2	9·3	12·2	2·5
3	1·44	4·2	3·0	9·1	11·9	2·0

IV **Effect of high carbon–vanadium ratio on hardness of HSS**

If we look at cutting speed against stock removal (Fig.III), the ideal material becomes the vertical line indicating a constant amount of stock removal independent of speed.

One concludes from the papers, in which there is very little mention of developments of better high-speed steels, that we are not likely to see very much improvement in the cutting characteristics of conventional high-speed steels, although some hope is held out by Dulis of high-alloy high-speed steels by the powder-metal route. One thing that the powder-metal route does bring back to mind is the fact that so many carbide tools are being used as either throwaway tips or inserts, and in this field, high-speed steel seems to have languished. The availability of atomized powder would presumably lend itself to direct sintering provided the oxygen content problem can be solved.

Alternatively, is there not a possibility for precision casting such tips? Both routes do allow the possibility of higher alloy compositions.

In conclusion, I hope I have not only given you some idea of what is in these very excellent papers, but have also brought out some areas where discusssion may be fruitful.

Mr J. Tidlund (Fagersta Bruks AB, Sweden)
I would like to make a brief contribution to what has been said about hot hardness in HSS, particularly by Kirk *et al.* We consider hot hardness as a most important feature for cutting material. In principal, it is the strength and hardness of the cutting edge at high temperature that determine the cutting ability.

I would like to show two examples of how composition affects the hot hardness. It is well known that a high carbon–

vanadium ratio gives a very hard martensite in HSS, if retained austenite can be avoided. This was done in the experiment shown in Fig.IV by lowering the vanadium content.

The hardest steel at room temperature remains hardest up to about 550°C. Above this temperature all modifications are just equal. A probable explanation is that in this case the martensite hardness determines the hot hardness up to about 550°C, because below this temperature the agglomeration and the constitution of precipitates is not changed if the steel has been tempered over this temperature.

If cobalt is added one also gets a higher room temperature hardness (Fig.V). In this case, however, a greater hot hardness remains even above 550°C. Cobalt has a solution-hardening effect on martensite and also a retarding effect on agglomeration of secondary carbides.

My conclusion would be that room-temperature hardness does not tell us everything about cutting ability. Hot hardness tells us a lot more, but I am sorry to say it might not give the whole truth.

Dr P. Hellman (Stora Kopparberg, Sweden)
I would like to comment first about hot hardness, and specifically about the figures given by Kirk *et al.* in their paper.

In their paper, if I read it correctly, it is said that T1 has a better red hardness than M1 and M2, and also a very much better red hardness than M42. This seems difficult to believe, especially in the light of what we have just heard.

Much has already been said, and more will, I suppose, be said, about the properties of high-speed steel determining tool life, but these tools have to be produced. In other words, properties like machinability and grindability of high-speed steel are very important properties.

With regard to sulphur additions, Mr Child said that the Germans add only 0·06% sulphur and that should be an advantage. In my opinion, one should be very careful when

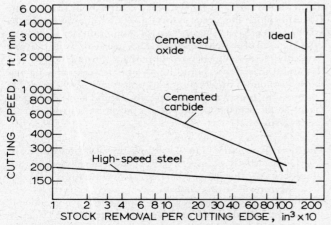

III **Modern cutting-tool materials approach the ideal concept of a constant amount of stock removal at any speed**

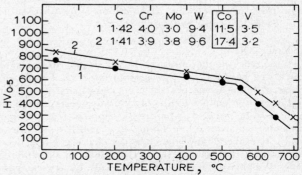

	C	Cr	Mo	W	Co	V
1	1·42	4·0	3·0	9·4	11·5	3·5
2	1·41	3·9	3·8	9·6	17·4	3·2

V **Effect of high Co content on hardness of HSS**

judging the properties of a high-speed steel from the sulphur content, that is the percentage of sulphur present in HSS. Other factors are very important, primarily the size of the sulphides in the steel, and their distribution. In fact, the size, number, and distribution of the sulphides in the steel are more important than the weight percentage of sulphur.

I have been studying grindability by a new method, which has recently been published in *Jernkontorets Annaler*, 1969, **153**, 583, and would like to give some of the results. First, the grindability of some standard high-speed steels, T1, M2, M7, and M35: from the grindability measurements we do, we give the results in the form of a grindability number. This grindability number is proportional to the volume of steel ground off until the abrasive grains have reached a certain predetermined wear level. Therefore, a high grindability number means better grindability. In this scale, T1 has a grindability number of 7·4; M2 is slightly lower at 7·2; M7 is very much lower at 3·8, and M35 is 6·6. M2 and M35 are identical except for the cobalt addition in M35. I think that these grindability numbers correlate very well with experience in practice.

When adding sulphur, we get sulphides in the steel. Ordinarily, it is a matter of a rather large sulphur addition, i.e. sulphurized steels with something like 0·1% sulphur. The sulphur is added to improve the machinability, but grindability is also improved. If, however, you can add sulphur in such a way that you get very small sulphide inclusions, then a moderate sulphur addition is sufficient to give even better grindability. If you add 0·04–0·05% sulphur to M2 and do it in the proper way, then you get an M2R with a grindability of 15 as compared to a grindability of 12 for the resulphurized steel M2S. The explanation for this is simple. In M2R with a rather low sulphur addition, the number of sulphide inclusions in the steel is larger than in M2S where the size of the sulphide inclusions is larger. With less than half the sulphur addition, the grindability is better. This is rather easy to understand. Considering the number of cutting edges and the sizes of these cutting edges involved in grinding, it is easily understood that, to improve grindability, the number of sulphides should be large and the sulphides themselves may well be small. Also, if we do the same with M7, that is M7R, the grindability number will be 6·1. The grindability of M7, therefore, may be improved up to levels in the neighbourhood of T1 and M2.

Sulphur is important for other properties too. We all know that in resulphurized steels, there is a tendency to a decreased toughness in the transverse direction, but this is not due to the 0·1% sulphur in itself; it is due to the size of the sulphide inclusions. If a high-speed steel is made with the very small sulphide inclusions, then nothing will happen to the toughness; that is, in these R grades containing 0·04–0·05% sulphur, each property is the same as in the low-sulphur parent steel except for improvements in grindability and, to some extent, in machinability.

I would like to make a very short comment on the British Standard. The British Standard contains 15 high-speed steels. For example, T1, M1, and M2 are used for roughly the same purposes. There may be rather small differences between them in some cases. My question is, why is M7 not included in the British Standard?

Mr Kirk
With reference to the questions raised relative to hot hardness *v.* red hardness and the position of M42, I must emphasize that the properties shown in bar-chart form in Fig. 2 represent the factual results of laboratory tests carried out on each of the steels, 'red hardness' being taken as the room temperature hardness remaining after triple tempering at 650°C. The overall appearance of the chart is, I feel, a fair reflection of the expected service behaviour of the different grades. We have not found hot hardness always as meaningful as red hardness in comparing different grades of steel,

although as Mr Tidlund has done, it may be possible to show some significance in hot hardness relative to alloy variation in one basic composition.

I agree that hot hardness, representing the resistance to deformation of tool material at cutting temperatures must be of importance, but in many instances cutting operations are of an intermittent nature and therefore resistance to tempering is highly significant. Furthermore we regard the response to tempering at 650°C as a useful measure of the rate of overaging, that is, the rate of agglomeration of secondary carbides, and we feel this factor is of particular importance.

As Dr Hellmann has remarked, the low red-hardness value shown on our chart for M42 seems at first sight rather anomalous, but nevertheless this was the result obtained. Some confirmation is shown in Table 3 which contains values for hot hardness of certain grades, and again we have M42 giving the lowest value at 650°C, showing 290 HV as against 330 for T1 and 385 for M2. These results are again a little surprising, but nevertheless factual. It is important to remember, however, that M42 was designed for cutting the harder materials, usually at slow speeds. We believe here that the development of a cutting edge with high hardness and adequate toughness is more important than the ability to resist high temperatures. I feel that one must put this steel in its correct place in machining requirements as compared with T4 and T6 which, as experience has confirmed, are to be preferred for really rapid metal removal. The necessity for high tungsten with high cobalt where the best red hardness is required is well illustrated in Fig.2.

Passing on to the subject of resulphurized grades, I was very interested in what Dr Hellmann had to say about the effect of relatively low sulphur levels on grindability. However, the requirement for resulphurized high-speed steels in this country has arisen from the advent of the unground form hob, wherein the hobmaker wishes to generate a form by a machining operation which produces a sufficiently good surface finish to put straight into service after heat treatment without the need for grinding. It is for this reason that sulphur levels in M2 of 0·1% and over have been the vogue. I am encouraged by what Mr Füssl has reported relative to the use of 0·06% sulphur levels. I have contended for a considerable time that a heat-treatment operation before form relieving is a desirable procedure, and if it is shown possible to operate with a lower sulphur level, then the problems of the steelmaker in supplying such grades should be alleviated with some economic benefit.

The question has been raised as to why M7 is not in the British Standard. This is a relatively recent introduction, not in very great demand or availability in this country, and was for this reason excluded. A great deal of discussion was centred around the steels to go into the standard, and those were included which generally are in more popular demand. I agree, nevertheless, that this is one of the steels deserving further attention, and indeed work is already in hand in the UK to explore this grade more fully. The question must be asked, however, whether in view of the popularity of M2 the introduction of a new steel is really justified. Does anyone have any factual evidence to support its introduction? I feel that the future trend should be towards fewer grades having more economic production and improved service to the customer in mind, and that a really careful look should be taken at all aspects before new grades are introduced as standard materials.

Mr R. G. Kennedy (The Cleveland Twist Drill Company). Answering Mr Kirk about M7, it was developed in our laboratory, but this is not why I am answering his remarks.

In the rotary cutting tool industry in my country, a type similar to M7, designated M10W, and M7, have practically supplanted all other types, except for some M1 and a small amount of M2. In the tap industry, there is still quite a bit of M1 used, but in the drill industry M7 and M10 with a

VI   Ausformed M2 high-speed steel punch

high carbon content are becoming predominant. M7 is being used not because it is any easier to make for the steelmakers, but, as far as the users are concerned, because the performance of the tools is the answer.

I would very much like to see M7 put in your list. I was on the ASTM Tool Steel Committee and I have brought a list of the tool steels included in their latest publication, for your information. I would like your committee at least to consider inclusion of M7 and possibly high-carbon M10.

As a general comment, the papers at this meeting appear to me to contain useful information for users and producers of tool steels. I would hope that a future meeting would include information on vital factors affecting tool performance such as residual stresses introduced into tool surface by heat treatment and grinding and assessment of grinding damage.

Dr M. G. Cockcroft (NEL)
Mr Child has already referred to the mention by Tomlinson and Beard of thermomechanical treatment. The BISRA work, I understand, refers to the high-temperature version of TMT; I suggest that the low-temperature version, ausforming, should at least be considered. We do not have any information so far on the cutting performance of an ausformed high-speed steel, but we have been interested in a job that is of more immediate application from our point of view; this is the production of ausformed high-speed steel for cold-forming applications, particularly cold extrusion. Figure VI shows a billet of M2 high-speed steel which is austenitized, quenched to about 600°C, and then put in the extrusion chamber. First, the billet is upset to fill the chamber, and it is then forward extruded in the normal way and ejected. Figure VI shows the product as extruded and after sand-blasting. After slitting off the lightly deformed front end the product is ground to size. The amount ground off is about 10 thou all over, except for a little more in the radius between the head and the shank. Forming close to size is important. If ausforming is to get anywhere, it is necessary to restrict the amount of machining that has to be done on the material. The punches of ausformed M2 steel have not fully been examined and tested, but the hardness after triple tempering is in, or a little above, the normal range of 850–930 HV.

Much of the development work was done on a lower alloy 5%Cr steel, H13, and this has shown the anticipated increases in yield, tensile, and hardness without a decrease in ductility. As yet, we do not know very much about our ausformed high-speed steel, still less do we know anything about its cutting performance but, in the current phrase, we are cautiously optimistic.

Mr A. J. Fenner (Ministry of Technology)
I would like to comment generally on the Ministry of Tech-

nology and tool steels. The Ministry is interested in supporting the BSI committee which is responsible for the specification for tool steels, and it also has a general interest in the development of tool steels. In consequence, two programmes of experimental work in this field have been planned.

The first of these is designed to obtain performance data on the M2 and T1 grades which has been requested by the BSI committee. The purpose of this would be to assess how far the molybdenum grades can replace tungsten grades in UK usage to save expensive imports of tungsten.

The second is a larger, collaborative programme on the application of thermomechanical treatments in the manufacture of cutting tools with appropriate sections of the work undertaken in four different laboratories.

The objectives here would be two-fold. First, to investigate the feasibility of using thermomechanical treatments to develop tool steels of lower alloy content than those currently used, and second to endeavour to improve tool life for the current grades by those treatments.

The total cost of this programme is about £110000, of which the Ministry is prepared to find 50%. I understand that currently there is difficulty in getting industry to provide the other half of the cost, although the possible annual savings which might result should amply repay any outlay.

Dr G. Barrow (University of Manchester Institute of Science and Technology)
I would like to make some comments regarding the use of ausformed cutting tools. We did some preliminary cutting tests about two years ago. At that time no ausformed high-speed steels were available so we tried cutting with some low-alloy steels and H13. The performance of these tools was, as expected, inferior to T1 HSS. However, H13 gave 60% of the tool life of T1 when machining En1a and performed very well on brass at low cutting temperatures. From these preliminary tests we feel that there is some potential with ausformed cutting tools, but in the first instance only a small expenditure is justified. In the next few months we hope to undertake tests on M50 and M2 ausformed steels. It is felt that the main application will be in intermittent cutting operations and small tools (taps and drills).

With reference to the comments on the machinability of ausformed steel I should like to say that we have been working on this topic for nearly three years. I agree that any ausforming process should eliminate as much machining as possible, since we have experienced considerable problems in machining ausformed steels under roughing conditions.

Mr F. M. Karlsen (London and Scandinavian Metallurgical Co.)
I would like to remark on Mr Fenner's statement that the Ministry of Technology is going to do some work on molybdenum high-speed steels to see whether they can replace the tungsten high-speed steels, and so save foreign exchange.

First, I think the reasons given are wrong, and the position in the rest of the world is that molybdenum high-speed steels seem to be coming into use more and more.

To give saving of expensive tungsten imports as one of his reasons is crystal-gazing. First, there is potentially quite a lot of tungsten ore available in this country which could be exploited if only the right people could be found to do this. Second, there is no guarantee that molybdenum (which is imported) may not double or treble in price, as it did about five years ago. Third, there is the problem of using vanadium in this type of steel, and everybody knows what has happened to vanadium in the last year.

I would suggest that this type of investigation should be based primarily on technical forecasting rather than on dubious commercial forecasting of raw material prices.

Dr D. J. Latham (BISRA, the Corporate Laboratories, BSC)
I would like to add one or two comments relating to the

Therefore some doubt lies in our minds as to the validity of the tests outlined in Tables 6 and 7. The predominant mode of failure appears to be tool wear. However, it is not expected that carbide banding would appreciably affect wear resistance. Our experience has been that such raw material defects are responsible for major tool breakage and local cutting edge chipping. Indeed, we would suggest that the most significant result in Table 6 is that five drills of VDE rating 5d suffered from chipping, whereas only one drill of VDE rating 2L encountered such problems.

We feel that more significant information would be forthcoming from torsion tests of finished drills and heavy-duty breakthrough drilling tests, where cutting-edge chipping predominated as the mode of failure.

Catastrophic tool failure has always been most undesirable, and in view of the increasing use today of numerically controlled and other automatic machine tools, where such failure may have disastrous results, we view with some alarm any suggestion, however tentative it may be, from the high-speed steel manufacturers that carbide banding is not important.

In this connexion it should be mentioned that we, as tool manufacturers, would welcome the widespread commercial availability of ESR material. Our experience has been that in comparison to good quality air-melted high-speed steel only marginal improvements in tool performance are likely, but by far the most important benefit is the improvement in consistency throughout the ingot and hence from bar to bar and tool to tool.

The ever increasing, but nevertheless understandable, demands of the engineering industry for predictable and repeatable tool performance must be met by all the means at our disposal, and improvement in the quality and consistency of high-speed steel raw material supplies must be obtained.

Finally, Mr Child, in his summary of the papers presented, referred to the lack of a paper on the performance of cast tools. My company has recently carried out a programme evaluating the performance of cast toolbits on continuous turning. The material was 5–6–2 high-speed steel with varying additions of zirconium, and while we found that increasing amounts of zirconium improved the performance, the best performance was only about 50% that of a toolbit produced from forged and rolled material.

Mr Kirk

I am grateful to Mr Caisley for his various comments representing a user viewpoint on some of the more controversial aspects of the paper. I feel that drill test results of the type shown in Tables 6 and 7 can only be taken at their face value. While a greater incidence of chipping is indicated for the material with the inferior carbide distribution, the overall lives, as indicated, do not show any significant variation, and it is not possible to draw positive conclusions. In the absence to date of conclusive evidence that better carbide distribution is of really over-riding significance, we felt the data to be worth including. It is not our intention to infer that good carbide distribution is not desirable, but to suggest that the matter be kept in true perspective. In the view of the steelmaker there are many other factors in tool manufacture such as heat treatment and grinding, also in the manner of use which are of at least equal significance in determining the tool life obtained.

As many of those present are aware, considerable strides have been made in this country during recent years in improving the quality of high-speed steel bars. Mr Caisley's reference to the desirability of electroslag-refined material is especially appreciated, and the data shown in Fig. 15 in the paper exhibits quite clearly the improvement in consistency of drill life which may thus be attained. Considerable ESR capacity is available in the UK, but if a customer specifies electroslag-refined material, he usually asks for special test requirements, and inevitably some element of premium must be involved. On the other hand there are manufacturing areas in which it may be possible to utilize the process to internal advantage, and the steelmaker will continually try to do this.

In the paper we show in Fig. 14 the microstructure of 7¼in diameter T1 steel produced by both the air-melting route and by electroslag refining, and suggest that the improvement due to ESR is overemphasized because the 16in square air-melted ingot size was perhaps a little too large for T1. Figure IX represents the same kind of approach in the case of M2, wherein the use of a 16in square air-melted ingot is acceptable, and compares 3½in square material produced from such an ingot with a similar product orginating from a 17in diameter ESR ingot. A significant advantage in carbide distribution exists for the ESR material, that shown probably equalling the best found in the product of the bottom end of the air-melted ingot.

Figure IX also, I believe, answers a point of concern which has been expressed in the past, that ESR material tends to be coarser at the outside than normally. As will be seen, this is not necessarily the case.

Mr W. Fairhurst (Climax Molybdenum Co.)

I think I must answer the rather unusual challenge raised by one of the speakers, namely that the molybdenum grades of high-speed steel should be asked to compete technically on equal terms with the tungsten grades and, secondly, that he felt there was some doubt in relation to the supply of raw material which may affect the situation.

From the user's point of view, there is only one criterion for the selection of tool materials and that is the most economical removal of metal consistent with the production of a satisfactory component.

If one looks at the range of machining operations carried out by high-speed steel in this country, a vast proportion is met by either T1 or M2. In terms of technical performance, I do not think there is very much differential between the two. The minor differences are really negligible. The big difference between T1 and M2 is in relation to cost. Considering black bar costs, I think at present it would work out, on a volume to volume basis, that the M2 grade is 30% cheaper than T1, and this really is the incentive in going towards the molybdenum grade. It is a question of economy.

In relation to supplies of raw materials, all I would say in that regard is that we have in hand at the moment one of the biggest mining projects in North America. Its aim is to go into production in the mid 1970's at a rate of around 50 m. lb of molybdenum per year. The investment in this alone will be about $200 m. To put this 50 m. lb per year into context, about 50% of the free world's requirements for molybdenum in the past has come from one main source, the Climax mine in America, which has a production capacity of around 56 m. lb a year. We are doing our best to maintain continuity of supply, and if I were putting my money on the availability of materials, comparing tungsten with molybdenum, I know which one I would back.

WRITTEN CONTRIBUTION
A. E. Longley (Rolls-Royce Ltd)

With reference to the paper on 'High-speed steel technology: the manufacturers' viewpoint', I would welcome the author's views on the following points.

Now that fracture-toughness testing has reached the stage of a proposed British Standard specification, would it not be feasible to consider elevated temperature fracture-toughness testing as a means of evaluating high-speed cutting steels?

It would be interesting to compare the elevated temperature fracture-toughness properties of air-melted v. electroslag refined cutting steels.

A possible correlation of elevated temperature fracture-toughness with tool life would seem most desirable.

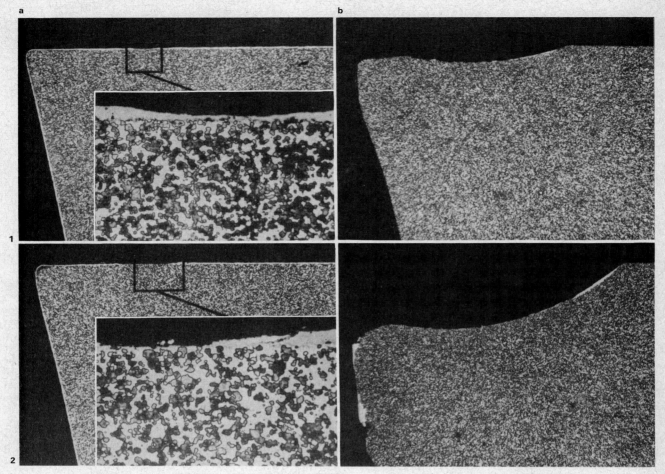

a GC 125, ×66 and ×535; *b* S4 (ISO P30), ×66; *1* after 1·5 min; *2* after 2·5 min; material Sandvik 17C (approx. SAE 1095); hardness 275 HB; cutting speed 180 m/min; feed 0·3 mm/rev; depth of cut 2 mm

XI   Comparison of actual wear generated on a GC 125 and a standard S4 cutting edge

points out that such protection is not afforded by formulations containing tungsten as the only carbide, but is afforded when titanium or tantalum are present, so that the formation of a stable layer owes much to the chemical reactions between titanium and the inclusions materials. I would like to speculate on the nature of these interactions, based on practical work carried out by Dr Singh and myself, and shortly to be published.

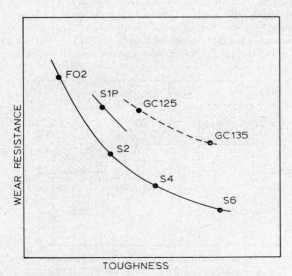

XII   Comparative wear resistance–toughness relationships of some Sandvik Coromant metal-cutting grades

I can begin by observing that since carbide-oxide compounds are not readily formed, the most probable form of interaction is that between the oxides present in the steel (including, incidentally, iron oxides formed directly by reaction with air at the cutting temperatures attained) and the oxide present as a result of degradation of carbide by environmental conditions.

That such oxides are likely to form when a carbide is exposed to a normal atmosphere is well known, and one measure of the propensity to do so is the ratio of the free energies of formation of the oxide and carbide. Where this is high, the tendency to degrade to oxide will be correspondingly high. Figures for the materials that I wish to discuss (at 280°K) are as follows:

	Oxide–carbide stability ratio	Oxide lattice
Aluminium	19	stoichiometric
Tungsten	7·4	stoichiometric
Titanium	2·1	10% O_2 deficient
Zirconium	3·3	20% O_2 deficient

Consideration of these ratios alone might lead one to expect that aluminium and tungsten would give good protective layers. Mr Lardner has indicated that tungsten gives no protection and our work on the interactions of iron and very fine alumina in conditions where atmospheric oxygen was absent showed that no detectable interaction product was formed. It should be noted that the oxides of both these metals are stoichiometric.

Turning next to titanium, we found (by X-ray back-reflection and powder diffraction) that an oxide compound

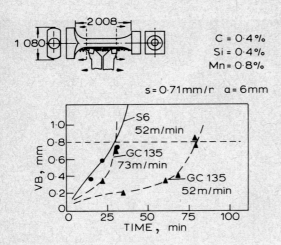

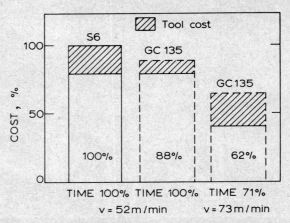

XIII   Data, comparative results, and economies achieved when machining a forged connecting rod with GC 135 instead of S6

Fe_2TiO_5 was very readily formed, and we suggest that the reason for this ready formation lies in the non-stoichiometry of the TiO_2 lattice. Since O^{2-} is a very much larger ion than Zr^{4+}, the oxygen forms a reference framework for diffusion, which will normally have about 10% 'holes' in place of oxygen ions. Further, in conditions where some oxide is already present before vacuum or hydrogen sintering (which is likely to be the case in normal hardmetal tool production) the low-oxygen partial pressures during the sintering would produce a still greater degree of oxygen deficiency in the titania lattice so that the diffusion rate will be further enhanced.

Now it so happens that the mechanism of formation of these oxide compounds has been extensively studied by chemists. It turns out that the mechanism involves the formation of molecular oxygen by:

$$2O^{2-} \rightarrow O_2 + 4\epsilon$$

Build up of oxygen gas pressure at the interface leads to microcracking and ultimately to spalling off of the protective layer.

It might then be supposed that an even better protective layer could be formed in a situation where this mechanism was absent. Zirconium provides such a case. The ZrO_2 lattice is even more oxygen deficient than that of TiO_2 and our work showed that this resulted in interaction effect at least an order of magnitude greater than just described with TiO_2. In this case, the most probable oxide compound to be formed in the machining of steel containing silicate inclusions is $ZrO_2 \cdot SiO_2$ in a silicate solid solution. This should form very rapidly, and being free from molecular oxygen, would be less likely to spall away and so afford better protection in machining.

My suggestion is therefore that a case could be made, on grounds of reduced wear in ferrous machining, for trying zirconium carbide as an alternative to titanium carbide. For the zirconium materials the basic carbide stability is somewhat lower and the oxide–carbide stability ratio is somewhat higher than for titanium, so that the protective layers will be rather more readily formed. An additional bonus accrues from the fact that zirconium carbide stability actually *increases* with temperature increase (a rare phenomenon) whereas that of titanium carbide decreases in the usual way. The magnitude is such that the oxide–carbide stability ratio for zirconium is down to about 1·7 at 1000°C, so that the extent of carbide degradation to oxide will be self-limiting with temperature rise, affording some degree of protection in the overload conditions which must occasionally occur in this type of machining.

Mr C. P. Mills (ASEA (GB) Ltd)

I work in the British Division of ASEA. Among the wide range of products that we produce, mainly in the electrical field, is the Quintus range of high-pressure equipment. This includes hydrostatic extrusion, powder presses, isostatic compaction, cold forging, rubber presses, and deep forming. I should like to give an account of the high-pressure application of hot isostatic compaction, which I think should be of interest to a number of companies represented here.

Most of you will be aware of the possibility of applying this technique to the manufacture of sintered carbides, ceramics, and alloy steels made from powder. At present, the powder is cold compacted and vacuum sintered separately afterwards; by hot isostatic compaction, sintering under pressure is available which enables control of the metallurgical structure of the component being made, in particular

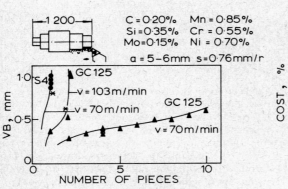

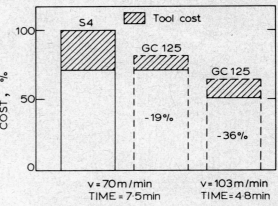

XIV   Data, comparative results, and economies achieved when machining a heavy tractor axle with GC 125 instead of S4

XV   Operating configuration for a 1 600 ton hot isostatic press

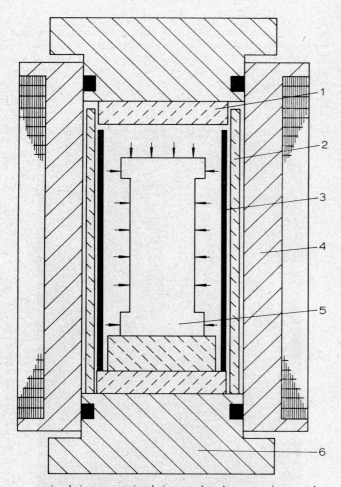

1 insulation cover; *2* insulation mantle; *3* heater; *4* wire-wound vessel; *5* hot-pressed item; *6* cover

XVI   **Schematic section through a vessel**

control of grain growth which is desirable in many instances.

Another facet is microporosity of carbide components which can be reduced in certain instances. We have found that the wear rates of some components due to this hot isostatic compaction are reduced by a factor of 8.

Figure XV shows the design philosophy behind ASEA's isostatic vessel design, and shows the general configuration. The isostatic vessel shown here is a cylinder, which is in essence a simple tube wire-wound with steel wire. The end caps are simple plugs and are light. There are no threads associated with this design of plugging at all and stress raisers on this score are not there. The end load resulting from pressure within the vessel is taken out by the frame on to the two yokes at either end. The frame is also wire-wound around its periphery.

This enables a compact design which, of course, has good fatigue properties as well.

Accessibility to the equipment is achieved by moving the trolley out of the way, the vessel itself being fixed to the stand.

Since it is hot isostatic compaction, the vessel itself has a mantle around it, a water cooling jacket, and we supply extra cooling to this. Since there are no threads on the end plugs, there is no axial stress on the chamber which is good from a safety point of view.

This principle is applied to both hot and cold vessels. We make a range of cold vessels. Figure XVI shows the schematic arrangement for the pressure vessel insulation, the heater element, and the end caps. The furnace has been specially developed to operate under an inert gas at high pressure and temperature and still provide accurate temperature control. Most work to date has been done with helium as the pressure medium. We have chosen argon for two principal reasons. Argon is low in cost compared with helium. The hot conductivity of helium is much higher than that of argon at high pressure. Consequently, argon promotes a more even temperature on the workpiece. The main objection to argon has been its pronounced tendency to convect. With ASEA's furnace design, this problem has been overcome and less than half of the heat losses from the furnace are caused by convection.

The furnace is built up of five main parts: the insulation mantle and the top forming a gas-tight shell, and no overall convection from inside to outside of the insulation mantle is allowed. Only holes for pressure equalization are provided. Inside the insulation mantle, gas convection is permitted and desired. There, the convection helps to create uniform temperature in the work space, and effectively promotes heat transfer from the heaters to the workpiece. The heaters are of the resistance type and divided into independently controlled zones. The desired temperature can be regulated within very narrow limits, and we reckon $\pm 1\%$ at temperature, which is, in the standard equipment, $1450°C$, and we have an option to go to $1750°C$. The heating equipment is thyristor controlled.

Figure XVII shows the system of supplying the gas to the vessel, with the bottles, gas filter, compressor, cooler, and pressure gauge. We allow the gas to flow into the vessel and then it is pumped with a diaphragm-type pump to bring it up to pressure, and the heating of the vessel is used to increase the pressure finally.

On decompression, the gas is allowed to pass straight out back into the bottles through filters and possibly the cooler, and the pump is used again and it goes back through the cooler and into the bottles. The loss of gas after each cycle is therefore quite small.

Figure XVIII shows the general layout of the two main parts. The unit as such is a package. The frame can run along on rails. This is a simple design and can be motorized. There is a pneumatic arrangement for raising and lowering the bottom plug and hence used for loading (Fig. XIX).

Just to give some idea of this particular model, the normal

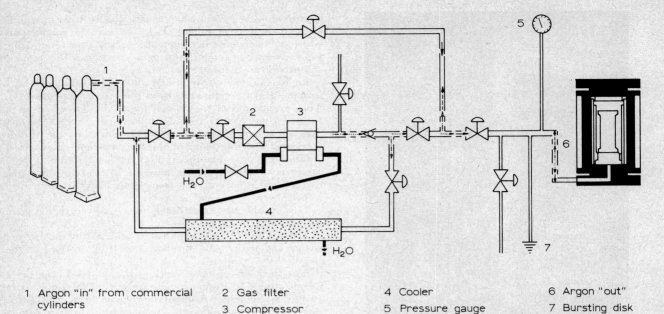

1 Argon "in" from commercial cylinders
2 Gas filter
3 Compressor
4 Cooler
5 Pressure gauge
6 Argon "out"
7 Bursting disk

——▶ Gas flow by compression until pressure is balanced
········▶ Gas pumping to a desired pressure
----▶ Gas flow by decompression until pressure is balanced
—·—·▶ Gas pumping by decompression

XVII   Flow diagram for gas equipment

maximum temperature is 1450°C with option for 1750°C. The pump pressure is 700 atm but there are options for 1000, 1600, and 2000 atm. This particular model is probably more used for development work. On a production job, preheating would be used probably in the furnaces up to 1000°C before bottom loading into the vessel, and in order to get the pressure of 2000 atm, a higher pumping power would be required because the heating would not be so effective.

For the one we have seen, the useful inner diameter is 6·3 in, the useful inner height 19·7in. If we go to 1750°C, this is reduced due to increased thickness of insulation, to 5·5in and 17·7in. Thus today we are offering a hot isostatic system on the lines of that which we have just seen.

The applications which we foresee are good, and in particu-lar for (a) sintered carbides for large pieces and assymetric shaped pieces, and improved wear properties of certain items; (b) high or super alloys made from powder in large pieces where segregation may be a problem; (c) ceramics and cera-mic-to-metal bonding.

XVIII   Side elevation of a press showing the bottom loading arrangement

XIX   End view of a press showing the wire-wound frame and lateral position of the loader

WRITTEN CONTRIBUTION
Mr B. Pegg (Rolls-Royce Ltd)

In view of some of the comments made during this discussion suggesting that the large users of sintered tool materials are passive in their activities, this is strongly refuted and the following comments are given in defence of the statement.

Sintered tool materials, especially tungsten carbide types, are the major tool materials in current use. With the advent of numerically controlled machine tools, the thoughts of such a machine tool being dependent on a 5*s* piece of carbide has not gone unnoticed. In this respect however, we are not provided with much assistance from the carbide manufacturers to enable us to select the right grades of tool material. It is contended that these people have a responsibility to promote a usable tool selection system rather than rate his customer for competence and turnover, and then restrict the information supplied accordingly. Rolls-Royce spend a considerable amount of time and money on metal-cutting development and application. In this respect, it has taken us something like ten years to earn the respect of carbide manufacturers as to our capabilities, competence, and sincerity. So much for 'passive activities' and reciprocation of informa-

tion! In this respect it is only recently that we have managed to obtain reliable information from manufacturers on composite analysis and associated data so that we can attempt to classify carbides in such a manner that a tooling engineer can perform his activities in tool selection on a scientific basis.

Both ISO and BHMA carbide classification systems, used independently or in conjunction with one another, offer little hope for even initiating a metal-cutting operation, let alone analysing it. Literature researches reveal the inadequacy of much scientific work, in that the consistency of composition of the tool material is frequently neglected, with regard to its effect on the work in hand.

It is from this situation that it was hoped to make an appeal to all sintered tool material manufacturers, to make generally available data on grain size (average and range), a standardized hardness value (Vickers would appear to be completely satisfactory), together with a standard composition analysis and an indication of constituent form (i.e. solid solution, etc.) Maybe other details will show their importance, but at least publication of the information quoted will assist in bringing analysis of carbide tool application into the scientific category it deserves.

Selection and heat treatment of high-speed steels for drill manufacture

C. E. Pillinger and G. Huddy

The twist drill is primarily a metal-cutting tool which, in common with other types of tool, can give excellent results if properly used within its limitations. The standard general-purpose tool is necessarily something of a compromise in both its mechanical and metallurgical characteristics. For specific types of work there are available specially designed drills, and in some cases the nature of the work involved calls for steels with properties beyond those available from the basic high-speed steels.

HIGH-SPEED STEEL
The prime quality of high-speed steel can be regarded as its ability to retain a useful degree of hardness at temperatures of the order of 500°C: a property usually known as red hardness, and a related property of regaining the original room-temperature hardness provided the original tempering temperature has not been exceeded. It may not be appreciated that the energy released in the form of heat during metal-cutting operations can quite easily reach these sort of temperatures, even in the presence of a flood of coolant. High-speed steels are also characterized by the high degree of hardness that can be consistently attained. Until about ten years ago the usual range of maximum hardness available was 800–900 HV, but these limits have been extended by the latest generation of specialized steels aimed at maximum hardness, and with these 950 HV, can be readily attained and 1 000 HV is just possible in exceptional circumstances.

The remaining features of high-speed steels tend to be liabilities which are accepted in pursuance of the foregoing; thus a very high austenitizing temperature is required in the range 1 200–1 300°C. These high temperatures carry a penalty of a greater tendency to distortion and a much greater tendency to decarburization than the low-alloy tool steels which are austenitized in the range 800–900°C. Other short-comings are a rather higher hardness in the fully annealed condition, a greater tendency to segregation, and a high density, which implies that a greater weight of steel will be required to produce a given tool.

There are quite a number of high-speed steels in existence; the older series which were in universal use until about 1940 were based on tungsten as the major alloying element. Economic factors caused the replacement of all or part of the tungsten by molybdenum and have led to the adoption of molybdenum-based steels. This movement has been greatest in the North American continent where at least 90% of the high-speed steel manufactured is molybdenum based. In contrast to this the production in the USSR is entirely tungsten based.

Table 1 lists a number of compositions of high-speed steels.

M2 and T1 are the compositions most commonly used for standard drill production together with M1 and M10 in the USA. Where greater red hardness is required, there are available compositions basically M2 or T1 with 5% or 10%Co additions. Another variation, aimed at augmenting the wear resistance, increases the vanadium content together with the carbon content to increase the amount of the very hard vanadium carbide in the steel, resulting in analyses typified in T2, T3, M3, M4, and M7 steels. The T15 and M15 steels are a combination of these two lines of approach giving greater hardness, together with improved red hardness.

Indeed they provide these properties to a degree which can be an embarrassment to the toolmaker who is required to grind the fully heat-treated tool to shape with conventional grinding wheels. The latest improvement, M42, was arrived at by examining several series of alloys to find the combination which would give maximum hardness without undue brittleness, and with one eye on the grindability. As can be seen, the composition is basically M1, with the addition of cobalt and with increased carbon content but only slightly increased vanadium. The high hardness levels, around 950 HV, which this steel can achieve are very useful where tools are required to machine the newer heat-resisting alloys.

Thus there is great variety of high-speed steels to choose from, but naturally the standard steels are more readily available and are used for a very large part of production. As far as Great Britain is concerned the economics favour the use of M2 for general twist-drill manufacture. Where improved performance is required, although there is still a body of opinion favouring the use of tungsten-based steels, the modern approach is to use a molybdenum-based steel designed for additional red hardness, greater wear resistance, or maximum hardness.

QUALITY CONTROL
The users of any high-speed steel in appreciable quantities will find it necessary to establish some sort of specification, and the following is an account of some of the difficulties which may be encountered.

Tolerance on the various alloying elements may need to be specified: any single supplier will usually supply consistently within the limits of his own specification, but it is very rare to have only one source of supply, and one soon finds that each supplier has his own preferred range for each element. Thus if the user without further thought accepts the standard product from a number of suppliers, although each one will be allowing say a $\pm 0.03\%$ variation on a carbon figure, one may be basing this on 0.82%, another on 0.87%, and the user finds he is dealing with a carbon content varying

The main properties of high-speed steels are summarized, together with a survey of the development of the numerous types. The need for a specification is mentioned, and the main points which need to be controlled are outlined together with an indication of some of the areas where agreement between user and supplier may be needed. Heat-treatment requirements and limitations are discussed, and mention is made of steam tempering and liquid nitriding.

621.951.4.025:669.14.018.252.3

The authors are with the Sheffield Twist Drill and Steel Co. Ltd

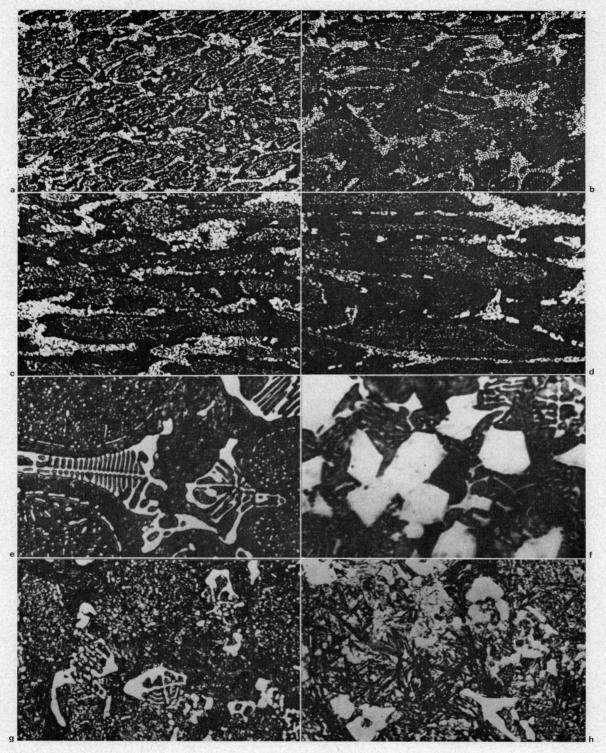

a cellular remnants; *b* dendritic remnants; *c* hookiness in forged blanks; *d* hookiness in rolled bar; *e* Chinese scripts; *f* coarse carbides;
g Chinese script remnants; *h* residual austenite associated with alloy-rich areas; *a–d* ×100; *e–h* ×400

3 Undesirable carbide formation

Working on the premise that as the section increases the structure will progressively depart from ideal it is necessary to devise a material standard which takes this into account and shows pictorially the carbide distributions that cannot be tolerated in a given section (Fig.3). In so far as the shape and form of carbides are concerned the remnants of the original eutectic cast structure cannot be tolerated, and yet it is surprising to find that often after much hot working remnants of 'Chinese scripts' are often found to persist. 'Hook' formations of carbides are often much in evidence

in the larger blanks and it is appreciated that with the upset forging method, some compromise has to be agreed between the amount of upset and danger of bursting; therefore some hookiness is often tolerated. With rolled bar or with small section blanks, the carbides are expected to be fairly well distributed and to be of spherical form and with these, the presence of large carbides of irregular shape form the subject of rejection. With the larger size of blanks, far more tolerance has to be exercised although segregation of carbides into enlarged areas is frowned upon since this can and does represent

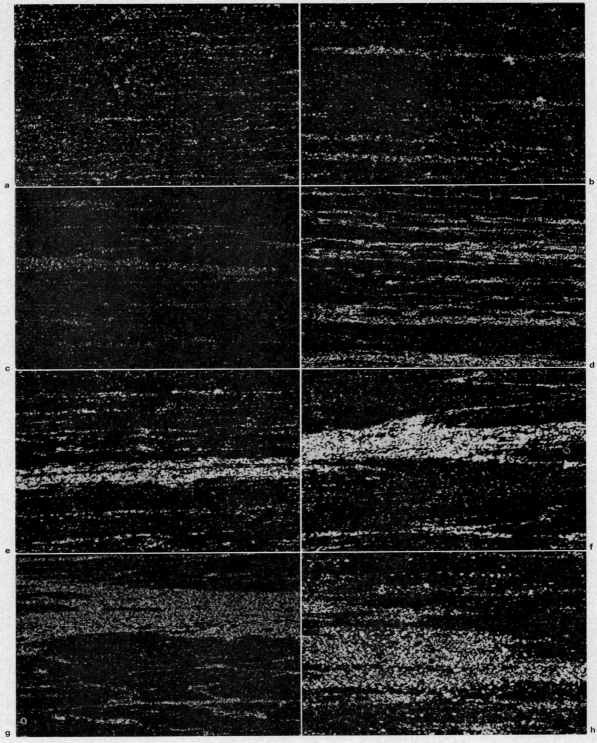

a grade 0; *b* grade 1; *c* grade 2; *d* grade 3; *e* grade 4; *f* grade 5; *g* grade 6; *h* grade 7

4   Carbide classification in bar stock   ×400

a segregation of alloys and during hardening this coring results in alloy-rich austenite which is difficult to transform.

For the unground type of hob, material in bar form is preferred to individually upset blanks since the general quality is usually superior, it is more consistent and its hardening behaviour more predictable. It has been argued of course, that in bar the carbides must necessarily be grossly elongated and unidirectional, but while this means that the grain flow runs across the hob tooth it does not appear to affect the cutting performance. There have been isolated

instances however, where abuse in grinding during resharpening has caused cracks to develop through and along the carbide stringers.

In the past, it has not always been easy to impose stringent standards on high-speed steel suppliers because they were able to sell most of their output against relatively loose acceptance standards. There have been times when they seemed to be suspiciously anxious to please, and these seemed to coincide with activity on the part of American or Continental steelmakers in developing new quality standards

Table 2 Relative grindability of 'super-hard' high-speed steels (Tarasov)

AISI type	Grindability index*	Hardness, RC
T15	0·6	66
M41	1·0	68–69
M42	5·6	68–69
M43	2·2	68–69
M44	0·95	68–69
M1	8·0	65

* A higher number denotes better grindability

Table 3 Composition specifications for M2 and T1 steels

Element	M2S (6–5–2+S) Min., %	M2S (6–5–2+S) Max., %	M2 (6–5–2) Min., %	M2 (6–5–2) Max., %	T1 (18–4–1) Min., %	T1 (18–4–1) Max., %
Carbon	0·80	0·88	0·80	0·88	0·72	0·80
Chromium	3·75	4·50	3·75	4·50	3·75	4·50
Tungsten	6·00	6·75	6·00	6·75	17·50	18·50
Molybdenum	4·75	5·25	4·75	5·25	..	1·00
Vanadium	1·75	2·15	1·75	2·15	1·00	1·40
Sulphur	0·11	0·15	..	0·05	..	0·05
Phosphorus	..	0·035	..	0·035	..	0·035
Manganese	..	0·45	..	0·45	..	0·45
Silicon	..	0·45	..	0·45	..	0·45

Table 4 Maximum allowable decarburization (hot-rolled bar, as received)

Bar diameter, in	Depth of total and partial decarburization, in
Up to ½	0·015
½–1	0·020
1–2	0·030
2–3	0·040
3–4	0·050
4–5	0·055
5–6	0·060
6–7	0·065

After rough machining, surfaces must be free from total and partial decarburization

but wherever such claims proved to be overconfident the British producers tended to revert to complacency. It must be said in their favour however, that in recent years the re-grouping of companies and cooperation with BISRA has led to serious attempts to produce higher qualities of high-speed steel, and at present certain of our suppliers are able to produce qualities comparable with anything available overseas.

STANDARD SPECIFICATIONS
National standards, material handbooks and suppliers' brochures show a bewildering array of high-speed steels. These are classified either as grade symbols or as proprietary brands originating from some standard conventional composition range, modified by raising or lowering the original main alloy contents, increasing the carbon and vanadium contents, or adding other expensive alloys such as cobalt in order that improved life, greater hardness, improved resistance to abrasion, and superior hot hardness values could be claimed. With the majority of such claims, however, there is surprisingly no quantitative data supplied that would enable the user to assist the economics of such more expensive grades. Undoubtedly these grades have their advantages where difficult materials are required to be machined but in the main any so-called advantages are outweighed by increased manufacturing difficulties. Table 2 shows how an attempt to produce improved wear-resisting grades has resulted in deterioration in grindability.

Experience in the tool manufacturing trade has demonstrated over the years that the two grades of overall highest merit are the 18–4–1 and the 6–5–2, and although the former has retained its popularity, it is now being gradually replaced by the molybdenum-steel grade. Perhaps the reason for the general reluctance to use molybdenum-base steels is due to unhappy experiences in the early days by the automobile manufacturers where decarburization during hardening resulted in poor life of the finished ground hobs. With modern manufacturing systems, the 6–5–2 grade has shown itself to be easy to machine and to be consistent in heat treatment and the prejudice against this grade is fast disappearing.

At the present time, the bulk of hobs is made in the 18–4–1 and 6–5–2 grades, the proportion being about equal, and consequently where the customer does not ask for a particular steel to be used for his tools, he will receive them in one of these grades; consequently most of the users specifications will cover these two grades only. Any such standards will lay down the form of blank, i.e. whether an upset forged blank or bar and composition will be restricted to relatively narrow ranges (Table 3). For blanks, there will be the familiar section covering freedom from defects and in addition, there will be a section outlining the maximum allowable decarburization (Table 4). Since the blanks will cover a considerable section range, there will be carbide segregation charts and network classifications and reference will be made to these in the section dealing with the permitted segregation which will be tolerated in any particular section range (Fig. 4 and Table 5). All such material is ordered from approved suppliers and it could well be that while a number of suppliers are included for small section blanks the number to be approved for the heavy blanks will be few.

Of course, such standards are constantly being modified in the light of steelmaking techniques and at any given moment of time they merely represent the highest commercial quality obtainable on the market in production quantities and with the most reasonable delivery times.

It is realized that at the present time this system can only operate as far as the supplier is concerned if he is willing to select ingots to a stringent quality and employs a reasonable discard system. Even then it means that users not purchasing steels to such careful quality specifications are liable to be supplied with grades of somewhat inferior quality.

One factor not covered by the present steel standards but which is of vital importance to the manufacturer is that of hardenability and hardness potential, and this no doubt will feature in future standards. At the present time the hob maker sets his heat-treatment cycles so as to produce parts having a final hardness level of 64–65-RC but he is only able to obtain this occasionally. Certain casts are found to be of high hardness potential material, i.e. 68 HRC, and when this happens in a mass-production line one can imagine the effects of bottlenecks arising in the finishing grinding section

Table 5 Permitted segregation

Diameter	Bar stock, at any position in the bar	
Up to and including 1in	Grade 1 maximum	
Above 1in and including 2in	Grade 2, isolated 3 tolerated	
	At a distance from the periphery	
	⅛ × o.d.	¼ × o.d.
Above 2in and including 3½in	Grade 2 to 3 Isolated 4 tolerated at 3in and above	Grades 3 to 4 Isolated 5 tolerated at 3in and above
Above 3½in and including 5in	Grades 3, generally isolated 4, and sparse networks tolerated	Grade 4, generally isolated 5, and knotted networks tolerated
Above 5in and including 8in*	Grades 3 to 4 knotted networks tolerated	Grades 4 to 5 knotted networks tolerated
Diameter	Forged blanks	
Up to 3½in	Carbides well broken down, relatively free from segregation. No evidence of networks, eutectic remnants, coring etc.	
Over 3½in but less than 5in	Some degree of segregation and traces of fine networks will be permitted, but no evidence of eutectic remnants, or coring etc.	
Over 5in but less than 8in	Some degree of segregation permitted, but no thick networks or cored structures	
Over 8in	To be free from coring and cellular networks, to be supplied only by named suppliers	

* In bar sizes in excess of 6in diameter occasional patches of dense network will be tolerated; some degree of coring will be acceptable, but continuous, cellular, dendritic structures must not be present

Tool steels and their application to milling cutters

A. L. Horne

Although a number of new metal-removal techniques have been introduced and developed over the last decade, it is accepted that machining by metal cutting will remain for a considerable time the principal metal-removal process. Hence there is a continuing need for an informed approach to the selection of the correct cutting medium, and of tool design, so as to ensure that machining costs as related to unit cost are kept to a minimum.

This objective can only be realized by close cooperation between user and supplier, and also, in the case of new projects, with the machine-tool manufacturer, at the pre-machine design stage. All relevant information at the pre-production planning stage should be available to the tool supplier to ensure, as far as possible, that the tools are suitable for the operations to be performed. The basic information required by the tool manufacturer is as follows:

(i) component details including material specifications and operation to be performed
(ii) production rate required
(iii) machine and fixture details
(iv) services available to maintain tooling
(v) minimum desirable tool life.

Component details and operation to be performed
This should include full information on component material, whether forging or casting, etc., the amount to be removed, and full details of component condition, i.e. details of any machining before the operation under consideration which has a bearing on the operation to be performed.

Production rate required
This has an important bearing on the selection of the cutting medium. It should be stated whether small-batch, large-batch, or continuous production is envisaged. The number of components required per hour or per shift should be indicated.

Machine and fixture details
Full details of the machine to be used should be supplied, giving the type of machine, e.g. whether it is a standard type, transfer machine, or numerically controlled, as this can be an important factor in tool selection. In the case of tooling for numerically controlled machines it is also necessary for the tool manufacturer to close down the overall tool tolerances to guarantee the maintenance of datums. Fixture details are desirable since they show how the component is held, enabling the tool designer to avoid any obstructions.

Services available to maintain tooling
This information is particularly important where high-volume production at high feed rates is involved since this may indicate the necessity for modern automatic cutter grinding machines to be available for tool servicing. Should this type of machine not be available it could influence the selection of the type of tool to be employed. In this case and also where very limited servicing facilities were available to the user, the tool designer might be influenced in the direction of indexable throwaway-type tooling, in suitable cases.

Minimum desirable tool life
A realistic estimate should be given by the user. This is usually determined with regard to the most convenient time to change tools, ideally between shifts, and servicing facilities available.

Machining can still be said to be more an art than a science. The tool supplier cannot guarantee that the initial choice of tool or tool material will be the ultimate. This can only be finalized by following up when in production, but if the foregoing information is available to the tool supplier the initial choice has a far greater possibility of being the correct one.

This paper presents the manufacturer's rather than the user's point of view. It may be argued that there should be no difference in approach since the objective, an efficient cutting tool, is the same. It should be emphasized, however, that cutting-tool manufacturers operate in a highly competitive field. Price and delivery are frequently the most important criteria, from the user's point of view, at the time of purchase. It is only when the tool arrives on the shop floor that the other factors become all-important.

In practice, therefore, sheer economic necessity tends to restrict the manufacturer's choice to tool materials of known quality which possess good wear-resistance characteristics, adequate toughness, offer no undue heat-treatment or grinding problems, and which are competitively priced and readily available.

SOLID HSS TOOLS
The principal cutting-tool manufacturers hold large stocks of a wide range of standard HSS tools enabling them to offer a quick delivery at competitive prices resulting from quantity production (see Table 1).

For this type of tool we believe the choice is between T1 (18–4–1) and M2 (6–5–2). Present trends indicate that M2 is perhaps replacing T1 as first choice for standard tools, since the early difficulties of decarburization have been overcome with improved manufacturing techniques, and, perhaps most important, M2 is cheaper while giving equal or better performance than T1. It appears likely that M2 will eventually virtually supersede T1.

For special tools, particularly those which are required to machine higher tensile steels or the 'exotic' materials used in the aerospace industry, and also tools used in the high-volume production plants in the automotive industry, the choice is between T15 (5Co–5V–12/13W) and M15 (5Co–5V–6·5W–3·5Mo).

There is a wide range of tool materials available to cutting-tool manufacturers. This paper sets out the various factors which motivate a cutting-tool manufacturer in the selection of materials. It suggests that cost, quality, and availability are perhaps the major factors in the selection of materials for standard cutting tools. It points out that different criteria are employed when selecting tool materials for special tools, discusses some of the alternative materials, and touches on various treatments which are thought to have a bearing on performance.

621.914.025 : 669.14.018.252.3

The author is with Richard Lloyd Ltd

Table 1 Composition of HSS grades

Grade	W, %	Cr, %	V, %	Co, %	Mo, %	C, %
T1	18	4	1	..	..	0·75
T6 (type)	20	4·5	1·5	12	..	0·80
T15	12	4	5	5	..	1·50
M1	1·5	4	1	..	8	0·80
M2	6	4	2	..	5	0·85
M7	1·75	4	2	..	8·75	1·00
M15	6·5	4	5	..	3·5	1·50
M42	1·5	3·75	1·25	8	9·5	1·00

These are approximate figures and are given as a general guide only

Both these types of tool steels have, during the last ten years, gained wide acceptance in industry. M15 may also, in the long term, replace T15 as first choice since it is cheaper than T15 and easier to grind, which has resulted in its being the preferred tool steel for high-performance form tools and ground thread taps.

It might be thought, from the foregoing, that manufacturers are imposing their own preferences on users and choosing to ignore a wide range of other tool steels but this is not the case since virtually all tool steels which become available are investigated by our metallurgists, and field trials are carried out in addition to trials in our own works, and the results studied as to their present or future implications. In most instances the results do not justify a change in policy and in those cases where good results are achieved it is often found that either the price is prohibitive or the material is not readily available.

It is evident that it would be uneconomic to hold adequate stocks of a wide range of tool steels, hence the need for preferred standards, but limited stocks of a number of other tool steels are held by Richard Lloyd Ltd where a regular call-off occurs as a result of field trials, usually in connexion with a special problem or customer preference.

OTHER HSS TOOL MATERIALS FROM METALLURGICAL VIEWPOINT

Physical properties, e.g. torque value and Izod impact values, are not easy to obtain and evaluate. The cutting properties, e.g. red hardness, are easy to evaluate since prolonged trials will determine this for anyone to see. Over the last four or five years we have focused our attention on M42 and M43 steels and it is thought they hold advantages over the conventional steels. Every steel user is looking for one HSS type that would perform every operation, under adverse conditions, with 100% efficiency, at less than basic cost and no risk of any shortage. It is sad that no such material exists nor is likely to in the near future. It remains for us to strive for consistently higher quality materials, to a controlled grain size, the absolute minimum of segregated carbides, and consistently high quality of craftsmanship in design and manufacture of the tools.

Recent trends in the manufacture of tool steels have varied from vacuum-melted steels to the resurrection of the electro-slag remelt process. The alleged advantages of cleaner tool steels particularly applied to the high-speed steel types are not definitely established yet. From a steelmaker's point of view the latter process would reduce the amount of scrap produced, but from a steel user's view the advantages are obscure. Doubtless this process has advantages if we were discussing the new ball-bearing steels and those used in the jet materials.

HAMMERED AND CUT-OFF BLANKS

A few comments here would be perhaps appropriate on a subject which is very old. In theory, a properly hammered blank should be one where the molecular structure has been broken down and rearranged in such a fashion as to minimize any distortion during subsequent heat treatment. A cut-off blank taken from a normally rolled or forged bar would in theory become more sensitive to distortion. If however a double upset forging operation is introduced to produce the bar then it is more than probable that this would create conditions similar to a forged blank. A proper balance could be drawn on the relative costs of these operations, saving of material, and the tool design before deciding on which operation is better than the other.

SURFACE TREATMENTS

For certain applications, it is advantageous for HSS tools to be given a surface hardness greater than that obtained by conventional hardening and tempering, with the object of increasing the useful life of the tool. The various treatments are as follows:

Nitriding
This process is well established, where the tools are immersed in a bath containing molten cyanide salts at a temperature of about 560°C, and nitrogen and possible carbon, liberated from the salt, diffuse into the surface of the tools, forming alloy nitrides.

Sulphidizing
Sulphidizing is similar to nitriding in respect of equipment and temperature, the essential difference being that the salt contains sulphur compounds as well as cyanide. This enables simultaneous diffusion of nitrogen, sulphur, and possible carbon into the surface of the tools.

The immersion times for both the above treatments can be varied to suit requirements, but for HSS tools, it is usually of the order of 10–15 min. This ensures that no softening of the matrix occurs and also that the layer depth and hardness is not excessive. With these short treatment times hardnesses up to 1000 HV and depths up to 0·001in are obtained. Because of the hardness of the layer, operations involving vibration or heavy intermittent cutting are not thought to be suitable for surface-treated tools.

Advantages are as follows: the improved surface hardness obtained by nitriding gives the tool increased resistance to chip failure; sulphidizing improves resistance to scuffing, and also, the likelihood of material pick-up is reduced because of lower frictional forces between tool and workpiece.

Steam tempering
In this process an extremely thin but adherent film of oxide is produced by heating the tools in a steam atmosphere. It is suggested that this layer will minimize the possibility of galling of the tool with the surface of the metal being cut. This is explained to a certain extent on the basis that the oxide layer will retain the lubricant at the cutting edge of the tool. Unlike most other surface treatments there is no increase in surface hardness of the tool, but increased life is usually obtained due to the reduction of frictional effects.

Chromium plating
Chromium plating is used to provide a hard surface with very low frictional characteristics to certain types of tools. Its main advantage is in minimizing the tendency of the tool to load or pick up material from the surface of the work. Only thin layers (0·001in) are applied, as heavy deposits tend to reduce the toughness of the tool. It must also be noted that the hard deposit becomes soft on heating, so that it is of no advantage to plate tools that become heated in use.

Molybdenum disulphide
One of the many alternative methods of improving the working life of tools being investigated by toolmakers is the use of molybdenum disulphide. This is sprayed onto the tool and allowed to cure, and it is claimed that the life of the tool is increased by several hundred percent.

INSERTED-BLADE HSS TOOLS

Larger sizes of HSS tools are usually made in inserted-blade designs (*see* Figs.1 and 2). We would always recommend inserted-blade cutters in preference to solid HSS except

1 **Galtona inserted high-speed steel blade alternate-angle milling cutter**

3 **Inserted carbide-tipped blade face mill**

flexibility in application and also provides for the use of the latest developments in HSS tool material as they become available. Changes in tool geometry are easily effected in this type of tool. This is a great advantage where variations in work material are encountered.

where the cutter under consideration was too small in diameter or too narrow in width, or where very fine tooth spacing was felt to be essential. Inserted-blade cutters, in addition to being of proved efficiency, are also very economical since only the blades are made of high-priced material, enabling the highest quality tool material, if required, to be employed at a moderate increase in total cost of tool.

Stock cutters are fitted with T1 blades but complete stocks are also held in T15 material. The design makes for greater

2 **Galtona inserted high-speed steel blade face mill**

4 **Inserted carbide-tipped blade alternate-angle milling cutter**

5 Two designs of indexable throwaway-insert face mills

INSERTED-BLADE CEMENTED-CARBIDE TIPPED CUTTERS
Stocks are held incorporating a full range of carbide grades (Figs.3 and 4). They, also, are able to take advantage of developments in the carbide field as they arise. Changes in tool geometry are easily effected.

INDEXABLE THROWAWAY INSERT
Indexable throwaway inserts are stocked without tips, and the choice of carbide grade is made at the time of ordering. This reduces the stock holding and ensures the best grade for the job.

Indexable throwaway-insert cutters (see Fig.5) suffer from one major disadvantage when compared with inserted-blade cutters, since changes in tool geometry are not possible. This is fixed at time of manufacture and cannot be altered.

Table 2   Carbide grades and their applications

BHMA no.	ISO no.	Application
7–2–0	K01	Precision finishing of cast iron
6–4–1	K05	A fine-grained alloy of high abrasive resistance especially suited to the turning of chilled cast iron of maximum hardness
5–6–1	K10	Machining cast iron at high feed rates; general machining of non-ferrous metals, and certain reaming operations
4–8–0	K20	Roughing grade, machining under unfavourable conditions, cast iron and all non-ferrous materials, austenitic stainless and nimonics
3–9–0	K30	Roughing grade, intermittent cutting, tougher than TC6 but slightly less resistant to edge wear
2–9–0	K40	Very tough grade for any application where heavy shock is encountered; woodworking cutters, heading dies, etc.
6–2–4	M10	Machining of cast iron with long chipping characteristics, austenitic and manganese steels at high cutting speeds, and where cratering is experienced
7–2–9	P01	Finishing of steel; exact final machining operations at very high cutting speeds, precision boring, highly resistant to cratering
5–5–4	P05	Hard but reasonably strong grade for the finish machining of steel whenever severe edge wear is experienced. Can also be used for milling as an alternative to the more usual fine-grained TC25 alloy
3–3–6	P10	Steel turning and special milling applications at medium to high cutting speeds
3–2–4	P20	Steel turning and milling at medium cutting speeds; copy turning
3–4–4	P25	Roughing grade, medium to heavy work
2–7–6	P30	Heavy and intermittent cutting of steel; slow speeds

GUIDE TO CLASSIFICATION OF CARBIDE GRADES
Increasing demand throughout the world for a standardization of hardmetal grade classification, has led tool-steel manufacturers to adopt the British Hardmetals Association (BHMA) coding system. This system is based on the three principle properties which are of most interest to the user, as follows:

(i) wear resistance
(ii) shock resistance
(iii) crater resistance.

These three properties are qualified as hardness, transverse rupture strength, and volume of crater-resistant carbides. The properties, in the order given above, are classified by the figures 0–9, so that each figure defines a definite range of each property. Hardmetal grades for machining are thus classified in terms of three single figures, which relate to their hardness, transverse rupture strength, and crater resistance. For example, a grade with the highest hardness, low transverse rupture strength, and intermediate crater resistance, would be described as 9–1–5, whereas a grade of similar hardness and toughness, but with no crater resistance, would be 9–1–0, thus, the higher the value, the higher that characteristic. Classification of grades according to this system automatically distinguishes the steel-cutting grades from cast-iron grades, as the latter shows zero or low crater resistance. For instance, a user who requires a grade with slightly less or more wear resistance than the one being used, can select the appropriate grade, and be able to assess whether the change in wear resistance has been achieved at the expense of one of the other properties. Users can therefore choose grades more surely, and can better understand the differences between the various grades (see Table 2).

FUTURE TRENDS
As manufacturers, we believe that in the future great benefits will ensue from even closer control of materials by tool steel manufacturers, and also by strict control of incoming materials by cutting-tool manufacturers.

This should be greatly assisted in the UK by the publication of the proposed British Standard specifications for tool steels now in course of preparation by BSI.

It seems clear that the change-over from tungsten to molybdenum steels will be greatly accelerated.

The increasing use of automation and the requirements of the aerospace industry will result in an increasing demand for special tools, designed to perform a specific operation for the whole of their useful life, hence the need for the required information, specified at the beginning of this paper, which enables the tool designer to assess the requirements of the operation to be performed and select the cutting medium and type of tool best suited to the individual case.

Further developments in surface treatments, particularly in the field of hole-producing cutting tools, may yield beneficial results.

Finally, in connexion with possible future developments, we would offer the suggestion that, in our view, the greatest benefits in the metal-cutting field would be likely to come from closer control of the materials to be cut rather than from the cutting tool itself.

ACKNOWLEDGMENTS

The author wishes to express his thanks to the Director of Richard Lloyd Ltd for permission to publish this paper, and also to Mr S. J. Fishwick, works manager, and Mr C. Thatcher, metallurgist, who assisted in its preparation.

High-speed steel bar quality

N. H. McBroom and T. B. Smith

The increasing sophistication of metal-cutting, shaping, and forming processes is such that more and more reliability is being sought from the materials involved. This applies not only to those materials being shaped (e.g. free-cutting bar stock for machining) but also to those such as tool materials, a successful and predictable performance from which is essential for obtaining economically viable engineering products. While it has long been the custom to invoke specifications to control the quality of feedstock materials, it is only in recent years that any detailed attention has been paid to the quality of tool materials. It is increasingly common to find tools working at the limit of their performance and consequently it is important to ensure that the quality of tool materials is of a high standard. (In response to the demand for improved quality, a British Standard for tool steels is now being considered by industry.) This paper reviews some of the developments in the quality of high-speed steel bar material in the last few years.

HIGH-SPEED STEEL QUALITY

In the GKN group of companies, a detailed examination of the performance of high-speed steel in various operations from the machining of crankshafts to the cold extrusion of large complicated shapes has indicated those quality parameters which are of importance. These have been summarized by the authors elsewhere.[1] The most important aspects of quality are:

(i) correct steel structure, i.e. freedom from segregation and inclusions, to ensure a uniform load-bearing capacity

(ii) correct steel composition, to ensure uniformity of heat-treatment response from batch to batch and from tool room to tool room.

General quality

Because they lead to local weak spots, defects such as ingotism and segregation cannot be tolerated for high-duty applications. However, defects such as those present in the 50 mm diameter bar of M2 shown in Fig.1 are not infrequently encountered. The incidence of defects of this sort led to the introduction in some GKN companies of quality specifications backed up with inspection procedures for incoming high-speed steel. In a recent period, 4000 bars of high-speed steel have been inspected and of these 8·6% were rejected for quality defects. A breakdown of the causes for rejection is given in Table 1. While specific local requirements can account for some of the causes for rejection (e.g. the large fraction for underannealed material, 28%, is in part due to the need for a very low annealed hardness in certain companies practising cold hobbing) the overall rejection level is disturbing since it indicates a lack of reliable inspection on the part of the supplier.

The fraction of rejections for mixed grades, 8%, is particularly disturbing, since the consequences of mixed grades can be very serious. Continual vigilance on the part of manufacturers, hire workers, and stockholders as well as consumers is required to eliminate this source of trouble.

Decarburization still remains a problem and amounts to 16% of the total rejected material. In GKN, the limits set on the maximum depth of decarburization have been related to machining allowances used in tool rooms. Values are 0·4 mm on bars less than 25 mm diameter rising to 1 mm on bars greater than 40 mm diameter. If decarburization is present beyond these limits, it may lead to reduction in the surface hardness of the tool which is quite unacceptable.

Of the other causes for rejection, poor carbide distribution (10% of total rejects) is probably the most serious, since not only is it undesirable in its own right, it is also indicative of more general segregation and lack of deformation in the bar. The consequences of carbide segregation and means of overcoming it are discussed in a later section.

The evidence in Table 1 suggests that practice to maintain quality within reasonable limits requires modifying in certain plants. Equally, there is no doubt that in response to demand from users in the last few years, very considerable progress has been made in upgrading the quality of high-speed steel.

Heat-treatment response

An obvious requirement for a high-speed steel is that it shall respond predictably and consistently during heat treatment. Thus different tools heat treated in different tool rooms can be relied upon to achieve the same level of hardness. As well as achieving the required hardness in tools subject to intermittent loading, it is essential to maintain a fine grain size. Gill[2] has shown that a fine grain size increases plasticity and hence the ability to absorb energy before fracture, an important property in tools subjected to shock loading A study of the change in grain size during hardening of high-speed steels has shown that many behave in a similar fashion. At lower hardening temperatures, grain growth is restricted by the presence of undissolved carbide. As the temperature is increased and considerable carbide solution occurs, there is a sudden increase in grain size as illustrated in Fig.2, which shows the as-quenched Snyder-Graff[3] grain size (the number of grains intersecting a 5in axis at a magnification of $\times 1000$) plotted against hardening temperature for T1, M2, T42, and M42 high-speed steels. The hardening times at each temperature are also shown. It can be seen that grain size remains constant with increasing hardening temperature until a certain critical temperature is reached, and then increases abruptly. This critical temperature coincides with conventional hardening temperatures. It seems reasonable, therefore, to be able to harden good-quality material with a uniform distribution of the grain boundary restraining carbides without

The need to control the quality of high-speed steel used in cutting and forming operations is indicated. The most important aspects of quality are seen as ensuring a uniform structure free from segregation and a consistent response to heat treatment. Experience has shown that the defects most commonly encountered in commercial bars of high-speed steel are decarburization and poor carbide distribution. Instances of mixed grades are also prevalent. Evidence is presented to show that as well as achieving the required hardness during heat treatment, good-quality high-speed steel should retain a fine grain size. This improves mechanical properties. Developments designed to improve the quality of high-speed steel are reviewed. In this context, mention is made of the strengths and weaknesses of electroslag melting and upset forging. The view is expressed that the quality of high-speed steel has improved considerably in the last few years.

669.14.018.252.3-422.1:658.562

The authors are at the GKN Group Technological Centre

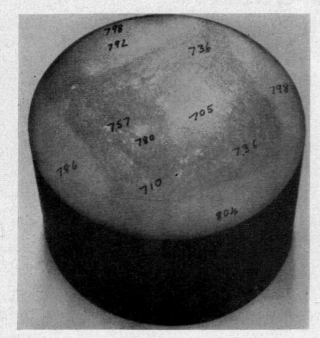

1 Segregation in 50 mm diameter M2 with superimposed hardness values

Table 1   Breakdown of rejections for quality defects

No. of bars inspected	4 000
Rejected for quality defects, %	8·6
Breakdown of causes for rejections, %:	
Cracked bars	5
Inclusions	9
Macrodefects	6·5
Chemical defects	0·5
Carbide segregation	10
Bars oversize	8
Mixed grades	8
Underannealing	28
Decarburization	16
Bars bent	9
Total	100

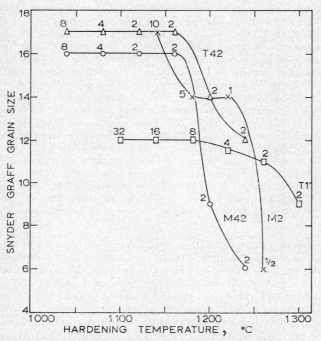

2 Increase in grain size with increasing hardening temperature; the numbers indicate hardening times in minutes

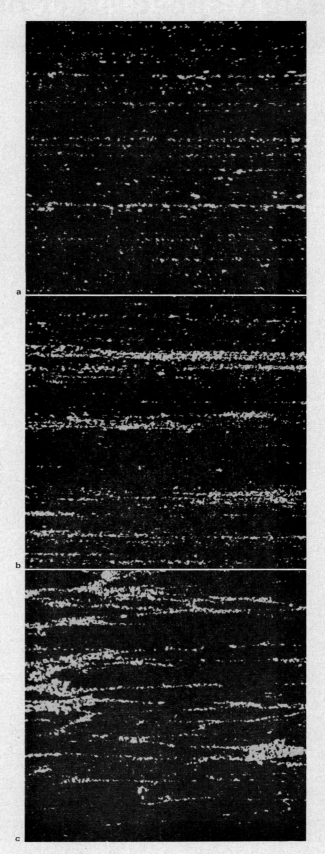

a 25 mm diameter; b 50 mm diameter; c 75 mm diameter

3 Maximum acceptable mid-radius carbide segregation in annealed M2   × 100

Table 2   Grain size limits attainable in high-speed steel at different bar diameters

Steel	Hardening temp., °C (2 min soak)	Max. mid-radius grain size (Snyder-Graff) after oil quench		
		<75 mm	75–125 mm	>125 mm
T1	1 260	12	10	8
M1	1 220	12	10	8
M2	1 220	12	10	8
M4	1 220	12	10	8
M15	1 220	12	10	8
M42	1 170	12	10	8

4 Electroslag remelted T1 material, mid-radius segregation in 25 mm diameter annealed bar   ×100

passing the critical point for grain coarsening. In this way the mechanical properties of the hardened steel are improved. Table 2 gives the maximum mid-radius Snyder-Graff grain size which should be present in T1, M1, M2, M4, M15, and M42 high-speed steel bars at <75, 75–125, and >125 mm diameter after quenching from the conventional hardening temperatures shown.

CARBIDE DISTRIBUTION

The adverse effects of poor carbide distribution have also been referred to by the authors in a previous paper.[1] Among

6 Electroslag remelted T1 material, mid-radius segregation in 125 mm diameter annealed bar   ×100

the consequences of segregation of the carbide phase are non-uniform transformation, non-uniform grain size, and a lowering of toughness. During hardening, carbide-rich areas will give rise to more alloy and carbon solution than in carbide-lean areas. A non-uniform structure is obtained very often with retained austenite associated with the carbide-rich areas. Such structures are undesirable in tools. A non-uniform distribution of the carbide phase can also give rise to local grain growth. In carbide-lean areas, there is little grain boundary restraint during hardening and considerable grain growth can occur. The most serious consequence of large carbide segregates is that they are an embrittling feature. They constitute a favourable crack-propagation route and their presence reduces the energy required for fracture.

The high incidence of poor carbide distribution as indicated in Table I and its associated effects are such that the problem is being vigorously tackled by the tool-steel industry. The nature of the problem is to disperse the segregation which occurs during the solidification of the original ingot. Conventionally, this is accomplished by applying a very heavy reduction in cross-section to the ingot during forging and hot-rolling operations. A deformation of above 95%

5 Electroslag remelted M2 material, mid-radius segregation in 175 mm diameter annealed bar   ×100

7 Electroslag remelted T1 material, mid-radius segregation in 75 mm diameter annealed bar   ×100

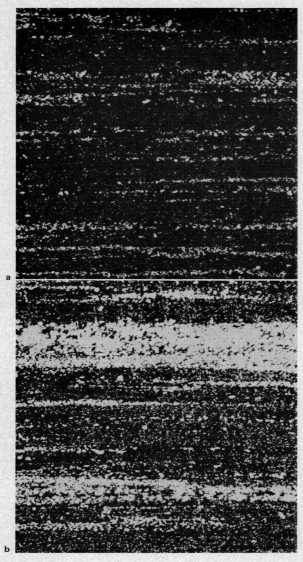

a ingot bottom position; b ingot top position

8 Conventional M2 material, mid-radius segregation in 50 mm diameter annealed bar   × 100

9 Upset-forged M2 material, mid-radius segregation in 110 mm diameter hardened bar   × 100

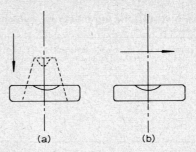

a upsetting direction; b rolling direction

10 Upsetting procedure, first method

is required to achieve the best results. It is for this reason that the best carbide distributions are seen in bars of the smallest diameter. For example, the maximum acceptable levels of segregation within GKN for M2 bars at 25, 50, and 75 mm diameter are shown in Fig. 3. To ensure the best results in larger bars, it is desirable to use larger ingots, and good manufacturers will match the bar stock size to the ingot size. However, by conventional manufacturing routes there is a limit to the degree of carbide dispersion that can be achieved in larger diameter bars. There have been several important developments to attempt to improve matters. These include electroslag remelting and upset forging.

Electroslag remelting

The electroslag remelting process is being increasingly applied to high-speed steel and the product is available to the customer at no extra charge due to the increase in yield obtained. Its success in minimizing carbide segregation lies in the very small freezing zone which exists during remelting. This limits the extent to which segregation in the remelted ingot can occur. There is no doubt that the process is capable of providing material which, after subsequent working, is of excellent quality. This is indicated, for example, by the structure of the 25 mm diameter bar of T1 shown in Fig. 4. However, there is some evidence that very close control over remelting conditions is required to ensure elimination of segregates. Figure 5 shows a 175 mm diameter bar of M2 material. The degree of segregation for this size of bar, which receives comparatively little deformation after remelting, is probably acceptable. The 125 mm diameter bar of T1 shown in Fig. 6 is not so acceptable however. After due allowance for the different character of T1 steel as opposed to M2, marked segregation for this size of bar is evident, suggesting that the consequences of a fairly large freezing zone during remelting have not been dispersed during subsequent deformation. This pattern of events can also be reflected in smaller diameter bars even after relatively large amounts of deformation. The 75 mm diameter bar of T1 shown in Fig. 7 is an example.

A further attractive feature of electroslag remelted material is the consistency of the product right through the ingot. Consistency in bars obtained from different ingot positions within a conventional ingot is not readily obtained. The sort

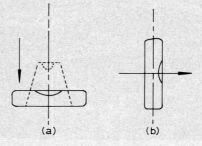

a upsetting direction; b rolling direction

11 Upsetting procedure, second method

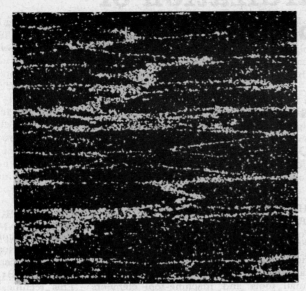

12 Upset-forged M2 material, mid-radius segregation in 200 mm
   diameter annealed bar   ×100

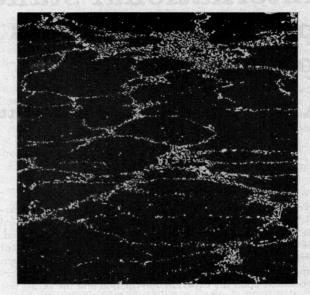

13 Upset-forged M2 material, mid-radius segregation in 200 mm
   diameter annealed bar   ×100

of deterioration in quality from the ingot bottom position
to the ingot top position in 50 mm diameter M2 is shown in
Fig. 8.

Upset forging

By applying a very heavy reduction to an ingot by upsetting
it into a flat bloom, it is possible to break up local carbide
segregates and also to partially disperse uniformly through
the bar the marked segregation which occurs in the ingot top
position. This technique can therefore increase both yield
and quality, and the latter can be very good as evidenced by
the 110 mm diameter bar of M2 shown in Fig. 9. (Figure 9
shows a hardened structure, while all other photomicrographs are of annealed structures.)

There are currently two schools of thought on the best
upsetting procedure. These are illustrated in Figs. 10 and 11.
In the first method (Fig. 10) the ingot is upset and is then
elongated by rolling it at 90° to the upsetting direction. In
the second method (Fig. 11) the ingot is upset and is then
worked along the upsetting axis. The first method is the most
commonly used and probably produces the best quality bar
material. Both methods suffer from the same limitation, however, in that whereas the forged structure can be very good at
positions remote from the original ingot top position, poor
carbide distribution is apt to persist where no real dispersion
of the ingot top segregate has occurred. Figures 12 and 13
illustrate this phenomenon for 200 mm diameter M2 material
upset by the second method. Figure 12 shows an area of very

good carbide distribution, whereas Fig. 13 shows remnants
of very large segregates which are relatively undisturbed.

This concentration of carbide in localized areas of large-
diameter upset-forged material can lead to grain growth in
areas devoid of grain-boundary restraint. Careful control of
heat-treatment conditions is therefore necessary to avoid
grain sizes greater than 8 as measured by the Snyder-Graff
method.

PROGRESS

It is pleasing to be able to record with due acknowledgment
the progress made in improving the quality of high-speed
steel in recent years, particularly in respect of large-diameter
bars. However, as machine tools and cold-forging machines
are called upon to shape more and more exotic materials, so
the demands made on high-speed steel will increase. Having
successfully faced one challenge, the industry must undoubtedly be prepared for a further one.

ACKNOWLEDGMENT

The authors wish to thank Mr F. D. Bridge, Head of Research Unit, GKN Group Technological Centre, for permission to publish this paper.

REFERENCES

1 T. B. SMITH AND N. H. MCBROOM: *Iron and Steel*, Special Issue
  (1968), 56
2 J. P. GILL: *Trans. ASM*, 1936, **24**, 735
3 R. W. SNYDER AND H. F. GRAFF: *Met. Prog.*, 1938, **33**, 377

Table 1 Reference materials:
Steel reference material

Hot-rolled medium-carbon steel corresponding to ISO/TC17 (sec. 692) 972 steel C45 ea

C	Si	Mn	S	P
0.42–0.50%	0.15–0.40%	0.50–0.80%	0.02–0.035%	0.035% max.

Maximum amounts, %
Ni = 0.20
Cr = 0.15
Mo = 0.05
V = 0.02
Cu = 0.20

Nitrogen
The nitrogen content should be 30–80 ppm by weight depending on the steelmaking source

Source
Open hearth or oxygen converter   30–60 ppm
Arc, single slag   40–80 ppm

Cast-iron reference material
ISO/R185 grade 25 hardness 200–220 HB

Table 2 Specification of standard high-speed steel tool material

C	Cr	Mo	V	W
0.8–0.85%	4–4.25%	4.75–5.25%	1.7–2.1%	6–6.5%

Si	Mn	P	S
0.10–0.40%	0.10–0.40%	0.030% Max.	0.030% max.

This corresponds to AISI M2 high-speed steel

medium-carbon steel corresponding to a draft ISO specification as shown in Table 1. Some delegates considered that the carbon content of this steel was too low, giving insufficient cratering effect on cemented-carbide tools. Moreover, it was claimed that in Germany it was difficult now to obtain this material without its having been given a special deoxidization treatment to improve its machinability. This material was, however, strongly supported on the grounds that it had been widely used in the past for machining tests, that it was used by the carbide-tool manufacturers in the USA for testing purposes and by CIRP as its reference material; it was thought that the adoption now of another reference material would result in much valuable data accumulated over the years being lost.

However, the ISO specification for this material was considered too loose for a reference material. A lower maximum sulphur content and tighter tolerances for nickel, chromium, molybdenum, vanadium, and cobalt have been specified as shown in Table 1. Special deoxidizing treatments used for the improvement of the machinability of such steels are to be excluded, and the reference steel must be deoxidized with aluminium with a minimum aluminium content of 0.01%. The nitrogen content is also to be specified as indicated.

Because of strong pleas from individual delegations, it was decided to leave to national standards organizations the specification of a secondary national reference steel material if they wished.

The reference steel is to be used after a normalization heat treatment producing a hardness within the range 180–200 HV, and in test bars having a minimum initial diameter of 100 mm. Because of the sensitivity of cutting-tool life and tool wear rates to the metallurgical condition of the workpiece material, it was considered necessary, whenever materials other than the reference materials were being machined, to specify not only the grade of material, its chemical composition, and physical properties, but to require also the reporting of information on its microstructure and its heat treatment, whenever possible. A proposal to require even greater quantitative information on the metallurgical condition of all work materials used in machining tests, e.g. grain size, inclusion count, was not accepted, though it was conceded that as much information as possible should be provided. Photomicrographs of the work material are to be provided.

The permitted hardness range of each reference material is to be specified and the hardness of each work material is to be determined over the complete cross-section of one end of each test bar. The cutting tests are only to be conducted on each bar within the range of work diameters over which the hardness falls within the specified hardness limits.

STANDARD REFERENCE TOOL MATERIALS

The working group, in considering the choice of standard reference cutting-tool materials, used similar principles to those used in selecting the reference work materials. It was agreed that in all cutting tests in which the tool material was not itself the test variable, or was not itself an important parameter, the investigation should be conducted with the appropriate one of the standard tool materials. Even under these exceptions, however, it was considered desirable that tests should also be conducted on the standard reference material for comparative purposes.

As when specifying a reference work material, it would ideally be desirable to have a single standard tool material for reference and calibration purposes. Such a material would have to be reproducible, readily available, reasonably cheap, provide uniform and constant wear characteristics from sample to sample, and be capable of machining all work materials. Such an ideal is impossible to achieve, and the working group has decided to recommend the specification of three standard tool materials: one high-speed steel and two cemented-carbide materials; of the latter, one should be of the steel-cutting type and the other being for machining cast-iron or non-ferrous materials.

Initially, a high-speed steel of the 18–4–1 type was considered, since this was thought to be widely available and widely used in the past for machining test work. Although its performance as a tool material was now considered to be relatively poor in comparison with more modern tool steels, this appeared not to vitiate its value as a reference material. Although the UK supported this view, it was claimed that in certain countries there was actual difficulty in obtaining 18–4–1 high-speed steels. The working group therefore agreed to adopt the steel specified in Table 2 which was claimed to be more readily available and to have a better performance. A study group was established by the working group to specify the appropriate heat treatment of this material for test purposes.

The working group was aware that substantial differences in the wear characteristics of cemented carbides of nominally the same composition had been observed in machining tests. Moreover, manufacturers of cemented-carbide tool materials were reluctant to provide information on their composition and manufacture. Many members of the working group believed that, even if all the information on the composition and manufacturing methods were available on a particular cemented carbide, it would be difficult, if not impossible, to manufacture samples of this material in the different countries which would prove to have similar wear characteristics. Others pressed the view that it was a bad scientific principle to use in standardization work a reference or standard material, the composition and manufacture of which were unknown; not even an attempt to reproduce the standard material would then be possible.

In order to try and overcome the practical difficulties of obtaining a reference material, the Swedish Standards Association have arranged with the cemented-carbide manufacturers in Sweden for the use of three standard grades of cemented carbide as standard reference materials, these grades being of the type P10, P30, and K20. By random selection of manufacturers and grades, three manufacturers had been asked to make one each of these grades in the form of standard tool inserts. After manufacture, all three types of inserts had been sent to one factory to be finish ground in order to obviate possible differences in grinding technique. These prepared standard inserts are now held in a tool bank by the Swedish Mekanforbund and may be drawn upon by companies and research institutes who wish to conduct machining investigations. Further batches of these materials can be made to correspond with the original batches, closely enough to make no appreciable difference to their wear resistance. Similar arrangements have been used in Sweden to develop high-speed steel inserts for use as standard reference tool materials. Germany had also developed some years ago a standard cemented-carbide material as a standard reference material. This however had been a non-commercial grade of carbide; it had low wear characteristics and was used only for short-time testing purposes. This national tool bank had started operating in 1956 but it had since been discontinued. They had found, like others, that substantial differences occurred in the performance of the standard material obtained from different manufacturers.

In the working group, it was suggested that some of the difficulties of specifying a reference material could be overcome by establishing an international tool bank, possibly operated by ISO/TC29 itself, and making use of the standard reference tool materials now available in Sweden. The objection that the manufacturing details of these materials were unknown still remained, and a number of delegations, including the Swedish delegation, undertook to discover whether or not their respective cemented-carbide manufacturing industries would be prepared to collaborate in producing standard reference materials and disclosing their composition and manufacturing methods in order to enable these to be produced on an international basis. Already, however, certain countries, including the USA, have already been using the Swedish standard cemented-carbide inserts for tool testing purposes.

The working group has now accepted in principle the establishment, at least for international calibration purposes, of an international bank of two or three sintered carbide reference tool materials, and in addition, substandard reference tool materials at national level. All these would, in the long run, be designated without the ISO/R513–1966 designation. However, as much information as possible would be given on each of these reference tool materials: chemical composition, grain size, and manufacturing information. It was to be understood that the international bank would supply the reference tools to any organization or private company which requested them or alternatively to banks at national level according to the preference of individual countries, and to provide assistance to users and national bodies to help them in establishing the substandard reference materials.

Pending the establishment of an international bank, the working group has agreed to make use of existing material banks (presently the Swedish bank, and perhaps the German bank) as the source of sintered-carbide reference tool materials. It wishes to encourage cooperation so that agreement can be reached on reference tool materials for use in specific ranges of application. However, it is to request Working Group 9 of ISO/TC29 actively to pursue the possibility of arriving at better standardization of the details of the existing application grades of sintered carbide defined in R513, for the purpose of eventually making possible the use of these application grades as reference tool materials.

The Swedish delegation has emphasized that Sweden has no wish to provide the material, or to have the control over, an international bank. Sweden would be pleased however to offer the facilities of their standard reference materials to any country wishing to use them.

Clamped insert tools are always to be used under the draft recommendation as reference tools for cemented-carbide materials, because it was claimed by the USA delegation that these had been found more reliable than brazed cutting tools. The working group however, rejected a proposal to specify

Table 3   Standard values for tool-in-hand angles

Cutting-tool material	Normal rake γ deg.	Normal clearance α deg.	Cutting-edge inclination λ deg.	Cutting-edge angle κ deg.	Included angle ε deg.
High-speed steel	25	8	0	75	90
Sintered carbide	+6 −6	5 6	0 −6	75	90
Ceramic	−6	6	−6	75	90

clamped insert high-speed steel tools as the reference tools for this material; the implication of this decision is that such tools should not be used in the standard machining test.

CUTTING-TOOL GEOMETRY

Methods of defining the geometry of cutting tools, both to enable them to be manufactured, e.g. the 'tool-in-hand' angles, and to establish the effective working angles when the cutting tool is set up in the machine tool, e.g. the 'tool-in-use' angles, have been under investigation for a number of years by Working Group 20 of ISO/TC29 on cutting-tool nomenclature. This work has resulted in the realization that the precise specification of cutting-tool geometry requires both new and complex definitions. This working group has now almost completed its first main proposal for an ISO recommendation on this subject, and Working Group 22 on machining tests has agreed that the new cutting-tool nomenclature should be incorporated into its own work. This has been achieved by means of a measure of common membership in the two groups, and will also explain some unfamiliar terms and definitions in the recommendation being proposed by Working Group 20.

The actual choice of values for standard cutting-tool angles in the standard machining test has, however, caused some difficulties due to significantly different practices having grown up in the respective countries and been established in the national standards. Moreover, with the advent of throwaway tooling, a number of the cutting-tool angles were in effect already standardized through the work of Working Group 9 of TC29. Though seriously tempted to accept a number of alternative standard values in the ISO recommendation, to enable certain countries to continue to use their own national standards, the working group has eventually decided to take into account the balance of usage in the various countries and to establish only one set of standard cutting-tool angles in their draft recommendation. Only in the case of unusually difficult-to-machine alloys, where it is impossible to utilize the standard tool geometry, are departures from the standard geometry to be permitted.

The standard tool-in-hand angles provisionally agreed by the working group are shown in Table 3, wherein it will be noticed that the angles specified for rake and clearance are the 'normal rake' and the 'normal clearance', though further consultation with Working Group 9 on this point is to be undertaken. The setting of the cutting tool in the machine for the standard test, and the permitted tolerance affecting this, are to be specified and defined as shown in the Appendix.

STANDARD CUTTING CONDITIONS

In selecting the standard cutting conditions for the test, there was on the one hand a desire to reduce the number of conditions, as regards feed and depth of cut, used in the standard machining test to a minimum, in common with the general philosophy of the working group to quote a single standard reference condition for each parameter; on the other hand, there was a need to ensure that the machining test would provide results representative of machining practice. Proposals before the working group therefore ranged from a proposal to have only one set of cutting conditions for the standard test, to a proposal for a very wide and extensive range of

Comparative assessment testing of cutting tools

A. Moore

Some ten years ago it became the policy at Hawker Siddeley Aviation (Manchester) to have all cutting tools coming into the factory submitted for quality tests by technical personnel. The management had been aware that the original procedure of leaving tool selection to the discretion of either buyers or shop floor supervisors had many imperfections. In particular, the former personnel were often motivated to purchase on cost alone without due consideration of technical merits, while the shop floor supervisors were often swayed by the personality of the salesman or opinion of the operators who were, in general, reluctant to change tools with which they were familiar. The answer to this dilemma came with the setting up of a machinability laboratory within the engineering research department. This was a facility equipped with standard and special machine tools under the control of technically trained labour. The management therefore instructed that the laboratory be made responsible for the development of standard testing techniques for cutting tools, and thereafter the assessing of the quality of such tools coming into the factory.

DEVELOPMENT OF STANDARD TEST PROCEDURES

Our first move was to examine work already done in this field, and naturally the British Standards on cutting tools were reviewed. Unfortunately this was unrewarding, since in only two standards were test procedures suggested, namely BS 1919:1967 on hacksaw blades and BS328:1959 on HSS twist drills. Subsequent use of these methods showed that of the two, only the hacksaw blade test procedures was severe enough to enable comparative assessments to be made. In the absence of suitable specifications in the British Standards, it was decided to go ahead and develop our own test procedures.

During this development phase many methods were assessed and discarded. Initially, accelerated test procedures were used, the basic aim being to reduce the time required to gather comparative data; these tests took one of three forms: high cutting speed tests, i.e. using a cutting speed far above that recommended for the particular tool material/work material combination; accelerated wear tests, whereby test material renowned for its abrasive-wear properties was selected to induce rapid flank wear, and finally, mathematically extrapolated results, i.e. the test was run until a number of points had been plotted on the tool-wear graph. The curve was then extrapolated by means of the method of least squares in order to obtain the figure for the surface area machined at the tool-life endpoint.

Each of these techniques was shown to have severe limitations. The performance of tools at artificially high cutting speeds bore no relationship to their performance at the optimum speeds. Similarly, the data obtained on the abrasive work materials was applicable to these materials only. With mathematical extrapolation there was always the problem of not knowing where the slope of the wear curve was liable to change.

In addition to these test methods, comparative tests were also made using cutting force measurement as the basis for comparison; this method was eventually discarded when it was shown that the magnitude of the forces developed during cutting did not necessarily bear any relationship to the tool life achieved.

Stemming from this experience we have now one single premise when considering standard test procedures and that is to simulate as closely as possible during the laboratory trials the cutting conditions under which the tool is to be used on production. The material or materials upon which the tools are to be used are selected as test material and assessment of performance is based principally on tool life, although surface finish produced is also considered when applicable.

STANDARD TEST PROCEDURES
Single-point lathe tools

The machinability laboratory is equipped with a 18×48in centre lathe driven by a 25 hp variable-speed motor. All initial turning tests are done on this machine. A 4–6in diameter bar of the test material is obtained when possible, about 2 ft long. The bar is centred at one end and mounted between a three-jaw chuck and the tailstock rotating centre. If the bar is in the rough-forged state then the surface layer is removed before commencement of tests. To enable sufficient points to be recorded on the tool-wear curve, the bar is divided into sections each about 100 in² with the aid of a $\frac{1}{8}$in wide parting-off tool.

The tools under test are ground, using standard equipment, to the geometry required. All tool angles are measured and recorded. For test purposes the tools are mounted in a quick-change tool post, care being taken to ensure that the unsupported length is identical for all tools under test.

A test consists of turning a section of the bar. The quick-change tool post is then removed from the machine and the wear measured with a travelling microscope as shown in Fig.1. This procedure is repeated until the tool-life end point is reached. The figure used for the end point varies from 0·015–0·030in max. wear land, the former figure generally being used on regrindable tools while the latter is adopted for throwaway-insert tooling.

All data is recorded on a results sheet, an example of which is shown in Table 1, provision is also made for plotting the tool-wear curve.

FACE-MILLING AND END-MILLING TESTS

All face- and end-milling tests are carried out on a 15 hp vertical milling machine. The size of test block is selected so that one pass would give a suitable increment in relation to the

The requirement of standard test procedures for the comparative assessment of cutting tools is discussed. The development of these test procedures at Hawker Siddeley (Manchester) is described. The methods adopted for testing turning tools, milling cutters, drills, reamers, taps, hacksaw blades, and engineers' files are outlined. The paper is concluded with examples of the use of standard tool-testing procedures. These examples include comparative testing of similar tools from various manufacturers to establish the company supplier and the comparative assessment of several grades of high-speed steel.

621.91.025:539.538

The author is with Hawker Siddeley Aviation Ltd (Manchester)

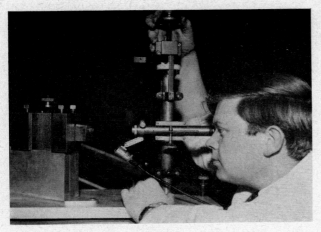

1 **Travelling microscope for measuring wear**

tool wear–surface area machined graph. With face mills the width of test block selected is made equal to $\frac{2}{3}$ of the cutter diameter. A test would then consist of a number of passes across the length of the block at the selected depth of cut. With end milling the type of cut can take one of three forms: a pocketing cut, i.e. when the radial depth of cut is equal to the diameter of the cutter; a profiling cut, where high-axial and low-radial depth of cut are employed, and a combination of both pocketing and profiling, see Fig.2. The type of cut used during a test will be related to that to which the cutter will be employed on production.

When the test cutters are *in situ* in the machine both axial and radial run-out of the teeth are measured; if either are found to exceed 0·001in adjustments are made until this specification is met.

To obtain a face upon which flank wear will develop, all cutters received for test are ground with a 45° corner angle. To measure the wear the cutter, complete with stub arbor or chuck, is released from the machine spindle. This assembly is then fitted into a universal workhead which is mounted on a table in conjunction with a travelling microscope. The

arrangement is shown in Fig.3. The tool-life end point used for these tests is generally not less than 0·020in max. wear land.

DRILLING, REAMING, AND TAPPING TESTS

General testing of drills, reamers, and taps is conducted on a 7 hp vertical drilling machine. The machine is fitted with a 24 × 12in work table capable of longitudinal and transverse movement under micrometer-indexed lead screws, this enables accurate pitching and alignment of holes to be achieved.

The test block used is a piece of actual production material upon which the test tools are intended for use. The faces of the block are milled parallel to each other and one face is then ground to reduce the risk of drill-point wander.

In general, drills and reamers are tested by producing through-holes equal in depth to 3 times the tool diameter. The production of blind holes $2D$ deep are used for tap performance testing.

The drill diameters selected for test are, when possible, reaming or tapping drill sizes; thus, the holes produced can be used on subsequent reaming or tapping tests.

All tools are reground before testing commences. For drills the point geometry is accurately produced, with reamers the chamfer lead is reground while the taper lead is accurately refurbished on taps.

A test will consist of producing holes until the tool-life end point is reached; this, in the case of drills and taps, is when there are visual or audible signs of cutting-edge failure, while with reamers it is when the wear land across the chamfer face exceeds 0·015in. All tools are removed periodically during the test for visual examination or in the case of reamers measurement of the wear land.

At the completion of the tests all holes produced are gauged with the appropriate plug gauges to ensure that the specified tolerances have been achieved.

HACKSAW BLADE TESTING

As previously stated, the test procedure adopted for testing hacksaw blades is that suggested in BS1919:1959. The test machine has a 6in stroke, and the cutting arm reciprocates at 60 cutting strokes/min.

Table 1 Test data record sheet

						Turning					
Material: Condition: Hardness: Initial bar size:						Tool no.: Front top rake: Front clearance: Plan approach: Nose radius:			Grade: Side top rake: Side clearance: Plan trail:		
Machine:						Cutting fluid:					
Pass no.	Speed, ft/min	rev/min	Feed, in/rev	Depth of cut, in	Bar dia., in	Length of pass, in	Surface area, in²	Flank wear, Average	Max.	Surface finish, μin CLA	
1											
2											
3											
4											
5											
6											
7											
8											
9											
10											
11											
12											
13											
14											
15											
16											
17											
18											
19											
20											

Test by: Date:

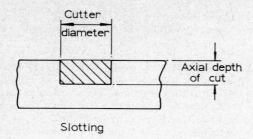

Slotting

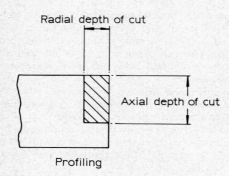

Profiling

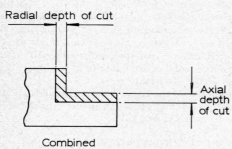

Combined

2   Types of cut used with end mills

The machine has a gravity feed which is controlled by loading the reciprocating arm so that, when it is at mid-stroke with the blade in the horizontal position, the total weight of the reciprocating arm is 12 lb. The test material, a medium-carbon steel heat treated to a specific hardness, is secured in a vice below the blade. The blades under test are made to cut a number of slices off the test material as continuously as possible; the slices being approximately 0·1in thick.

The time taken to cut each slice is recorded on a stroke counter. From these figures a graph is prepared from which wear rates can be compared.

FILE TESTING

The machine is of our own design. It is driven by a ½ hp motor at a rate of 60 strokes/min. The design is such that the file under test cuts only on the forward stroke when a constant load of 13 lbf is applied. The value of 13 lbf was selected from tests in which a number of fitters of various weights, ages, heights, etc. had filed a test block clamped to a force-measuring dynamometer. The downward thrust of each operator was measured and the average was found to be 13 lbf. On the backward stroke, the load is removed from the file by means of an arm controlled by a cam as shown in Fig.4.

The file operates in the horizontal plane and the length of stroke is 7in. The testpiece is weighed before being firmly clamped in a vice bolted to the table of the machine.

A test consists of cutting for 15 min after which the testpiece is removed and reweighed in order to establish the metal removal rate. The file is examined and cleaned with a file card before the above procedure is repeated. The test is

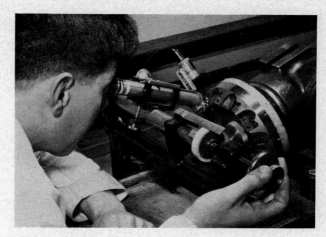

3   Universal workhead assembly mounted with travelling microscope

generally run for 5 h, by which time sufficient data has been accrued from which a comparative assessment can be made.

APPLICATIONS OF STANDARD TEST PROCEDURES

Having established standard test procedures it is possible to apply them to three general fields of testing. First, and primarily the function for which the tests were developed, they can be used for obtaining comparative data to enable a sound purchasing policy to be established for standard cutting tools. Second, having recorded performance data on the standard tooling in use within a factory, the tests can be used as a check to ensure that quality is being maintained. Finally, the procedures lend themselves readily to tool material selection testing. To illustrate these various functions of standard tool testing, a number of examples have been taken from the records at Hawker Siddeley (Manchester) and are detailed below.

Standard tool testing

To illustrate the procedure used, two examples have been taken. The first was to establish a supplier of HSS hacksaw blades, and the second to check the quality of an existing supplier's drills and to compare the performance obtained with alternative suppliers.

Hacksaw blade selection

The object of the investigations was to compare the performance of HSS hacksaw blades from four manufacturers, each of which had agreed terms with our purchasing department. The blade most commonly used in the factory is a 12in, 18 teeth/in, all-hard HSS type, therefore sample blades were obtained of each brand from stockists. The test procedure followed that suggested in BS1919:1967 and the blades were tested alternatively. To comply with the BS test a 18 teeth/in HSS blade must be capable of cutting off 36 pieces of the

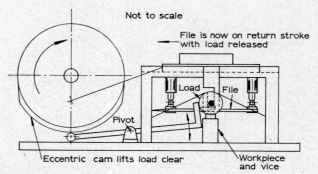

4   File-testing machine

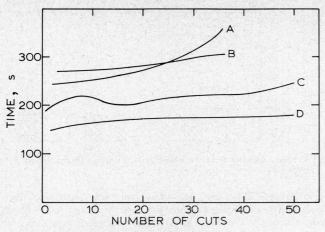

5   Tool-wear curves for hacksaw blades

Table 2   HSS tool bit compositions

| HSS grade | Identifying elements, % | | | | | |
	C	W	Mo	Cr	V	Co
T1	0·7	18	. .	4	1	3
T15	1·5	12·5	. .	4·75	5	5
A	0·8	18·25	0·75	4·8	1·25	5
M15	1·5	6·5	3	4·75	5	5
T6	0·82	21·75	. .	4·4	1·5	12
B	0·75	22	0·85	5	1	5·5
M42	1·07	1·5	9·5	3·75	1·2	8

test material while the maximum time for the final section should not exceed 7 min. It is considered that this figure does not represent a severe enough test and that a blade capable of just meeting this specification would be of poor quality.

The results of the tests in question are shown in Fig.5, where it will be seen that all blades easily passed the BS specified test. However, if the slopes of the curves are examined it can be noted that there is considerable differences in performance. Blade A for example wore very rapidly almost from the start of the test while blade B had a higher wear rate than either C or D. Although the wear rates of C and D were basically similar up to 40 cuts, it can be seen that after this point, blade C began to wear more rapidly. The differences in the initial cutting times were attributed to the ability of the blade material to keep a keen cutting edge. The result of this investigation was that the manufacturers of blade D were selected as suppliers.

TESTING OF HSS TWIST DRILLS

As a result of a number of premature drill failures on production, a comparative assessment was made of taper shank standard M2 HSS drills between our supplier at that time and two other well known small tool manufacturers. Since the failures had been occurring on 2½Ni–Cr–Mo steel, BS99, it was selected as the test material. The drill size used was 31/64in diameter and the holes produced were subsequently used for ½in diameter reamer tests. The speed and feed used had previously been established in laboratory trials.

Through-holes 1½in deep were drilled and the test on each drill was terminated when either:
  (i) the drill had completed 50 holes or
  (ii) drill failure occurred.
Each drill was removed periodically during the test and the point examined for wear.

The results of the tests are shown in Fig.6, where it will be seen that drills produced by manufacturer A had an average life of less than 10 holes while the average life of drills produced by the other two manufacturers was about 40.

The company had been purchasing drills from manufacturer A for some time on the basis that they were slightly cheaper than other brands; this was, therefore, a typical case of what on the surface appeared to be a sound buying policy, saving the company a few hundred pounds per year on capital expenditure, was in fact costing the company several thousand pounds per year in production breakdowns.

QUALITY CONTROL CHECKS

During the early 1960's there had been several suggestions from shop-floor supervision that the quality of small cutting tools varied and that production time on the factory floor often suffered because of this. The production management therefore proposed that an investigation be instigated to determine the consistency of the quality of small cutting tools by carrying out quality checks on tools received over a period of twelve months. A list was therefore prepared of small tools purchased during the preceding two years, and it was shown that the most commonly used tools were as follows: hacksaw blades, some 40000 annually; engineers files, 6500 annually; no.7 single-point HSS lathe tools, 2000 annually; and HSS twist drills from ⅛in–⅜in diameter, 27000 annually.

It was therefore proposed that the investigation be directed at these five types of tools and that a random sample of 3%, or 3 items on batches of less than 100, should be taken from each batch received. Test specifications were drawn up for each type of tool tested. The specification covered geometrical checks to the appropriate British Standard, cutting performance tests to standards previously set in the laboratory, and

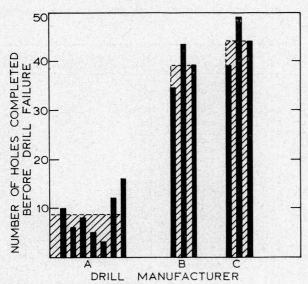

6   Drill life achieved on S99 steel

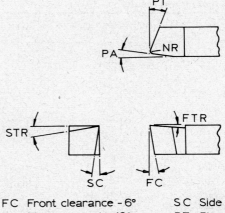

FC Front clearance - 6°   SC Side clearance - 6°
PA Plan approach - 10°   PT Plan trail - 5°
FTR Front top rake - 8°   STR Side top rake - 10°
NR Nose radius - 0·015in

7   Single-point tool geometry

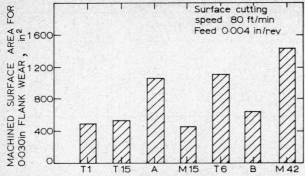

8   Average tool life of various HSS grades when turning S96 steel
     at standard feed rates

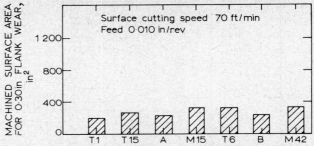

9   Average tool life of various HSS grades when turning S96 at
     high-feed rates

physical tests which consisted primarily of hardness tests to determine the heat-treatment control.

The result of this programme, during which some 500 small tools were tested at a cost of £1 500 to the company, was the recording of the failure of three batches of 18 teeth/in HSS hacksaw blades to meet the required cutting performance specifications. All items tested met the dimensional and physical requirement of the BS specifications. It was therefore concluded that the general quality of small tools was of a high standard and therefore did not warrant the level of testing made during the programme. Thereafter, quality checking was limited to tools receiving adverse comment from shop-floor personnel.

SELECTION OF TOOL MATERIAL

The work under this context has been basically concentrated on two types of tools, namely ½in square tool bits and end mills. The tool bits are used in substantial quantities on the automatic lathes at Manchester.

It was therefore decided to test a number of grades of HSS to determine the grade most suitable for use in this particular application.

The programme was divided into three stages: laboratory trials, shop-floor trials with uncontrolled geometry, and shop-floor trials with controlled geometry. The laboratory tests compared the seven grades of HSS shown in Table 2 and the tools were prepared as indicated in Fig.7. The tool angles used had been determined previously as optimum for S96 a 55–65 tons/in² Ni–Cr–Mo steel. This material had been selected for the test since it accounted for some 70% of the components machined on the automatics. Plain finish turning tests were conducted on the centre lathe in the laboratory. A test consisted of setting the tool bit on centre relative to the

spindle axis. The test bar which was 5in diameter×18in long had been sectioned into three 5in lengths. A depth of cut of 0·025in was taken and the bar machined in a series of 5in passes until 0·030in max. flank wear on the tool was obtained.

The results of the tests are set out in block form in Figs.8 and 9; each result shown is an average of at least two tests. If a significant difference in tool life was shown after two tests then a third test was made and the average of the three taken as the final result.

The results in Fig.8 represent the tool life achieved at generally recommended conditions of speed and feed. Figure 9 illustrates the effect of greatly increasing the feed rate; this was recommended by the manufacturer of the M15 HSS who claimed it gave optimum performance at this higher feed condition.

From these results it was decided that the three grades of HSS giving the highest performance under the generally recommended conditions should be submitted for shop floor trials. A fourth grade, the M15 HSS was also eventually included in the shop-floor trials, since the foreman in charge of the automatic section stated that in his experience this particular grade gave the best performance under the conditions which prevailed on the auto machines.

The procedure adopted for the first phase of the shop-floor trials was as follows: a suitable long running component in S96 steel was selected for the trials. The component was a bolt, and the test cut consisted of turning the shank diameter down to ⁵⁄₁₆in from ½in A/F hexagon bar. The tools were ground by the autosetter on an off-hand wheel. He was instructed to keep the tool geometries identical. The setting up and operation of the machine were carried out by the autosetter, while a technician observed and recorded the results.

Each test was concluded at the operator's discretion, i.e. when, in his opinion, surface finish was below that required,

Table 3   Phase 2 tool bit test results

HSS grade	Front to back rake, deg.	Side rake, deg.	Incl'd angle of 'V' deg.	Land width, in	Angle to groove to front, deg.	Tool wear, in	Machined surface area, in²	Aver. machined surface area, in²
T6	17	11	120	..	47	0·0228	74	200·6
T6	..	13	115	..	50	0·082	187	
T6	..	5	119	..	30	0·0688	273·8	
T6	40	9	102	0·004	48	0·0876	267·6	
M15	27	7	110	..	40	0·0313	107	169·6
M15	..	9	110	..	34	0·0655	181·4	
M15	Not recorded					0·0418	140·8	
M15	38	10	108	0·007	33	0·0586	247	
M15	52	3	113	0·0025	23	0·0218	172*	
A	32	10	119	0·007	26	0·074	117·7	154·7
A	Not recorded						122	
A	24	..	107		44	0·0535	216	
A	Not recorded						129·7	
A	25	8	111	..	43	..	175	
A	31	8	123	0·010	52	0·0426	168	
M42	Not recorded						203	210·1
M42	Not recorded						211·8	
M42	24	10	97	..	40	0·0778	162·5	
M42	39	8	107	0·008	35	0·083	242·2	
M42	28	10	108	0·007	47	0·065	212·3	
M42	54	14	118	0·0025	23	0·0245	219	

*Test incomplete

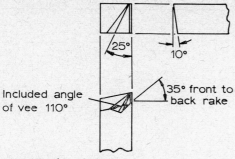

10 **Chip-stream tool geometry**

Included angle of vee 110°

35° front to back rake

Table 4 End mill HSS grades

HSS grade	Elements, % C	W	Mo	Cr	V	Co
T1	0·7	18	..	4	1	..
M15	1·5	6·5	3	4·75	5	5
T15	1·5	12·5	..	4·75	5	5
T4	0·75	18	..	4	1	5

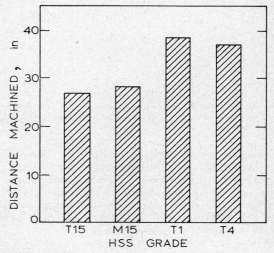

12 **End mill results**

or when dimensional stability of the component ceased. The number of components machined, tool wear, and the tool geometry were then noted by a technician.

The results of these tests, shown in Table 3, in general agreed with the laboratory trials, although the difference in performance was by no means as pronounced, and also the M15 tools performed substantially better on the autos than when used on the laboratory centre lathe. However, the most significant facts which emerged from these tests were:

   (i) the considerable scatter in tool life obtained for similar tools

   (ii) the enormous variation in tool geometry which was attributed to the use of off-hand grinding techniques.

The type of geometry employed for auto turning is shown in Fig.10; this is generally known as a chip-stream tool and is used in conjunction with a roller box. As can be seen, it is a complex geometry which is virtually impossible to repeat identically by off-hand grinding.

It was therefore decided to design and manufacture grinding fixtures in order to produce the tool geometry consistently. This was done and, after optimizing of the tool angles, a final set of tests were made. The results shown in Fig.11 indicated that the T6 grade of HSS gave the highest tool life of the three grades tested. The M15 and M42 grades of HSS gave similar results about 25% lower than the T6 grade. Grade A was not included in the final tests because of its relatively poor performance during phase 2 of the programme.

EVALUATION OF HSS GRADES FOR END MILLS

A second example of the use of standard tool testing for HSS grade selection concerned end mills. The company had begun using NC milling machines and, in order to optimize on metal removal rates, a series of grade selection tests were commenced. The work material in question was again the Ni–Cr–Mo steels, and for these tests the high-strength alloy BS99 (80–90 tons/in² UTS) was selected.

In the first set of tests four grades of 1in diameter end mills were assessed as shown in Table 4. Two types of cut were taken: a slotting cut where the axial depth of cut was made equal to ¼ of the cutter diameter, and a profiling cut, of ⅜in axial and ¼in radial depth of cut. The slotting was done at one set of conditions while with the profiling cut, two feed rates were used and both conventional and climb milling were attempted.

The slotting results shown in Fig.12 were inconclusive. It was considered that the reason for this was premature tool failure caused by the cutting speed of 47 ft/min being too high. The remaining profiling tests were therefore made at 38 ft/min. The results of this test, shown in Fig.13, clearly indicated that the T1 grade of HSS gave the superior performance on S99 steel. It was also shown that higher tool life can be expected when the climb-milling technique is used.

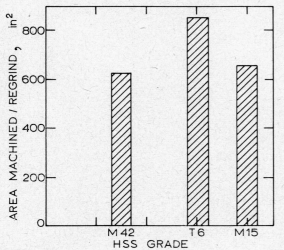

11 **HSS tool bit results**

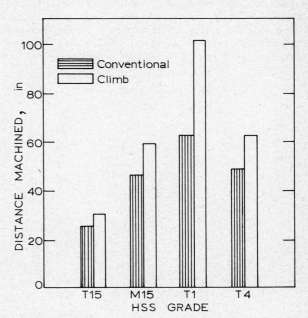

13 **End mill results**

Table 5 Effect of speed and feed on tool life

Approx. feeds, in/ton	Speed, ft/min 39	49	72	90	110	140	
0·0015						3·5in/min Chip burning	Optimum conditions obtainable
0·002							
0·0025					4·5in/min Chip burning		
0·003				4·5in/min Chip burning			
0·004							Optimum conditions not obtainable
0·0045			6in/min Low tool life of 120in travel				
0·0055							
0·0080		7in/min	Poor finish low tool life				
0·010	7in/min	Chips slightly burnt, table vibrations					
0·013	Small table vibrations						
0·016	Chips slightly burnt						

Material: S99 hand T (380–435 HV); cutter: 1¼in diameter M42; depth of cut: ⅛in radius; length of cut ⅝–1in; cutting fluid: synthetic soluble oil at 20:1 dilution

Subsequent to these tests the company used T1 end mills for cutting the Ni–Cr–Mo steels until earlier last year tests were made on a further grade of HSS, namely M42. The method adopted for this assessment differed slightly from the earlier tests in that accelerated testing was used initially to establish optimum conditions of feed and speed. Thereafter, the standard life test was adhered to.

The tests were made with a particular S99 component in mind in which profiling only was to be used. A pass at 0·125in radial depths of cut and 1·8in axial depth of cut was demanded. The following test parameter was required: the highest metal removal rate to give a tool life of 200in length of cut. The tool-life end point was taken as 0·030in max. wear land.

The accelerated testing method adopted to determine the maximum metal removal rates was as follows: an estimated optimum speed was selected and the test commenced at the desired depth of cut at a low feed rate. As the cutter progressed along its length of cut, the feed rate was gradually stepped up an increment at a time until discoloration of the chips was observed. The increment below this feed rate was taken as the optimum feed value at that particular speed. This procedure was repeated for various speeds and from the accrued results the speed/feed combination giving the highest metal removal rate was selected. A tool-life test was then made at these conditions. If a tool life of less than 200in was achieved then further tests were made at lower speed/feed combinations until the desired tool life was obtained.

The results of the tests are shown in Table 5 where it can be seen that the conditions determined for the maximum metal removal rate were 49 ft/min and 7in/min feed rate. Higher feed rates were attainable at lower speeds but ma-chine-tool vibrations and component movement were experienced. These figures with M42 HSS end mills represent a metal removal rate four times higher than had previously been achieved with T1 HSS.

CONCLUSION

The ideal procedure for comparative tool assessment is to have controlled testing under production conditions. Unfortunately, this type of testing can only be successfully employed in establishments where long batch runs are the rule. Generally, in the airframe manufacturing industry long running components are the exception rather than the rule; therefore, reliable laboratory testing is required. It is the opinion of the author that if testing is done along the lines described in this paper the results can be confidently used to establish a sound purchasing policy for small tools. Other benefits which will accrue from this type of work are the recording of accurate data on metal removal rates, tool life, and surface finish which will prove invaluable to rate fixers, estimators, and production planners. There is, however, one pitfall that must be avoided and that is the tendency to read more into the results of a test than can be substantiated. It must always be borne in mind that the results obtained refer only to the type of tool tested on the particular test material selected for the conditions under which the test was made.

ACKNOWLEDGMENTS

The author wishes to express his thanks to the directors of Hawker Siddeley (Manchester) for permission to publish the paper and to his associates in the Engineering Research Department for their assistance in preparing the paper.

High-speed steels for cutting aero-engine materials

D. Milwain and J. M. Thompson

This paper deals with the integration of metallurgical principles into the system of selection and control of high-speed steel cutting tools, within the Aero Engine and Bristol Engine Divisions of Rolls-Royce Ltd.

The historical development dates back to the period before the interests of the two companies were merged, but comparison of the lines of development has shown that experiences were remarkably similar and where possible these have been combined. Although there was considerable investigation into both the choice of material identification systems and methods of hardness testing, this was only completed recently, but in order to present a clearer impression the AISI/BSI notation and Vickers units have been used throughout.

HISTORY

As long ago as the early 1960s it was realized that the high-speed steels then most favoured commercially (T1, T4, and T6) were inadequate for machining the Ni and Co base alloys used in gas-turbine production.

Plain turning and some milling operations have been investigated over the years by engineering departments and 'throwaway' cemented-carbide tips largely standardized. However, the position of special tools produced in high-speed steels presented a slightly different challenge, in that the properties of the cutting material could be varied much more in respect of composition and thermal treatments.

Initial laboratory involvement was through injuries caused to the operator by tool disintegration. This usually led to an investigation by the Factory Safety Officer involving a laboratory report on the tool to determine Company liability.

In the area of the company which is now the Bristol Engine Division this led to a combined metallurgical and purchase department approach to suppliers. It was possible only to invoke 'common law' rights regarding grossly bad defects, such as extreme central segregation, cracks, or inclusions. This was later enlarged, when new alloys were introduced, to three single-page specifications, and was superseded by the new Rolls-Royce Ltd commercial supply specifications for high-speed steels.

The process of involvement did not proceed as quickly in the Aero Engine Division and did not really get under way until the high-C high-V alloy T15 was offered by manufacturers as an alloy capable of increased hardness (900 HV) over existing alloys.

In both divisions tools were ordered as a result of manufacturers claims, and although initial troubles were experienced due to faulty heat treatment and/or softening due to grinding, the material became for a time the preferred tool material for cutting Ni base alloys. Later a similar alloy, M15, was introduced which also proved successful.

The biggest drawback with the two alloys, T15 and M15, was that grinding of the accurate or fine forms required for some broaches was extremely difficult because of the large wear-resistant carbide content (high in V). Despite this, large numbers of side and face cutters were produced and in the area of the company that is now the Bristol Engine Division, these were covered by customer–manufacturer agreements.

A further new alloy (T42) was offered and taken up. This was similar to M15 and T15 with a lower C and V content but a higher Co content. Performance on milling Nimonic alloy 105 turbine disks with T42 was of the order of 6–7 times that of T1 under identical conditions. This was superior to T15 and at least as good as M15 (but with slightly better grindability). The performance of T42 on broaching was disappointing due to a lack of toughness at >900 HV.

A comparison of T1 and M15/T15 for broaching showed that while T1 gave a reasonable performance at 850 HV and above, on the occasions when the hardness dropped to 800 HV or less, the performance was negligible. While the materials M15 and T15 gave the desired improvement at hardness values of 850 HV and above, their low grindability made both manufacture and maintenance an expensive business. The problem was so acute that some intricate forms were found impossible to produce.

This difficulty was not solved until the arrival of M42 which was put forward as having the performance of T15, M15, and T42 with the grindability of T1. It was certainly capable of very high hardness and in many cases, high performance, but until the heat treatment was thoroughly investigated (over a period of two years) it was troubled by chipping problems. A number of heat treatments have now been defined in order to cover different applications and these are recorded in the Rolls-Royce specification CSS10/M42. The specification quoted and the corresponding ones for M2, T1, M15, and T15 were not raised until after the two companies merged and represent the combined experience of the Aero Engine and Bristol Engine Divisions. This was derived from failure investigation, laboratory and production trials, and meetings with tool and tool-material manufacturers and other tool users.

MATERIAL QUALITY REQUIREMENTS

The main raw material controls found to be necessary were identification (AISI system adopted), standards for composition ranges, central segregation, carbide distribution, material cleanliness, decarburization, and heat-treatment response. These requirements were considered in part with an informal tool user group and the joint experience is embodied in the Rolls-Royce Ltd commercial supply specification

The historical changes are described in the high-speed steel alloys selected within the Aero Engine Division and what is now the Bristol Engine Division of Rolls-Royce Ltd. The material quality requirements for high-speed steel cutting materials are indicated, and the Rolls-Royce specifications covering both material quality and heat treatment are introduced. The opinion is expressed that heat treatments and hardness values quoted by some manufacturers have not been sufficiently accurate for critical applications and that specific heat treatments, with tight hardness ranges, are required. The currently available alloys and their properties are reviewed and suggestions made for consolidating usage on a smaller number of alloys. The relationship of the metallurgical laboratories to the other departments involved in selection are briefly stated.

621.438:669.018.44:621.91.025:669.14.018.252.3

The authors are with Rolls-Royce Ltd

CSS10 for high-speed steels and the supplements CSS10/M2, T1, T4, M15, T15, and M42.

The principal faults eliminated by application of the specifications were poor tool performance due to low hardness values, and prevention of cracking at the roots of teeth or other highly stressed areas in tools manufactured in poor-quality material.

HEAT TREATMENT

From the multitude of different recommendations issued by manufacturers, specific data has been obtained by fixing hardening times and temperatures for salt-bath treatments and tempering times and number of tempers. The hardening temperatures were obtained by using standard-sized samples (now $\frac{1}{2} \times \frac{1}{2} \times 1\frac{1}{2}$ in approx.), standard times which took account of possible decarburization, and a number of trials in the region of the local melting temperature. Where possible, the specified hardening temperature was fixed at 20–30°C below the local melting temperature of high-quality material. The latter was then specified as the 'overheating temperature'.

Tempering temperatures were then obtained using the standard hardening temperature and various tempering temperatures above and below the temperature recognized to produce maximum hardness. Usually hardness fall-off (with salt-hardened tools) for the first 10–15°C above peak hardness is fairly slow, and because toughness and stability were always at a premium, most minimum tempering temperatures were fixed at slightly above that required to produce peak hardness.

The number of tempers required was fixed by metallographic examination of samples having various numbers of tempers and from long-term results of crack testing hardened and tempered tools.

It was also found necessary to define a number of lower hardness ranges with some alloys to obtain increased toughness. Because these were for metal-cutting tools, the lower hardness was obtained by overtempering whereas, if the requirement had been for cold forming or similar applications, lower hardening temperatures would have been used.

HARDNESS TESTING

During investigation of both heat treatments and production problems, many discrepancies were found between hardness values quoted. Of the two suitable scales, HRC and HV, the former was the most widely used but was subject to variations of ± 1 HRC on a given test sample (there were even larger variations between readings obtained on different machines). This was partly due to the extreme accuracy required of the mechanical depth gauge head (i.e. 1 HRC represents a difference in depth of 0·002 mm [0·00008in]).

The variations obtained could not be resolved nor could a HRC–HV conversion chart be produced, so the HV scale was adopted as standard with hardness ranges of 50 HV (less than 2 HRC).

REVIEW OF COMPOSITIONS AND PROPERTIES

After an initial period of assessment it is now felt that the high-speed steels commonly available can be arranged in groups in order to reduce the problems of selection and make easier the determination of development requirements. The groups which seem logical are given in the Table 1, which also gives nominal compositions.

Group 1

The basic material in this group for many years has been T1 but, because of material cost, work has been carried out on many applications to replace it with either M2 or M1. Tools manufactured in M2 are tougher for the same hardness and appear to perform as well as tools in T1. The biggest problem appears to be that of mass producing tools in M2 where above normal amounts of grinding are required. The increased V content certainly increases grinding difficulties.

Table 1 Groups and nominal compositions of commonly available high-speed steels

Identity	C, %	W, %	Mo, %	V, %	Co, %	Cr, %
Group 1						
M1	0·80	1·5	8·5	1·2		4·0
M2	0·85	6·3	5·0	1·9		4·0
M2A	0·90	6·3	5·0	1·9		4·0
T1	0·75	18·0		1·2		4·0
Group 2						
M3–1	1·05	6·0	6·0	2·2		4·0
M3–2	1·15	6·0	6·0	3·0		4·0
M4	1·30	6·0	4·5	4·0		4·5
T14–4–4	1·25	13·5		4·0		4·5
Group 3						
M15	1·55	6·5	3·0	5·0	5·0	4·5
T15	1·55	12·5		5·0	5·0	4·5
Group 4						
M33	0·9	1·5	9·5	1·2	8·0	3·8
M42	1·07	1·5	9·5	1·2	8·0	3·8
M34	0·9	1·5	8·5	1·9	8·0	3·8
M36A	1·0	6·3	5·0	1·9	8·0	4·0
M35	0·85	6·3	5·0	1·9	5·0	4·0
T4	0·75	18·0		1·2	5·0	4·0
Group 5						
M44	1·15	5·0	6·3	2·3	12·0	4·2
T42	1·25	10·0	3·0	3·0	10·0	4·2
T6	0·80	20·0		1·5	12·0	4·2

M1 has been adopted by some manufacturers for tools like taps but no conclusive evidence is available as yet that M1 gives as good a performance at the same hardness as T1.

Assuming that the price of W will remain sufficiently high to give the advantage to the Mo containing high-speed steels the development of a basic high-speed steel containing only 1–1·3V to replace both M1 and M2 would reduce the selection problems.

From the gas-turbine industries' point of view, the most important application is on broaches for Ni base alloys where a minimum hardness of 850 HV is required and this is being pursued using M2 containing higher than usual C content (0·85–0·92%) i.e. M2A. This is a viable proposition because broaches are usually ordered in relatively small quantities and thus a poorer grindability can be tolerated.

However, a basic high-speed steel containing say 7W–6Mo–1·2V or 9Mo–3W–1.2V capable of achieving 850 HV min. would be of great interest particularly in view of the fact that the high-V M2 appears to have a lower hardenability than T1. This has been reported on tools which have to be air hardened because of capacity or plant breakdown problems. This group of alloys are amenable to salt-bath nitriding at temperatures of 480–550°C and a case of 0·0001–0·0004in has proved extremely useful on drills and taps for Ni base alloys.

Group 2

The basic British alloy in this group was the 14W–4Cr–4V alloy (no AISI equivalent) which is reported to have been used considerably at Derby, before the introduction of T15 and M15 to achieve higher hardness values. It is felt that a critical evaluation of this group of high-V high-C high-speed steels could be of considerable value. The range of V contents follows on from that of M2 in the order M3–1, M3–2, M4 (14W–4Cr–4V).

Group 3

This group and group 2 are characterized by increased wear resistance in proportion to C-balanced V content. Unfortunately, the difficulty in producing tools in the material is also increased, both machining and grinding being made more difficult. The increase in overall cost per tool, over say T1, can be 5–100% depending on the amount of grinding required and the limits that must be achieved. The usual increase is between 30 and 50% and it follows that this increase in price must be matched by a corresponding increase in performance to justify selection.

Group 4

M33 and M42 alloys contain a higher than usual C content

in relationship to the C-balancing elements (particularly V) but nevertheless contain more carbides than lower C alloys.

The increased C content together with the Co content increase the hardening response of the material M33 which will heat treat to 870–930 HV at 1 180°C hardening temperature; the high-C M42 will heat treat to 920–1 020 HV at the same hardening temperature, but is more often used in the range 890–940 HV or 850–900 HV.

M34 and M36A are similar alloys to the above, but have slightly higher V contents (1·5–2·0V) and are restricted to slightly lower hardnesses than the previous two alloys. They will give similar performances providing heat treatment and hardness levels are matched to the application and are similar to those which would have been used for M33 or M42.

One of the difficulties which arise with alloys of increased C content is that of sensitivity to overheating on hardening, in this respect M42 gives most problems.

An alloy which does not strictly belong to this group is M35 but it is well worth a mention here since it is a medium-Co alloy capable of taking an increased C content to give higher hardness capability and is at present being evaluated at Derby against other alloys in the group. T4 is now considered to be obsolete for virtually all applications.

Group 5

These three are the most highly alloyed of the high-speed steels as regards Co content and in consequence their toughness is very much reduced at comparative hardness levels, when compared with other high-speed steel alloys.

The traditional alloy is T6 and this has given extremely good performances on high-speed machining, where the tool form is solid and gives good support to the cutting edge, particularly on turning.

The usual hardness is in the region of 850 HV but this has been increased on occasions to 900 HV for some applications where toughness is not of paramount importance.

The other two alloys, M44 and T42, are slightly tougher than T6 and can be heat treated to hardness values of up to 1 000 HV with acceptable toughness.

To obtain maximum hardness, the C content may have to be specially controlled.

The more interesting alloy of the two is M44 since its grindability is better than T42, though both alloys would only be recommended for high-speed applications (preferably on small tools where the increased material cost would be small).

The alloys recommended for engineering use are now M2, as the basic alloy, and M42 as the second choice if the performance of M2 is insufficient (although M15 may be used if the cost–performance ratio can justify this).

In group 5, no alloy is at present recommended on Aero Engine Division drawings but T42 is being used occasionally by both Divisions.

Further work is required to establish the economic boundary between M42 and T42/M44 particularly where the grindability of M44 could affect the issue.

SUMMARY

The metallurgical investigation of high-speed steels has now reached the stage where the properties required or available in any specific tool can be estimated and controlled with a moderate degree of accuracy. A great deal of emphasis is placed on matching heat treatments and hardness levels of the various tools to the application. This can be critical when machining Ti alloys at 300–400 HB and Ni base alloys at 300–450 HB, both of which are prone to work hardening.

The experience related here has been brought together in the Rolls-Royce commercial supply specifications for high-speed steels: CSS10 and the supplements CSS10/M2, CSS10/M42, etc.; but it should be emphasized that in cutting aero-engine materials, there are also the problems of tool geometry and machining rates, etc. These are dealt with by production engineering departments with which there is very close cooperation.

The metallurgical laboratories have contributed considerably to the present selection system which is organized by the tool design standards department and operated by draughtsmen in conjunction with shop cutting-tool and production engineers.

APPENDIX

Table A1 Heat treatment of high speed steel currently in use at Rolls-Royce (AED, BED) Ltd

Material	Hardness range, HV	Heat treatment	Hardening temp., °C (salt)	Tempers, °C (salt)
M2A	835–900	1	1 220	3 × 550 (broaches) 2 × 560 (general)
	790–850	2	1 200	1 × 560, 1 × 580
T1	820–900	1	1 270	2 × 560
	790–850	2	1 250	1 × 560, 1 × 580
M15	870–930	1	1 200	3 × 550
	850–900	2	1 200	2 × 550, 1 × 580
T15	870–930	1	1 220	3 × 550
	850–900	2	1 220	2 × 550, 1 × 580
M42	890–940	6	1 180	3 × 550, 1 × 565
	850–900	2	1 180	3 × 550, 1 × 580
	940–1 000	4	1 195	4 × 540
			Air hardening	
			Gas muffle	Gas muffle
	900–960	7	1 195	4 × 530
	870–930	8	1 195	4 × 550

A1 M42 broach, used after normal regrinding period   ×5

A2 M42 broach, material too hard for application   × 5

A3 M42 broach, failure as a result of interruption in the coolant flow while cutting the nickel-base material   ×5

A6 T1 broach tooth, showing excessive carbide segregation leading to failure   ×3

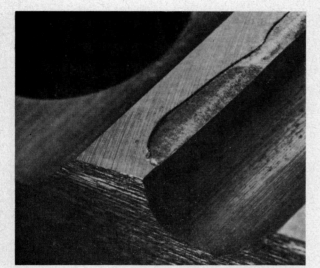

A4 M42 broach, defoliation of broach material due to abuse in grinding   ×5

A5 T1 bar samples (1 × 1½ in) rejected for poor carbide distribution   ×2

Selection and quality assurance of tool materials used in the machining of earthmoving components

R. Dickson

The development of higher strength metals to meet today's rugged requirements in the earthmoving industry and the need to meet high production demands has resulted in a more critical appraisal of cutting-tool materials to be made over the last decade. More emphasis is laid on tool metallurgy than ever before and the contents of the paper deal primarily with tool selection and quality assurance from a metallurgical point of view.

TOOL-MATERIAL SELECTION

Selection of the basic material for high-speed steel tools is based on four main application requirements: abrasion resistance, toughness, red hardness, and grindability. The final choice of chemistry depends on which of the foregoing are most desirable and necessary. The vast majority of tools purchased fall into two composition ranges, tungsten grades and tungsten–molybdenum grades. It is claimed that the molybdenum steels have better retention of hardness at elevated temperatures, although no evidence of this has been found in practice to date when comparing AISI M2 and T2 steels. Abrasion resistance and grindability characteristics have also been found to be similar with molybdenum and tungsten steels.

Toughness with molybdenum steels is generally superior to tungsten steels (Fig.1). It is desirable to standardize for general-purpose tools of all description on M2-type material. It is not always possible to follow this through since some tool manufacturers will not use molybdenum-based steels and others cannot always obtain it. This is a situation which seems to have improved in the last few years.

Caterpillar Tractor Company Ltd use tungsten cemented carbides extensively for machining cast iron, steel, and non-ferrous materials in the ranges 130–900 sft/min. These carbides are generally stronger than the titanium grades and ceramics, but more brittle than high-speed steel. Titanium carbide grades are used for machining steel in range 500–1 500 sft/min. In general two major grades of tungsten carbides are used: (*a*) abrasive-resistant grades for cast iron and non-ferrous materials; (*b*) crater-resistant grades for machining steel. In selecting carbide grades, irrespective of the type of machining problem, the following approach is used:

1 The type of wear the tool has to combat is determined. If it is the abrasive type of wear associated with non-ferrous or cast-iron materials, a straight tungsten-cobalt grade (C2, C3, C4) will be selected. If the wear is caused by cratering, seizing, welding, or galling such as would be encountered in machining steels, a grade containing titanium or tantallum (C5, C7, or C8) would be selected.

2 The strength level of the carbide that is required to withstand the cut condition is determined. The softer the carbide the greater the strength, but this strength is obtained with a corresponding sacrifice in tool life. For maximum tool life the hardest grade that will withstand the cut condition without chipping or breaking should be applied.

Pure titanium grade carbides and ceramics have even less strength than the C4 to C8 grade carbides and are most successfully applied to high-velocity finishing operations. These materials should be engineered for special applications only.

METALLURGICAL QUALITY CONTROL AND SAMPLING OF HIGH-SPEED STEELS AND CEMENTED-CARBIDE TOOLS

Specifications for high-speed steel and cemented carbides have been drawn up by Caterpillar after many years of cooperation with American toolmaking companies and SAE, ASTM, and AISI committees on tool properties. In addition, the company has developed their own investigation programme, where tools are tested for tool life and shock-resistance properties, coincident with metallurgical examination. Typical specification sheets for high-speed steel and cemented carbides are shown below:

Molybdenum-base high-speed steel, M1

Scope This specification covers the essential applications, requirements, and heat treatment of M1 material.

Requirements

Chemical composition (similar to AISI M1)		
Carbon	0·78–0·84
Manganese	0·20–0·40
Sulphur	0·020 max.
Silicon	0·20–0·40
Chromium	3·75–4·25
Vanadium	0·90–1·30
Tungsten	1·25–1·75
Molybdenum	8·20–9·20

Microstructure (after heat treatment) Specimens for micro-examination shall be taken on a longitudinal section and shall show a structure of tempered martensite, free from retained austenite and injurious inclusions. The Snyder-Graff grain size average shall be not less than 10 up to a 3in round or less than 9 over a 3in round. The carbide distribution

Development of higher strength metals to meet today's rugged requirements and the need for higher production has resulted in a more critical appraisal of cutting-tool materials. The metallurgy of these tools is receiving more attention than ever before. This paper deals with tool selection and quality assurance, at the writer's company, from a metallurgical point of view. The choice of steel is based on what usage the tool is intended for and properties considered in selection are discussed. The system of quality assurance employed by Caterpillar Tractor Company Ltd is discussed, both from the point of view of initial approval of a tool-supplier's material and from audit sampling of approved suppliers' material at the receiving stage in the user plant. Typical reasons for rejection and indications of lack of control by some tool suppliers are highlighted by actual examples of tools cut up for audit checking purposes.

669.018.25+669.14.018.252.3:658.562

The author is with Caterpillar Tractor Co. Ltd

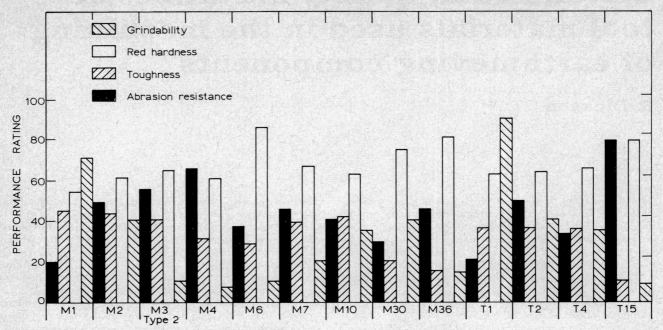

1 **Qualitative comparison of properties of high-speed steel**

and network shall not exceed that shown in MS4100. The carbide size shall be typically 2–4 with no carbides exceeding 6.

Cemented carbide steel roughing grade

Scope This specification, together with the general specification, covers the essential application and requirements of a cemented carbide composed of tungsten carbide, titanium carbide, tantallum carbide, columbium carbide, and cobalt of 91·0 HRA hardness.

Application This material is used for roughing steel with cut conditions including scale, cast surfaces, and heavy interruptions. It has good resistance to thermal deformation, good crater resistance and excellent shock resistance.

Requirements Composition in wt-%: (*a*) tungsten carbide 70·0–75·0; (*b*) titanium carbide 6·0–8·0; (*c*) tantallum carbide and columbium carbide 10·0–12·0; (*d*) cobalt 8·5–9·5.

Hardness Nominal 91·0 HRA, range 90·6–91·4 HRA (ASTM B294–54T).

Density 12·4–12·8 g/cm³ (ASTM B311–58).

Apparent porosity Average shall be better than A2–B2–C2. (ASTM B276–54).

Microstructure Grain size shall be within the following ranges: (*a*) tungsten carbide grains 1–3 μm; (*b*) recrystallized tungsten carbide grains 4–10 μm; (*c*) titanium carbide, tantallum carbide, columbium carbide, and combined carbide grains 2–8 μm.

There shall not be a consistent occurrence of recrystallized tungsten carbide or combined carbide grains over 10 μm or cobalt lakes over 10 μm. The carbide shall not exhibit the 'rapid etch' phase indicating carbon deficiency.

The product shall give acceptable performance in a straight turning test, an interrupted cut test, and in shop operations to assure good tool life, excellent shock resistance, and resistance to edge chipping.

Before a tool from a new source of supply or a new grade from an existing supplier is tested for machining characteristics it is approved metallurgically. Inferior-quality high-speed steel tools can be detected by metallurgical inspection of the critical characteristics of hardness and microstructure. Cemented carbides likewise can be detected, but in this field the possibilities for introducing errors or variables into the metallurgical examination and testing procedures are many. The extremely high hardness of cemented carbides necessitates the use of diamond for machining operations. All cutting and grinding must be carefully performed under coolant to prevent the formation of microcracks from overheating. Errors in hardness checking can be easily introduced by non-parallelism of the faces of the pieces being checked because of the shallow penetration of the indenter, and errors in density can arise from non-removal of surface imperfections. Only qualified personnel with a good knowledge of the significance of the metallurgical examination carry out initial evaluation tests.

Initial metallurgical approval of new high-speed steel tools is based on chemistry, microstructure, and hardness. Microstructure is examined at ×100 and ×500 unetched for sulphide size and distribution. The specimen is then etched and examined at ×100 for carbide distribution and network, and at ×500 for retained austenite and carbide size. The etchant is normally 10% Nital. Sulphide distribution, carbide particle size, and carbide distribution are checked on longitudinal sections and carbide network on transverse sections. Snyder-Graff grain size is measured at ×1000.

Cemented-carbide tools are initially approved similarly to high-speed steel tools, the physical requirements checked being hardness, density, and microstructure. Hardness is determined on a Rockwell A scale after accurate grinding using a 220 grit size diamond grinding wheel. Surfaces between anvil and indenter must be absolutely parallel to ensure accuracy of hardness readings and the accuracy of the testing machine is checked with tungsten carbide test blocks.

Density is determined by means of the immersion method. Microstructure examination covers apparent porosity on the unetched polished specimen and identification of phases present after etching in freshly prepared 1:1 potassium hydroxide–potassium ferricyanide solution (10% by weight). A typical approved sheet is shown in Table 1. When the tool, high-speed steel or carbide, has passed the above tests the supplier becomes tentatively approved for that particular grade, subject to satisfactory shop trials. After initial qualification, the next ten shipments of the new tool are tested metallurgically and if found to be correct the supplier is put on the approved list.

Routine tests on accepted grades are made at the receiving station at predetermined frequencies and controlled by computer. In addition to sampling of incoming batches, all tools giving poor shop life are submitted for metallurgical examination as part of the routine investigation to determine the reason for premature failure.

Table 1 C6 grade general-purpose steel

Property	Findings	Acceptance level	Remarks
Hardness, HRA	93		
Specific gravity	11·8		
Apparent porosity	A₁, C₁	A₂ B₂ C₂ max.	
Micro α1 gr	1–2 U	1–3 U	
α2 gr	3–8 U	4–10 U	Accepted
WC/Tic Sol Soln	1–8 U	10 U max.	
Cobalt	1 U	10 U max.	
η-phase	None	None	

TYPICAL REASONS FOR REJECTION OF CUTTING TOOLS

Metallurgical causes for rejecting high-speed steel cutting tools can be split into two basic causes:

   (i) faulty raw material

   (ii) faulty heat treatment.

Faulty steel

Premature tool failure when related to faulty steels has been found generally to fall into six categories:

1 *Voids* These are normally the result of 'piping' or 'bursting'. Pipe is caused by insufficient cropping of the ingot, bursts can usually be attributed to centre breaks caused during rolling or forging and are characterized by a rough intergranular fracture.

2 *Tears* These are transverse defects caused by continued working of the steel at too low a temperature.

3 *Poor carbide distribution* (Fig.2) shows the microstructure obtained on a face and chamfer tool, made in M2 material which shattered almost immediately it was put in service. The excessive carbide network visible at × 100 magnification is due to insufficient hot work at the forging stage. This invariably leads to sudden brittle failure along the planes of the massed carbides. Figure 3 shows a micrograph of a tap at × 5 magnification showing excessive carbide distribution.

4 *Poor sulphide distribution* This defect, caused by improper melting procedure and/or rolling practice, results in edge chipping or crumbling along the strung-out sulphide inclusions.

5 *Bar stock instead of forging* The use of bar stock instead of forging for large-diameter tools such as gash cutters or shaper cutters can lead to sudden tool failure. Failure in this instance is again due to insufficient hot work being done on rolling large-diameter bars to break up massive carbides. Forging also offers the obvious advantage of producing

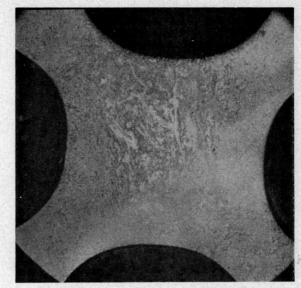

3   **Section of tap showing excessive carbide distribution**   ×5

preferential flow lines with improved toughness and shock resistance.

6 *Wrong material* This is a defect which has had a higher incidence than any of the foregoing. The reasons vary from obvious poor stock control at the toolmaker, as in the case of a two-flute drill made from mild steel, to using an alternative process and material without due regard to service requirements. Figure 4 is an example of this; the material designed for the tool holder is 1½%Ni–Cr–Mo En24 material through hardened. As can be seen from Fig.4, the holder was supplied in mild steel case hardened. Apart from the fact that the core properties were completely inadequate for the job, the holder has a pin hole in it which was also cased and would have acted as an excellent notch if the holder had been put in service.

Faulty heat treatment

Incorrect heat treatment or poor control of the heat-treatment operations account for many premature failures in cutting tools. Typical defects produced in heat treatment are as follows:

1 *Overheating* This is the most common heat-treatment

2   **Excessive carbide network due to insufficient hot work at the forging stage**   ×100

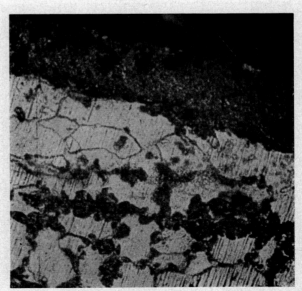

4   **Tool holder, mild steel case hardened**   ×500

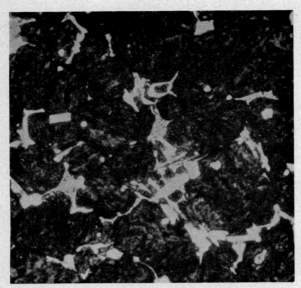

5   Coalescence of carbides and grain coarsening due to overheating   ×1000

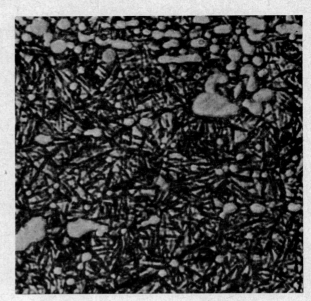

7   Typical structure obtained by single tempering   ×1000

defect found on high-speed steel tools. Figure 5 shows a structure found on a drill. Carbides have started to coalesce and grains to enlarge with the formation of cellular carbides and eutectic envelopes at the grain boundaries. The hardness on this drill was 68 HRC which is typical of this type of structure, and it could be expected that the supplier on routine hardness testing during the process would have been able to catch this defect with nothing more sophisticated than a Rockwell machine.

2   *Underheating* Although not as common as overheating this defect has been found on a few occasions. Figure 6 shows a hob which has been underheated showing the typical areas of ferrite and transformation products along with tempered martensite and carbides. This defect results in poor tool life.

3   *Single temper* Figure 7 is of a gear hob which shows the effect of single tempering. Multitempering transforms the retained austenite and tempers the resulting martensite. This hob was low in hardness and should have been picked up by the supplier before delivery.

4   Other heat-treatment defects such as carbon enrichment or decarburization of the surface layers cause premature failure and reduce tool life. Quench cracks, frequently

attributed to poor heat or quench techniques may also be caused by poor tool design. Sharp corners, undercuts, and drastic section changes, which look good on a drawing board, can cause endless problems in subsequent heat-treatment operations.

With cemented-carbide tools, rejections are due either to segregation of the constituents at the mixing and milling stage or improper sintering.

Tungsten carbide occurs in two forms, fine α_1 grains and larger α_2 grains. Much larger α_2 grains, greater than 10 μm in diameter, often occur as a result of grain growth or recrystallization during sintering. Excessive numbers or segregation of these large grains is undesirable for the cutting-tool grade of cemented carbides.

ϵ-phase (Fig.8) is a detrimental condition caused by precipitation of carbon on cracks, caused by mechanical stresses in the grains. These high stress plains which appear as strain lines on a micrograph are lines of weakness which result in premature failure.

In complex carbide tools, incomplete diffusion of titanium carbide in the tungsten titanium solid solution causes premature failure.

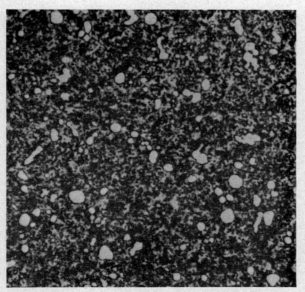

6   Typical areas of ferrite and transformation products associated with underheating   ×500

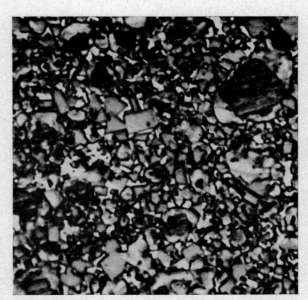

8   Transgranular strain lines caused by ϵ-phase precipitation   ×1000

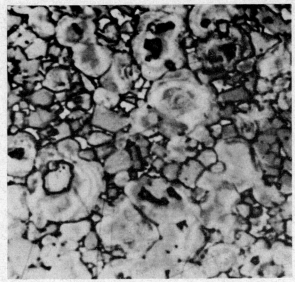

9   Large single crystals which reduce cutting properties   ×1000

11   Cutting-edge failure due to excessive cobalt 'laking'

Good cutting performance with cemented-carbide tools is dependent on the presence of properly formed chains. The disruption of these chains into large rounded single crystals (Fig.9) results in decreased cutting properties.

Uneven distribution of the cobalt binder (Fig.10) through the matrix resulting in excessive 'laking' is undesirable, as it creates brittle areas devoid of binder material resulting in failure of the cutting edge as shown in Fig.11.

η-phase, due to low carbon content and characterized by high hardness and brittleness, reduces tool life. Figure 12 is typical of η-phase found in a throwaway insert.

A- and B-type porosity depend upon carbon absorption by the cobalt phase. C-type porosity indicates the presence of free carbon in the initial powder and points to lack of control in mixing. C-type porosity establishes planes of brittleness which causes premature failure when it occurs near a cutting edge (Figs. 13–14).

Evidence obtained from the testing and approval programme has indicated that many problems associated with metal cutting are due to metallurgical deficiences in the tools themselves. This is a fact which for many years has been overlooked, solutions to problems of poor tool life being

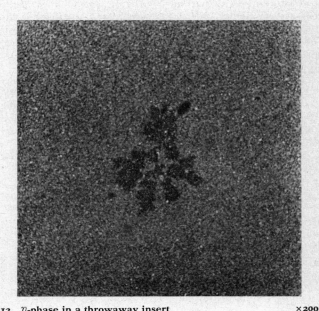

12 η-phase in a throwaway insert   ×200

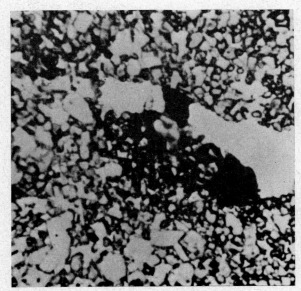

10   Uneven distribution of cobalt binder   ×1000

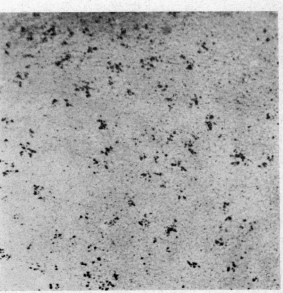

13   Excessive C-type porosity   ×200

14   Cutting-edge failure due to excessive *C*-type porosity

sought for, in the main, in the areas of tool geometry and the metal being cut. Many problems in the past which were solved by changing geometry and base metal, may well have been solved more economically and in a better technical sense by critical appraisal of tool metallurgy. The current and future needs to machine ever harder and tougher materials at ever higher speeds has accentuated still further the need for more attention to be paid to metallurgical considerations when designing tools, and for closer metallurgical process control of raw material and heat treatment to ensure consistent quality.

It has been our experience that it pays to deal with suppliers who have their own laboratory facilities and process-control techniques. This of necessity eliminates a number of the smaller suppliers, who buy high-speed steel material, hoping that it conforms to the specification and that it will respond normally to heat treatment.

At the start of the programme on carbide tools it was discovered that there were twenty-seven sources of supply. Less than half of these suppliers actually manufactured the carbide, the balance purchased sintered carbides from large manufacturing companies and made the tools up from these. This made consistent tool life a lottery and evaluation an impossibility since variation in performance within even one shipment was inevitable.

It is now Caterpillar Tractor Co. Ltd's practice to buy carbide tools only from sources which have the facilities for the manufacture of carbide. This has resulted in a marked improvement in consistency and performance.

It has been said that tool users play too passive a role when dealing with tool suppliers' shortcomings, and rely too greatly on what the supplier says his tools will do. Whether this criticism is justifiable or not depends to a great extent on what side of the fence one is standing on.

From the users' point of view, the majority are small to medium sized companies who do not have and cannot afford the facilities necessary to develop their own tool specifications. These companies must rely on the knowledge and integrity of the tool supplier's representative for guidance on the best tool for the job. This situation could be improved by tool suppliers issuing specifications detailing minimum properties of the tools they make. This is common practice on the Continent and in America, but few British tool suppliers issue specifications for the benefit of customers.

The blame has also often been laid at the door of the user in that he does not communicate his problems to the supplier, and because of this inability to convey the necessary information, the supplier cannot do too much to improve general tool quality to meet present day demands. It has been the experience of the author, whose company's practice is it to contact the tool supplier to discuss each and every rejection, that the problem conveyed to the local representative gets back to the toolmaker's establishment, but too often seems to die there.

There is an evident need for better communications on both sides. A good starting point from the user's point of view would be the issuing, by the supplier, of specifications covering the minimum properties to be expected from each grade of tool steel and cemented carbides.

ACKNOWLEDGMENTS
The author wishes to express his thanks to Caterpillar Tractor Co. Ltd for permission to publish this paper and to acknowledge the assistance of his colleagues in its preparation.

Cutting tools in mass-production machining of auto parts

N. P. Riley

In presenting this brief paper it is not the intention of the writer to deal in depth with any one topic of automobile component manufacture, but rather to outline the Ford Motor Co. approach to the problems of performance, purchase, and housekeeping of cutting tools.

The broad approach to cutting-tool materials and applications are well laid out in a paper 'Metallurgical development and quality control of automotive tooling material' prepared by W. J. L. Woodhouse, then of the Ford Motor Co., for the conference on Machinability in 1965. In this present paper it is hoped that by comparing the current approach with the conditions existing five years ago useful lessons may be drawn by personnel both within and without the company.

Rereading the 1965 paper, one is beset with mixed feelings. For instance, the answer to the plea for a British Standard for tools, although apparently then imminent, has yet to come to fruition in an entirely acceptable form. It is noted that the British Standard for tool steels forms the basis of a paper at this conference and its contents are awaited with interest. One topic of which the original conception was outlined in 1965 may be reviewed with some satisfaction within the company: the continued progress toward standardization and rationalization of cemented carbide for cutting application. It is this company's cemented-carbide cutting-tool standard which forms the basis of this paper. It is considered justifiable to concentrate on this subject for it is felt that much progress has been made and that further progress can be made towards a more ideal system for the specification of cemented carbide over the range of applications within the motor industry.

In general, the main aim of the Ford Motor Co. in its approach to tooling is to rationalize and to standardize tooling based on an acceptable performance level, rather than to aim for an ultimate in performance for individual applications. It is felt that, should the latter approach be the main criterion of the company, the tool-producing industry cannot necessarily produce the consistency to maintain that approach economically. Some of these inconsistencies will be mentioned later.

The following system of standardization has been based on surveys of production trials over a period of years, and to a certain extent the standard tends to lag behind the continuing developments and changes of equipment. Hence an essential part of the system is the constant revision and updating by all parties concerned. Tool engineering departments coordinate individual tooling trials enlisting the aid of purchase and laboratory personnel where necessary to establish actual tooling costs or to evaluate material queries. When deciding the most economic tooling, the tool engineering department will of necessity take into account such items as initial tool cost, cost and number of regrinds, setting, and machine idle times. These aspects will, of course, be additional to overriding factors such as the speed of the production time and associated tool changes. Hence to a certain extent the tooling trends reported here reflect only those conditions prevailing within the Ford Motor Co., and may not necessarily appertain to other organizations. However, it is perhaps not inconceivable that other industries applying similar reasoning may arrive at similar conclusions which would further arouse the desire for a reduced number of marketed grades each with a more versatile application approach. Improved economics and quality may well accrue and, perchance, even a national cutting-material standard arise.

CEMENTED CARBIDES FOR CUTTING APPLICATIONS

The current system of specifying cemented carbides is summarized in Table 1. It will be noted that cutting applications are divided into five groups under the separate headings of cast iron plus non-ferrous materials and steels. These grades are based on Co–WC or Co–WC–TiC–TaC plus additional carbides, the exception being the G9 grade which is based on the Ford 4J material developed in the USA. Typical photomicrographs of the various grades are shown in Figs. 1 and 2.

Some 90% of the cemented carbide for cutting applications is estimated to be purchased by means of the system outlined in Table 1. An early estimate of the usage of the various categories is shown in Table 2, and although of interest it is not exactly representative of the current situation, for which figures are not yet available. The main divergence between Table 2 and current practice is the increased usage of the G9 groups. (G9 grades are mostly purchased from the USA, and although found to out-perform conventional carbides on some application, the TiC materials may not in certain cases be used due to the economics of import duty etc.; some UK development of titanium grades could be of advantage here.)

Table 2 also excludes brazed tool and bladed items, most of which are purchased to G1, G2, G5, and G6 grades; some G4 and G9 grades are also brazed.

Group G1
In this group which is based on 6 wt-% Co and a medium- to

The system adopted by the Ford Motor Co. for the specification and purchase of cemented carbide is outlined. In this system, the tool engineering department have identified ten major groups of cutting applications, each of which is depicted by a typical composition and hardness of cemented carbide. With the exception of one group of applications, these cemented carbides are based on cobalt binder with varying amounts of tungsten titanium and tantalum carbides and are supplied to a Ford approved chart by five to six suppliers. The carbides within the individual groups, as supplied by the different suppliers, are considered interchangeable when operated with the grade system. Some cemented carbide defects are noted and observations are made on the disturbing level of geometrical and finish complaints. Some metallurgical defects are cited but do not feature as significantly. Brief comments are made concerning the usage and trends in cemented carbide and other cutting materials, for instance laminates, coated carbide, and high-speed steels. It is anticipated that by extending the role of TiC cutting tools, the number of individual application groups could be further reduced from the present level of ten.

621.431 : 669.018.95 : 661.878.621

The author is with the Ford Motor Co. Ltd

Table 1   Summary of the 10 Ford cutting grades, indicating typical application, hardness, and composition by volume

Composition by vol.-%		Co	WC	Additional carbides	Hardness HV30
Cast iron and non-ferrous materials					
G1	Roughing cuts (heavy and interrupted)	10	90	*	1 500–1 600
G2	General purpose (reasonable surface and roughing cuts)	10	90	*	1 565–1 685
G3	Light finishing (general purpose machining)	8·5	91·5	*	1 640–1 760
G3A	Light finishing (general purpose special operations)	8·5	70	21·5	1 750 min
G4	Precision finishing (exact final operations)	6	94	..	1 750 mm
Steels					
G5	Roughing cuts (heavy and interrupted)	13	54	33	1 450–1 550
G6	General purpose (reasonable surface and rough cuts	11	56	33	1 500–1 600
G7	Light finishing (general purpose machining)	11	35	54	1 500–1 600
G8	Precision finishing (exact final operations)	7	35	40	1 700 min
G9	Precision finishing (high cutting speeds)	7	53	Nickel/molybdenum/titanium carbide	

* May contain up to 5% TaC or other carbides

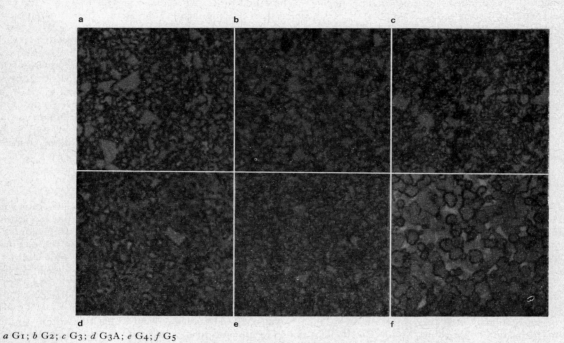

a G1; *b* G2; *c* G3; *d* G3A; *e* G4; *f* G5

1   Photomicrographs of various cemented-carbide grades

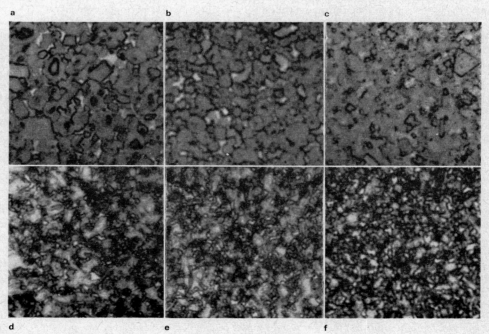

a G6; *b* G7; *c* G8; *d* G9–4J 30 HV; *e* 5H 30 HV; *f* 7G 30 HV

2   Photomicrographs of various cemented-carbide grades

Table 2   The percentage of throwaway tips purchased in each Ford grade (figures derived from an early survey)

Cast iron and non-ferrous machining			Steel machining		
Ford grade	G1	1%	Ford grade	G5	15%
	G2	18%		G6	33%
	G3	1%		G7	16·5%
	G3A	12%		G8	less than 0·5%
	G4	2·5%		G9	0·5%

coarse-size tungsten carbide, most applications are found in the brazed tools for, say, boring cylinder block cast iron, and similar materials. This group has sometimes been found to have insufficient wear resistance for the intended use and the jobs have been upgraded. There is a general tendency for a diminishing demand for this grade.

Group G2

This group is based on 6 wt-% or lower cobalt content often with grain-controlling carbides, and in general a medium-size tungsten carbide is maintained.

The G2 group is one of the more commonly used carbides having a wide range of applications from spot facers, counter bores, hobs, and form cutters, to brazed and bladed milling cutters. It is predominantly used on grey or alloy irons. Again instances arise where the application requires upgrading to utilize a higher wear-resisting grade, for example in the machining of aluminium piston materials where G3 or G3A grades are now specified.

Group G3

The majority of this group are found to be the 5–6 wt-% cobalt binder types having a medium-fine tungsten carbide grain size. Though also a diminishingly utilized grade it is still maintained for milling blades and brazed tooling application requiring higher wear resistance than G2.

Group G3A

The G3A group was introduced as a result of developments in the carbide industry to produce cemented carbide of high wear resistance for high cutting speeds, while still retaining a measure of crater resistance. Its composition is based on the G3 type in which a proportion of the tungsten carbide is replaced by titanium carbide, and to a lesser extent tantalum carbide. The G3A microstructure is often typified by a fine-grain tungsten carbide and a fine- to medium-sized solid-solution crystal.

In the range of application formerly undertaken by non-brazed tools of G3 and G4 groups G3A has often been preferred, causing a reduction in the usage of these categories. To a certain extent, further expansion of the usage of this grade is limited by inconsistencies in manufacture and availability. An additional problem has been the poor finish of some G3A inserts when applied to finishing operation, for instance poor insert finish was experienced in the final boring of crankshaft main bearing bores and seriously jeopardized production.

Group G4

The G4 composition reverts to a Co–WC composite often employing 4·5 wt-%Co or less. Sometimes grain-controlling carbide is additionally employed to maintain a fine WC grain size. The main application is one of fine boring either as a brazed insert or throwaway tip, although other applications such as gun reamers exists for machining cast-iron components. As stated previously, the range of applications for G4 are receding and are being replaced by G3A. To some extent the TiC range may find future application in this category.

Group G5

The first of the steel-cutting grades has a composition based

on approximately 10 wt-%Co 70 wt-%WC and the balance Ti and Ta carbides. Typical microstructures show a predominance of solid-solution crystal together with some recrystallized WC in a well dispersed cobalt binder. The general carbide sizes are medium or medium-coarse.

The G5 material is generally used to carry out initial rough-turning operations on forgings. As the standard of forging tolerance improves, the requirements tend to transfer from toughness to wear resistance. G5 materials are also used in the brazed form as milling cutters and heavy- and interrupted-cut turning tools. It is fair to say that within this group some anomalies of composition and performance have to come to light over long-term evaluation, and steps are in hand with various suppliers to modify manufacture. Some steps have been taken already, such as refinement in grain size and additional titanium carbide addition to improve crater resistance. Further efforts by the tool engineering department continue to improve predictability of performance in this group.

Group G6

The G6 group have lower cobalt contents than G5, of the order of 9 wt-%, and a proportional increase in titanium and tantalum carbides to give a greater area of solid-solution crystals in the typical microstructure. This grade finds considerable application for general-purpose machining in the form of throwaway inserts or brazed tools.

Group G7

Another high-usage group is the light-finishing grade counterpart to G6. Its composition is approximately 10 wt-%Co, 60 wt-%WC with the balance being made up of Ti and Ta carbides to give a predominantly solid-solution structure medium-coarse in grain size; some WC crystals are usually present.

G7 is used on the complete range of steel material employed within the company and is used particularly in fine finishing and boring where it may be in the form of a brazed insert.

Group G8

This group has a small number of suppliers and a limited range of applications, the composition approximating to 5 wt-%Co, 75 wt-%WC, with the balance being predominantly TiC. The microstructure is characterized by a low cobalt binder content and a predominating amount of solid-solution crystal. G8 is used as a brazed insert for fine boring and other finish applications at medium cutting speed.

Group G9

G9 is based on the Ford 4J carbide developed by Ford Motor Co's applied research department in the USA. The binder is a Ni–Mo composite and the carbide is Ti. Its microstructure is that of fine TiC crystals in a well dispersed binder. (The subject of the metallurgy and development is covered in detail by either papers in this volume. The G9 group is particularly successful on fine-finishing operations such as connecting-rod bore finishing at high cutting speeds. It is employed either as a throwaway tip, or brazed insert. Specially developed brazing techniques must be employed to minimize stress cracking of the tip.

Note

The usual cast-iron materials which are to be machined may range through the BS1452 grades 12–17 grey cast iron, BS2789 SNG 27/12, SNG 32/2 nodular cast iron, or alloy cast iron such as the camshaft irons. Steel components may be manufactured from one of a wide range of steels which include BS970 En8, En16, En18, En19, SAE 4028, SAE 8620, SAE 4630, SAE 4147, and carbon-manganese-boron steels. The heat-treatment condition could be 'normalized',

'normalized and tempered', or 'hardened and tempered'; this applies to gear box, transmission, and some engine components.

Derivation

Having outlined the typical cemented carbide and its application requirements it is perhaps worthwhile indicating briefly how such a system was derived. With the rapid increase in the usage of cemented-carbide cutting tools in the late 1950s and early 1960s a situation arose where many different carbides were being specified for individual applications throughout the company. After initial surveys and rationalization the number of cemented-carbide categories was identified as 18 in 1965. Further surveys and rationalization by the tool engineering department brought this figure down by almost 50%. At the same time a survey of composition and physical properties indicated a typical carbide for each group application. The composition by volume was chosen in an attempt to take into account the various preferred compositions of individual manufacturers. The volume percentage of the additional carbides (Ti and Ta) gives some indication of crater and wear resistance, while the binder content give some indication of the toughness. Metallurgical observations were subsequently confirmed by shop trials, under correct machining conditions.

Operation

The value of the specification system is obtained in its fullest operation throughout the Ford Motor Co. In conjunction with the specification, an approved chart has been derived which then enables the tool engineer to obtain an insert which will give an acceptable performance for a particular application.

Currently, the approved list shows 5 or 6 approved suppliers, so that for any one operation there are 5 or 6 commercially available cemented-carbide grade equivalents for that application.

The successful operation of the system relies heavily on the tool engineer who will, in the light of the variables involved in the individual application, designate the application with a grade category. He will be required to take into account component material, component condition, equipment availability, cutting speed and feed, speed of line, etc. under acceptable economic conditions. After this decision, an insert is chosen which may have been produced by any one of up to 6 manufacturers and will be anticipated to perform to a norm. Any discrepancies are first investigated within the designated grade, and second by trials with adjacent grades, e.g. less tough or tougher grade, or finally outside the system altogether.

The derivation and maintenance of the approved chart has its initial step with the buyer who will investigate the economics of bulk purchase of the proposed grade against existing suppliers. The second stage is normally to seek from the supplier technical information to identify material type and properties; vendor application definitions are not necessary in line with the company's conception of roughing grades etc. Metallurgical examinations are made to obtain a provisional evaluation of manufacturing quality, and may be supplemented by occasional visits to the suppliers premises to consider quality control features. If the cemented carbide thus appears to meet the economic and specification requirement sample inserts will be tested. These steps reduce the time involved in the expensive process of machine trials by eliminating the uneconomic and low-quality or unsuitable supplies before the performance testing stage.

Here again on the question of performance testing the tool engineer must take into account the varying conditions of the individual grade requirements since it is current practice to undertake performance testing as part of current production runs. The most and least arduous application will be selected within the group to be tested, the proposed against the existing 'mean' performer. If trials are acceptable within

maximum 10% deviation between the two materials, limited acceptance is given. Brazing trials will be undertaken where applicable. A successful review of six months of trials will give inclusion to the approved chart. Laboratory checks are made throughout these tests to ensure compliance with the standard requirements and establish potential quality.

Discussion

As stated earlier, not all cemented carbides purchased fit readily into the system outlined. Non-compliance may arise as a result of equipment purchased outside the area normally bounded by the approved list, or the existence of non-standard tooling which is not available in the grades desired. However, within the company it is recognized that a proportion of applications will give below par economics when machining is undertaken by inserts taken from the standard groups. Tooling for such application are considered separately on individual merits.

It could be argued that the Ford system is oversimplified, even inaccurate, and not worthy of the efforts made to maintain it. The prevalent feeling is that with the development of more versatile materials the future will see a wider use of the standard or modified standard grade system. It is also accepted that within each grade there is the odd occasion, say, a milling application, where one carbide will be preferred to the remainder in that group. At times reasons for these differences are established and may for instance be associated with equipment condition, tip finish differences, toughness variations due to grain size, or cobalt-content variation. Such a deviation may warrant a modification to the standard, or alternatively a grade may remain acceptable by virtue of the overall 10% performance discrepancy allowance.

Since it has been found essential to judge actual cutting performance entirely on production equipment, thereby ignoring results predicted from controlled machine trials, the alternative to rationalization is to evaluate every individual application, an obviously impractical and uneconomic approach. Bearing in mind the wide range of equipment and cutting conditions, and in the absence of any discriminating system of designating commercially available materials, a system of some kind is essential. In stating typical composition and hardness it is not intended to dictate developments to the carbide manufacturer, rather it suggests that material complying with the specification generally meets the needs of the application. That it does not inhibit development and cause the usage to stagnate is evidenced by the group 3A which was included after evaluation, although hitherto falling outside the stated requirements.

That the specification does not define all the factors which influence the suitability of a material for a cutting tool is apparent, such items as carbide particle size and porosity levels are omitted. At this stage insufficient evidence is available to justify a decision on these counts, similarly geometrical aspects other than broad dimensional requirements are not yet part of a company standard.

The nett result of the current specification and approval system is that the Ford Motor Co. is supplied by 5 or 6 competitive manufacturers whose carbides within the 10 groups are considered interchangeable. The realization of such a situation has led to the anticipation of closer collaboration with suppliers benefiting both customer and supplier alike. An example of such collaboration is an instance where a deficiency in crater resistance as compared with other members of group G5 has resulted in one supplier investigating his grade and proposing a modification.

Quality aspects

The aspect of quality of cemented-carbide inserts which cause most concern is undoubtedly that of deficiencies in geometry and finish. That this is not a problem confined to our industry is made obvious in discussion with colleagues in other large 'carbide user' industries. For fear of labouring

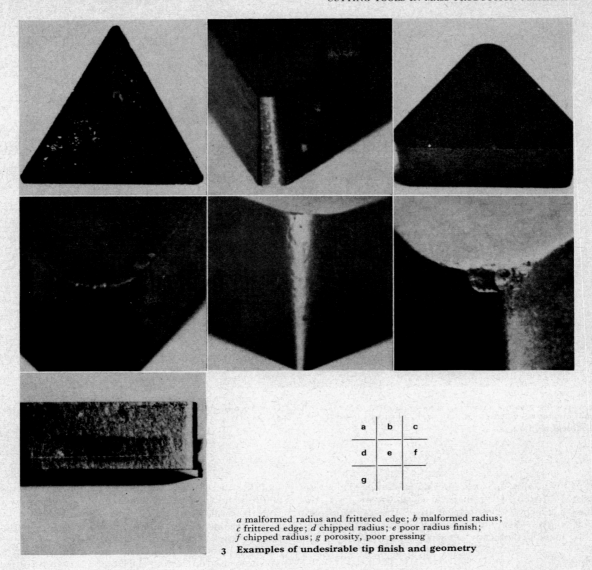

a malformed radius and frittered edge; *b* malformed radius;
c frittered edge; *d* chipped radius; *e* poor radius finish;
f chipped radius; *g* porosity, poor pressing

3 Examples of undesirable tip finish and geometry

an aspect which will surely be raised elsewhere in this conference the usual complaints are brought forward for reference.

Poorly blended nose radii are common, and due to their nature of origin can result in whole batches of tips being rejected. This defect results in high wear and difficulties in maintaining component finish. These problems have been experienced in operations such as the final crankshaft main bearing boring. Chipped radii also figure highly and have similar results to poorly blended radii. This defect is of random occurrence and therefore not as predictable on the production line.

Frittered edges, pressing defects, poor surface finish, and porosity are also fairly common each of which, depending on its degree of presence, will have detrimental effects on performance.

Edge reinforcement on steel-cutting grades is considered to be critical. The desired edge reinforcement is related to application, i.e. light or heavy duty, and cannot be interchanged. A few examples of undesirable tip finish and geometry are shown in Fig.3.

Metallurgical defects

Defects which are identified by metallographic examination occur to an apparently lesser extent. Again, some defects arising during sintering may be applicable to whole batches, whereas contamination due to pick-up may only occur in isolated instances.

On occasions when the cemented carbide has been faulted, and excessive tip wear has been investigated, the cause has most often been attributed to oversintering or high porosity levels (on occasions ASTM B276 A5–6). These complaints have normally been detected in groups G3A and G4 which requires fine grain size and a low cobalt content. An example of a G3A carbide which fell below hardness of 1750 HV minimum is shown in Fig.5. A similar below-specification hardness was noted on a G4 insert, though on this occasion the deficiency was attributed to excessive porosity (*see* Fig.4) and not oversintering. The presence of high amounts of graphite has also on occasions been found to reduce wear resistance in cast-iron cutting grades (*see* Fig.4). High graphite contents have occasionally been noted on some steel-cutting grades, usually G5, often with no adverse complaints of high wear. These defects usually apply to batches and should be detectable either by hardness or metallographic examination at source.

Insert material

Deficiency in performance due to edge breakdown has been attributed to a number of defects, the most common of which has been identified as a compacting or pressing defect. This fault has occurred most frequently in groups G3A. It has also been noted in G2 and G3 (*see* Fig.4). Gross porosity ASTM B276 B1 type has been thought on occasion to have led to edge breakdown. Another form of defect due to contamination is

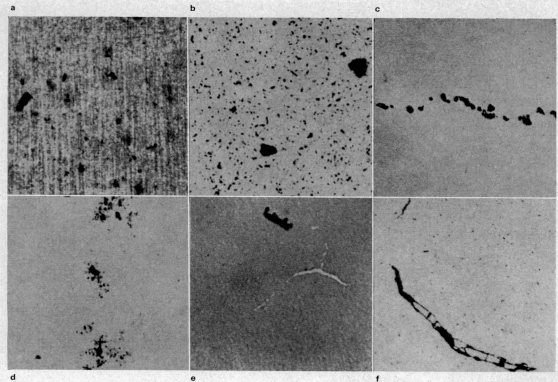

a gross porosity, G9 tip surface; *b* excessive porosity, G4 ×200; *c* lamella compaction defect, G9 ×25; *d* excessive graphite content, G3A ×200; *e* compaction defect, G3A ×200; *f* compaction defect, G3A ×200

4   Metallurgical defects

shown in Fig.4, although this time an isolated instance. An example of a batch of G4 material is shown in Fig.5, the structure of which is seen to have arisen due to inadequate sintering control. Again those defects applying to batches should be metallographically detectable at source.

Trends

Greater use is intended for multi-tipped heads as an aid to maintain accuracy and economics. This in turn will require greater versatility from cemented carbide to operate over a wider cutting speed range. These operations will also require the use of precision- and ultraprecision-ground tips which are not readily available in the UK. Improvements in machine rigidity will allow the use of less tough but higher wear-resisting inserts, and may require an upgrading of the specification level. The replacements of brazed tools by throwaway inserts will continue.

In 1965 there were seven cast-iron non-ferrous cutting grades and nine steel-cutting grades in addition to two grades identified as TiC groups. At the present time five cast-iron non-ferrous and five steel-cutting grades are listed, the latter including one TiC grade. Reference to Table 3 reveals the incentive to reduce the cast-iron grades from five to four, possibly utilizing G3A in place of G3 or G4 and the deletion of G8 from the steel-cutting grades. Investigations are under way to evaluate these possibilities. With the continuing development of the TiC grades further rationalization seems possible. Of the Ti grades developed within the company, 4J is well established as the fine-finishing grade G9. Two additional grades have been developed containing increasing 'binder contents', these grades are 5H and 7G which are proposed as general-purpose and rough-cutting grades respectively.

Trials employing the 5H TiC grade have been quite promising when compared with the current G6 application, and some impressive performance ratings were found to be possible. Success was also achieved with 7G, although it should be noted that a smaller number of tests were undertaken. It should also be stated that 5H and 7G specifications are not yet finalized, and development is being continued in the Ford applied research departments.

Although to date most trials employing titanium carbide have been used in cutting steel, some cast-iron cutting trials have been completed and a proposal has been made to apply the previously mentioned carbides in a similar manner to both steel and cast iron. Hence a schematic description of past and possible future trends is shown in Table 3. An acceleration of the development of TiC grades used within the company is likely to result from the recent manufacturing licences granted to a UK supplier.

The current high-usage steel-cutting grades G5, 6, and 7 are under continuing scrutiny and it is considered that developments and refinements of existing grades will give an improved predictability and consistency of performance.

OTHER CUTTING MATERIALS WITHIN THE COMPANY

Laminates

Laminates have been tried on cast-iron turning applications and compared to the Ford G3A carbides. Wear resistance has been comparable or slightly better. No economic advantage was gained, and a definite tendency to chipping noted. Further trials are planned with modified materials. It is possible that although currently outside the Ford standard, laminates may well fit in a similar manner to the titanium groups, as shown in Table 3.

Coated carbides

It is too early yet to give an accurate assessment. If the predicted potential is fulfilled their performance may be as versatile as the titanium grades. Initial trials have indicated incentive for further trials.

Table 3   Schematic layout of past and possible future trends

	1965	1970		Future
Cast-iron and	F21	G1		7G(G1, G2, G5, G6)
non-ferrous cutting	F20			
	F14	G2		
	F13			
	F12	G3	G3A	
	F10			
	F11	G4		
Steel cutting	F42	G5		5H(G2, G3A, G6, G7)
	F41			
	F43	G6		
	F40			4J(G2, G3A, G7, G9)
	F34	G7		
	F33			
	F32			
	F31	G8		
	F30			
	F60	G9		
	F61			

High-speed steels

In the period of review no startling revelations have occurred within the scope of high-speed steel-cutting tools. The same trends predicted in 1965 could be considered relevant now.

T1 and M2 steels

These two grades of high-speed steel still form the backbone of steel requirements for cutting tools. Although M2 is preferred by our company and many suppliers, the use of T1 persists due to a national demand with some tool manufacturers to supply this material. Perhaps the combined persuasion of the price increase in wolfram and the efforts of the Ministry of Technology will give impetus to the trend away from T1. Trials have continued on the comparison of these two materials with the reputedly successful M7. These trials have been confined to various forms of drill though the conclusions are generally identical, i.e. negligible difference in performance. In fact trials have been seriously impeded by the persistent delivery of tools showing overheating or over-soaking, and Snyder-Graff intercept grain size of 8–9 on drills of 0·5in diameter have been recorded.

M2 continues to be employed successfully on unground hobbing application where the resulphurized version is accepted interchangeably with M2, depending on the conditions prevailing at the tool makers.

NB Reported uses of solid cemented-carbide hobs possibly relate to a proposed gear-cutting application on a cast-iron camshaft auxiliary gear. These Continental tools, which are of the order of 3·5in diameter, apparently have no counterpart among UK manufacturers. A hob employing brazed cutting surfaces has been successfully cutting the phenolic timing gear on the V–4/6 range of engines.

At present the steel quality available for hobs is being investigated. A preliminary observation on this quality would be that, based on carbide segregation and distribution, no great progress has been achieved overall. Although electroslag-refined material has been proposed for this purpose, and excellent examples of stock have been examined, the view of the tool manufacturers contacted is that consistency has yet to be proved.

Although the isolated use of more highly alloyed steel such as the T15 type continues, their success has been somewhat variable. Brazed cemented-carbide drills have taken on some of the tasks once performed by solid high-speed steel tools, resulting in improved productivity.

The trend towards improved performance on that of T1 and M2 tools is to a limited degree being satisfied by the nitriding of these materials rather than a change in steel. It is anticipated that increasing numbers of tools will be nitrided in a similar manner to drills, on which nitriding is becoming almost standard practice.

Alternative steels

The search for alternative steels to obtain improved performance has received some impetus due to the company's desire

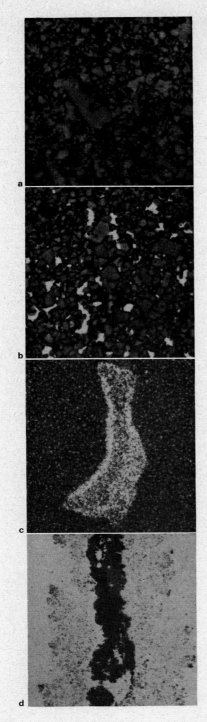

a coarse WC grains, G3A $\times$ 1 500; *b* cobalt segregation, G4 $\times$ 1 500; *c* segregation, G9 $\times$ 120; *d* inclusion, G3A $\times$ 200

5   Metallurgical defects

Ceramics

These materials do find applications with the company but these are confined to cast-iron turning operations such as is required on flywheels and brake drums. The number of applications has been reduced partly due to equipment and environmental problems, in one or two instances ceramics being successfully replaced by 4J finishing at 1 000 ft/min on cast iron. Further applications of ceramics are proposed for the high-speed final boring operations on cylinder blocks, a cutting of up to 1 500 ft/min is anticipated. Constant review of the situation is made as new materials are developed.

to rationalize practices in the associated Ford plants on the Continent and in the UK. It is noted that, in addition to the wide range of steels specified, those steels which are common to Continental and UK practice, for instance M2, are often found to have a high carbon content to meet the desired heat-treatment requirements. It is understood that the subject of another paper in this conference is the new DIN high-speed steel specification. An appraisal from the steel manufacturers as to the merits of the proposed British and DIN standards would be welcome.

Tooling has been, and is being purchased in this country for use in Ford Cologne plants in materials consistent with UK practice. It is hoped that performance data will become available which will be helpful in predicting future trends.

OBSERVATIONS AND CONCLUSIONS

Although this paper is intended as a general survey a number of firm conclusions and observations are possible on the evidence presented.

An aspect of tool material which should be given high priority is an improvement in the quality of insert finish. All cemented-carbide manufacturers should avail themselves of the necessary equipment and expertise to avoid the defects mentioned. It seems fruitless to expend considerable effort to produce a metallurgically sound insert only to be ruined by mishandling or maltreatment for, as we have observed, tip damage accounts for a far higher proportion of rejections than any other reason. Where possible, quality criteria should be explicitly included with the dimensional requirements of BS4193 or similar.

Regarding the metallurgical quality of cemented carbide, as stated previously proved instances of deficiencies are low. However, to the large consumer such as the motor industry the breakdown of the production line is exceedingly costly. It is not unreasonable for customers to expect those defects which apply to batch or powder mixes to have been isolated at source. One would assume that individual powder mixes or heat treatments should be examined. There is a case for the supplier to undertake more and more searching metallographic examination as some faults are apparently not detectable by the normal rapid non-destructive measures, for example the compacting defects shown.

On the question of standardization and rationalization, the customer would prefer to be confronted with fewer more versatile materials from which to choose. This observation applies particularly to cemented carbides. Suppliers are encouraged to pursue this end and possibly couple the results to a national material standard for carbide cutting materials.

The formulation of the material standard for cemented carbide is appreciated to be a complex problem. The large number of independent variables which make up a single machining operation does not lend itself to a readily available solution. The Ford internal approach reported here is based on an essentially practical level, though it is known that many other users are equally inquisitive of the problems of specification. Another approach which has been suggested elsewhere will involve an attempt to identify individual material characteristics and evaluate the effect of various properties on cutting applications. In the meantime the ISO system remains insufficiently selective while the BHMA system, though potentially more discriminating, has lost impact due partly, it is suggested, to misquoting or quoting a variable which has an inherently large scatter band in the results produced. The first stage of high-speed steel rationalization appears to be occurring, that is, a rationalization of terminology; the next stage is eagerly awaited.

Discussion: Views of users

Chairman: K. A. Ridal (BSC, Special Steels Division)

Mr A. I. W. Moore, the rapporteur, presented the following papers: '*Selection and heat treatment of high-speed steels for drill manufacture*', by C. E. Pillinger and G. Huddy; '*Material criteria for gear-cutting hobs*', by W. T. Chesters; '*Tool steels and their application to milling cutters*', by A. L. Horne; '*High-speed steel bar quality*', by N. H. McBroom and Dr T. B. Smith; '*International standardization of performance tests for machining processes*', by Prof. A. W. J. Chisholm and Dr E. J. Pattinson; '*Comparative assessment testing of cutting tools*', by A. Moore.

Mr A. I. W. Moore (Production Engineering Research Association)

My task is to introduce six papers representing in various ways the views of users of tool materials. Three of these papers are based on experience in the manufacture of cutting tools, namely HSS twist drills, HSS hobs, and HSS and carbide milling cutters. The authors of two further papers come from large companies which use tools, one of which represents experience in the maintenance of HSS quality, and the other, experience in testing cutting tools supplied for factory use. The sixth paper is concerned solely with standardization of machining performance tests and describes the steps being taken at international level to coordinate experience and knowledge for the ultimate benefit of all kinds of companies and organizations concerned with machining.

From the standpoint of HSS tool manufacture, the main conclusion is that, hitherto, the quality of HSS has not been controlled sufficiently well, and the resulting inconsistency has carried with it a high risk of premature failure, through inferior resistance to fracture. It goes without saying that a tool should not fracture before it has worn sufficiently, but when one considers the variety of tools made from HSS, there are clearly difficulties in agreeing on acceptable standards of performance. Users in this country feel that a reasonable first step towards solving this problem is the establishment of a British Standard for tool steels. Welcome though this will be, its immediate usefulness will probably be confined to reducing the incidence of premature fracture, because the knowledge on which the proposed British Standard is based has mainly arisen from metallurgical investigations of failed tools. Nevertheless, this is the best foundation we have, and as, metallurgically, our knowledge of the wear properties of different kinds of high-speed steels and different cutting tools becomes more extensive, I believe we shall eventually see the basic standard being supplemented by further standards relating to specific kinds of cutting tools, e.g. high-speed steels for large twist drills, small twist drills, for broaches, for lathe form tools, etc.

The three papers concerning high-speed steel tool manufacture are indicative of the kind of information which will be increasingly required in the future, and they provide examples of the different approach which is required for different machining operations.

In the case of hobs, Chesters points out the wide range of hob sizes involved, from $\frac{1}{2}$in diameter to 13in diameter, and weighing from a few ounces to 700 lb. Like any precision form tool, a hob must be of high quality and give consistent cutting life. Therefore, the material chosen must assure high accuracy, dimensional stability, surface finish, and wear resistance.

High-speed steels used for this application give rise to structures which are less than ideal, mainly due to segrega-

tion and size differences in the free carbides. Therefore, the hob manufacturer standardizes on those grades which provide the best compromise between ease of manufacture on one hand and hobbing performance on the other.

Chesters welcomes the imminent British Standard for different grades of high-speed steel, and then emphasizes the additional need of the tool manufacturer, consistent machinability of tool steels. He explains in some detail typical deficiencies of steels supplied to his company, but no evidence is presented which directly connects these deficiencies with the performance criteria he has himself specified, e.g. abrasion resistance, form accuracy, stability, etc. although it is stated that the direction of grain flow does not appear to affect cutting performance. The author does, however, point out a connexion between grinding abuse and the development of cracks through and along carbide stringers.

In this matter of avoiding deficiencies in steels, it seems to me that there is only one way of achieving this, the one Chesters himself favours, which is standard material specifications. This does not mean that all the problems will disappear overnight, but it does mean that users of high-speed steels will be able to decide for themselves whether or not a non-standard steel is justified for their end product. When drawing up standards of this kind, in the absence of direct evidence of the connexion between, for example, carbide structure and tool performance, the experience and guidance of people like the author is invaluable. However, I would request more information on the relationships with performance.

In the case of milling cutters, Horne opens his paper with a general review of the information which users of cutting tools should supply in order that the tool manufacturer may select the most appropriate high-speed steel for the application.

He then indicates generally those high-speed steels which are most commonly used by his company. This is mostly in line with the experience of other tool manufacturers, and it is interesting to note that the higher vanadium grades (T15, M15) are also commonly used. Apparently, grinding problems are now tolerated, or have been overcome, and in this connexion it would be helpful if either Horne or Chesters indicated the relative grindability of M15 on the basis of Fig.5 of Chesters' paper.

Horne does well to remind us that a single high-speed steel for all industrial applications is unlikely to be feasible, but it is nevertheless clear that he supports the idea of standardizing on a few basic high-speed steel grades. The T1 and M2 grades are obvious candidates, but Horne, along with other tool manufacturers, foresees a changeover from tungsten-based to molybdenum-based steel, and it would be interesting to hear his comments on the possibility of M2 replacing T1 for inserted-blade cutters.

In their paper on twist drill manufacture, Pillinger and Huddy outline the factors which distinguish the various main types of high-speed steels available, and the basis on which those for twist drill manufacture are selected. In this country, the economics generally favour the use of M2.

They then outline some of the problems in obtaining steels to the same basic specification from different suppliers, and also quality factors which need to be controlled. These include the tolerance on the alloying elements themselves, the quantity of impurity elements, dimensional stability, and avoidance of faulty annealing. They also give their views on the extent to which decarburization can be tolerated, and the problems of defining acceptable limits of segregation in relation to the size of bar and different types of high-speed steel.

The paper indicates well the skill and expertise which needs to be exercised by tool manufacturers in taking a particular basic high-speed steel and, through careful, and in some cases novel, heat-treatment procedures, producing the most desirable properties for the type of tool they are making. Although all cutting tools require maximum resistance to abrasive wear, the different basic shapes of tools, the range of sizes involved, and widely different work materials to be machined, give rise to problems of manufacture which require continued vigilance over heat treatment. The need for consistent quality of bar material is therefore unquestionable.

The authors conclude on possible future developments, and point to the need for a better understanding of the basic relationships between high-speed steel properties and cutting performance. This they feel will surely come after the present number of basic high-speed steel grades has been rationalized down to six or seven, or perhaps even as few as three. Different properties may then be achieved through different heat treatments and other treatments such as nitriding.

The paper by McBroom and Smith deals with the somewhat broader field of high-speed steels for machining and forming and, having regard for the widely different nature of machining and forming operations, their main interest is in maintenance of bar quality. They specify that the most important aspects of quality are:

(i) structure, for uniform load-bearing properties
(ii) composition, for inter-batch uniformity of heat treatment.

The authors summarize the defects found in 350 out of a total of 4000 bars inspected over a period of several years, and draw particular attention to the fact that nearly 30 of these bars were of a different grade to that specified. This problem is not confined to tool steels as we ourselves found when obtaining medium-carbon steels through stockists. We also found considerable variation in the interpretation of the term 'normalized'. Quite clearly, the compilation of standard specifications for steels of all kinds is extremely complex, and in view of the fact that many of the people involved in selecting specifications are not metallurgists, there is probably a communication problem. It seems to me that if more emphasis is placed in standard specifications on control of composition and other metallurgical factors than on properties and application data, selection by users will remain hazardous.

The authors provide very useful evidence of the need for adequate heat-treatment control to maintain maximum shock resistance of high-speed steel, and further evidence shows the way for improving the grain structure of high-speed steel. All this effort by user metallurgists emphasizes the need for high-speed steel manufacturers to improve the quality and consistency of their products. However, I think it would be true to say that most of the quality defects investigated are primarily concerned with avoiding premature fracture, whether the high-speed steel is to be used for cutting tools or forming tools.

Several users welcome the improvements being brought about by manufacturers but stress the need for continuation of their efforts. Before leaving this subject, I feel I should mention that in my view the real period of progress will come after the current problems of achieving consistent resistance to premature fracture have been overcome. Then, I feel, we shall see more attention given to the criterion which is of greatest interest to the user, namely wear resistance.

I am not sure whether it is significant that in only one of the four papers I have received is mention made of the performance testing of cutting tools themselves, as distinct from quality control of the high-speed steel. The exception is Chesters' paper about hob manufacture, and in it he describes a test in which a single-tooth fly cutter is used to simulate a tooth of a hob. Tests with this cutter enabled comparisons to be made of the performance of 13 high-speed steels. With T1 as reference, the best performance was obtained with M1 steel, giving a 20% increase in life.

In tests of this kind, it would be wrong to attempt to draw wider conclusions about the relative performance of the various high-speed steels tested, and the tests are only of value therefore as an indication of hobbing performance. Desirably, the test technique needs to be correlated with full hob life tests.

A few years ago PERA carried out an investigation into side and face milling cutter performance, and although the variables were investigated using a single-tooth fly cutting technique, use of this technique was continued only after we had carried out correlatory tests between single- and multiple-tooth cutters made from the same batch of high-speed steel. The results were perfectly satisfactory and showed that, providing proper steps are taken when designing experiments of this kind, a great deal of unnecessarily lengthy testing can be avoided.

The whole philosophy of machinability testing is very complex and the remaining two papers, by Moore and by Chisholm and Pattinson, both deal with this subject, but from quite different standpoints.

In the paper by Moore, an outline is given of the factors which motivated the setting up of his company's machinability laboratory, and the standard techniques developed for testing and assessing the quality of cutting tools used in the factory.

In studies of accelerated test procedures, the author confirmed the inadequacy of several well known procedures discarded by other investigators, including first, the use of artificially high cutting speeds, and second, measurement of cutting forces. He found, as others have done, that cutting performance at high speed bore no relationship to performance at the speeds normally used in practice, and also that the magnitude of forces developed during cutting did not necessarily bear any relationship to tool life. Consequently, assessment of tool performance is based mainly on the development of tool wear to a degree which corresponds to a full tool life test.

Details are given of five standard procedures for machining tests adopted by the author's company, as follows:

(i) single-point turning
(ii) face and end milling
(iii) drilling, reaming, and tapping
(iv) reciprocating power sawing
(v) filing.

I think perhaps the most important feature of the procedures is that they are based on machining operations actually carried out in the company, and care has been taken to relate the work materials, cutting conditions, type of cut, and test criteria to common specific production operations carried out in the company. This approach contrasts strongly with that adopted by other people wishing to carry out some kind of machining performance test. Very often a search naturally starts for the best single test, and eventually it is realized that there is no better alternative, so far as tool testing is concerned, than the one adopted by the author. However, in most cases, companies are reluctant to set up a machinability laboratory unless they can satisfy themselves that the costs will be offset by savings in production. In most cases, companies will not take positive steps to make sure that they are using the best cutting tools for their requirements, and will often rely on the sales efforts of tool manufacturers for improvements in their machining practice. However, in other cases, companies commission an organization like PERA to undertake investigations to determine the best type of cutting tool for a given application, and the basic approach we have always used for many years is that adopted by the author.

The author goes on to say that in addition to the primary function of establishing a sound purchasing policy for standard cutting tools, the testing procedures are used for periodic checking of cutting-tool quality and also to investigate the merits of different cutting-tool materials.

Examples are given of these different applications of cutting-tool testing. The first of these relates to selection from four companies of a supplier of hacksaw blades. The test showed considerable differences in wear performance and enabled a choice to be made. However, the author points out that all four makes of blade easily met the requirements of the test specified in the appropriate British Standard, and he observes that a blade just capable of meeting this test would be of poor quality.

Another application, comparison of the performance of M2 twist drills being used with those from two other suppliers, revealed that the cheaper drills already in use gave only one quarter the life of the other drills. Although it is perhaps a detail, I think we would be interested to know, if the author can tell us, in what way drills from the various manufacturers differed?

At one stage, it had been suggested that the variations in quality of small cutting tools were creating uncontrolled variations in production machining time, but a programme of work covering some 500 small tools showed that the general quality of small tools was of a high standard and that a formal quality-control procedure for cutting tools was not warranted.

Finally, the author describes the third application of cutting-tool testing; comparison of different types of high-speed steel tool bits for automatic lathes, and also end mills made from different types of high-speed steel. A feature of the tool bit tests was the introduction of shop floor trials following laboratory testing. His account of these has a very familiar ring and indicates the enormous complexities in this kind of testing.

First of all, it was found that the high-vanadium high-speed steel (M15) performed better in the shop floor trials than in the laboratory tests, and in the absence of any comment by the author I assume that no apparent explanation for this was found. Also, perhaps not surprisingly, there were considerable variations in both tool geometry and tool life in the shop floor trials, and my view is that unless one can be sure that complete control can be exercised in such trials, they are virtually worthless. First, if the machine is manned by an operator, his integrity must be assumed, and I can recall instances in which an operator has attempted to sabotage a test purely to protect his own personal interests. Second, uncontrolled variables, e.g. tool angles, must be kept to a minimum, and third, the test criteria must be quantitative not qualitative.

In the final set of tool bit tests during which the chip-stream geometry was controlled, it was found that the tungsten-based T6, with high cobalt, gave the longest tool life, but even so, it is interesting to note by comparing the author's Figs.6 and 9 that the chipstream tool life for T6 in terms of area machined was only about three quarters of that of the conventional turning tool. Perhaps the author could comment on this, and also on the apparently curious performance of M15.

In the end milling tests, it was found that the standard 18–4–1 (T1) high-speed steel gave longer tool life than T15, M15, and T4. However, another later test based on metal-removal rate for a specified minimum life showed that M42 achieved a metal-removal rate four times that of T1.

The author concludes with a warning about attempts to read more into the results of a test than can be substantiated, and expresses the view that it must always be borne in mind that the results obtained refer only to the conditions adopted for the test. However, in view of the fact that it is not feasible to optimize every practical machining operation in the manner

described, perhaps he would agree with me that, in the absence of specific tests, it is sometimes possible to make some deductions about machining performance from other test data.

The sixth paper is concerned with the international standardization of performance tests for machining processes and I think we are fortunate to have the opportunity of commenting on the various aspects of the proposals which are currently in preparation. However, before doing this, I should like to present some views for discussion.

As will be appreciated from my comments on the previous paper concerned with cutting-tool testing, I take the view that for cutting-tool testing by tool users, the best approach is the one presented in that paper, i.e. the test should be closely matched to production requirements. Similarly, if the cutting-tool manufacturer wishes to check the overall quality of completed cutting tools from time to time, basically the same approach is desirable. However, he is one step removed from production, and usually finds difficulty in deciding on a single test which represents what the majority of his customers do, and is thus acceptable to them. On the other hand, the manufacturer of the tool material is a further step removed from production so far as his cutting-tool manufacturer customers are concerned, and when he is producing, for example, a particular batch of high-speed steel or carbide, he may not even know whether the material will be used to make a lathe tool or a face milling cutter. Which, then, is the best single test for him? A further complication is that he also supplies direct to tool users making their own cutting tools, and I believe this increases the difficulties which face the material manufacturer in maintaining consistent quality, not only in respect of avoiding premature fracture, but also in terms of machining performance.

I think we therefore have to recognize that the likelihood of a single machinability test being devised which will be acceptable to everyone for virtually every purpose is remote, and the sooner this can be accepted the sooner we shall make more efficient use of our effort. It seems to me therefore that the standardization of performance tests for machining processes needs to be considered, not in isolation, but in terms of the use to which a standardized test will be put. Having dealt with the testing of cutting tools, it would appear that the other important requirement arises from the need to control the quality of various kinds of materials.

It does not matter whether the material is a tool material or a workpiece material, the need is the same, 'Is this batch of material I have just made the same as another so far as machining performance is concerned?'

I believe that a tool wear test is essential. Because very often the kind of machining operation to be performed by the tool material or on the work material cannot be predicted, it is sufficient if the kind of machining operation on which the test is based produces a conventional metal cutting chip. From the many alternatives available, there is little doubt that a single-point turning test of some kind is the least complicated and easiest to control. Desirably, the test needs to be of short duration and to consume a small amount of material, and this suggests a sensitive wear technique which would enable small increments of wear to be detected.

Some years ago, those of us seeking such a test were optimistic about using radioactive techniques to measure rates of tool wear, and although satisfactory techniques were devised, they have not really caught on because of the need to organize irradiation of tools well ahead of them being required, and also the fact that, overall, the tests were no quicker. In the meantime, optical techniques have been developed to a high degree, and at PERA we are able to measure increments of wear in the order of 10^{-4} in without difficulty. To do this, very great care in tool preparation is necessary, and we feel that the techniques we have developed represent the best compromise having regard to all the factors. However, I would not like to give the impression that we use this kind of test all the time. The type of test is chosen carefully in accordance with the requirements of the investigation, and the wear rate test I have mentioned is used when quality comparisons are required. For example, this kind of test would be used to measure variations in machinability of a given type of work material, either within one bar, between different bars within a batch, or between different batches.

One final comment I should make is that in using the wear rate test it is desirable to choose cutting conditions to represent the different conditions of severity and wear rates typically met in industry. For example; light, medium, and heavy cuts should be considered.

I feel we now have a background against which standardization of performance tests for machining processes can be judged. If we are to meet industry's total needs, we require a whole range of standard tests for cutting tools, and there is no reason why such a range cannot be devised right away from which the various user/manufacturer combinations can select specific tests as a basis for commercial supplies. It is not at all clear from the paper by Chisholm and Pattinson what function the proposed ISO test is intended to fulfil. Reference is made in the introduction to the difference between 'practical testing work' and 'research to elucidate the basic physical processes involved in the machining process'. Practical tests are stated to be difficult to reproduce and are often carried out under widely differing conditions so that results from different sources cannot be compared. I suppose one might add that research work is often carried out in some laboratories under idealized unrealistic conditions and is often of no practical value.

It seems to be accepted that the ISO proposals should be orientated towards a tool wear test, although of course there is no reason why other measurements such as cutting forces, power consumed, etc., cannot be made once standardization of conditions for a test have been established. However, as no mention is made of the use of other criteria for purposes of comparison, I would suggest that the ISO document should include a presentation of the consensus of opinion of the various contributing countries on this aspect of performance criteria.

Reference is made in the introduction to various developments such as the advent of new and difficult-to-machine materials, the general trend to automatic production, pre-planned machining, numerically controlled machining, etc., which have generated interest in the establishment of banks of machining data. This in my view has little to do with standardized machining tests and may add further to the confusion which already exists.

The authors describe the background to the establishment of ISO Working Group 22 in October 1967 on 'Unification of tool life cutting tests'. BSI also established Committee MEE179 on 'Tool life testing' to develop the British viewpoint. The paper is an account of the progress made of the work of the ISO group and the BSI committee.

Initially the ISO group were asked to prepare a new ISO recommendation on a standard machining test for use by scientific laboratories, but the group decided at the outset to include the requirements of companies and industrial laboratories.

The BSI committee soon found that doubts were being expressed concerning the need for control represented by the proposed ISO test procedure, but after discussion in the ISO group, the latter emphasized that the proposed control was regarded as minimal for an adequate degree of reliability. Nevertheless, the group agreed 'to consider in the future the possibility of developing a recommendation for a short-time inexpensive machining test to supplement the full testing method'.

The introductory note of the proposed ISO recommendation states that 'the proposals have been so framed that they can be directly applied in industrial testing and in research'. However, I have a feeling that the primary function of the proposals is to provide a common base for comparing the results of machining research in different laboratories, and that in view of the extensive degree of control proposed, it is unwise to claim that they will be suitable for industrial requirements. My own view is that proposals need to be drawn up for tests for specific purposes, e.g. quality control of machinability of a particular type of work material, quality control of wear resistance of a particular tool material, etc., and these should be treated as an entirely separate requirement.

So far as the details of the proposed standardization of the machining variables is concerned, the approach seems to be eminently suitable for scientific comparisons of results from different laboratories.

The two proposed standard reference work materials are a hot-rolled medium-carbon steel to a modified ISO specification and ISO Grade 25 cast iron which has a hardness of 200–225 HB. If required, a second reference steel material can be adopted by individual countries.

Three standard reference tool materials are proposed, molybdenum-based M2 high-speed steel, and cemented carbides for cutting steel and cast iron respectively. The tungsten-based T1 high-speed steel was considered but rejected in favour of M2. In view of the fact that no standardization of carbides by manufacturing variables is in existence it is proposed that companies and organizations should be supplied with carbide tips from an international tool bank.

Cutting-tool geometry is given for ceramic as well as high-speed steel and carbide tools. The main features are:

(i) an approach angle of 15°
(ii) for high-speed steel a normal rake of 25° (we feel this is much too high) and 8° clearance
(iii) for carbide, positive or negative rake of 6°, and 5° or 6° clearance
(iv) for ceramic, negative rake of 6° and 6° clearance.

Some difficulty was experienced in establishing standard cutting conditions, and the compromise between the desire for a single set of conditions on the one hand, and a very wide and extensive range on the other, has resulted in four standard conditions being specified for combinations of feed, depth of cut, and tool corner radius.

Details are not given in the paper of the extensive precautions in the test procedure covering the proportions and preparation of the workpiece, and the grinding of tools.

Chisholm and Pattinson deal at some length with the matter of tool life criteria, and in particular definition of the end point of tool life. These criteria are still under consideration by the ISO group, but it seems likely that for carbide and ceramic tools, both flank wear and crater wear will be adopted. For HSS tools, catastrophic failure is likely to be the criterion. In general, a cutting fluid is not to be used.

Dr P. Greenfield, the rapporteur, presented the following papers: '*High-speed steels for cutting aero-engine materials*', by D. Milwain and J. M. Thompson; '*Selection and quality assurance of tool materials used in the machining of earthmoving components*', by R. Dickson; '*Tool materials used for machining turbine components*', by G. J. Hill; '*Cutting tools in mass-production machining of auto parts*', by N. P. Riley.

Dr P. Greenfield (GEC Power Engineering Ltd)
I have four papers to review, from Ford, Caterpillar, GEC, and Rolls-Royce, a formidable cross-section of industry. The first is from Rolls-Royce, by Milwain and Thompson, and deals with the historical changes in high-speed steel utilization at the Bristol Engine Division and the Aero Engine Division of that company.

In the early sixties, T1, T4, and T6 were becoming inadequate for machining nickel and cobalt alloys used in gas turbines. Tools burst in production and were dangerous.

Purchase specifications covered only grossly bad defects such as extreme central segregation, cracks, and inclusions.

At the supplier's suggestion, the high-carbon high-vanadium alloy T15 was introduced as being harder at 900 HV than existing alloys. Later a similar alloy, M15, was introduced, also successfully. The biggest drawback with T15 and M15 was that grinding of fine forms on some broaches was very difficult, due to the high wear resistance associated with the high carbide content.

Later, the suppliers offered T42, lower in carbon and vanadium than T15 but higher in cobalt. In milling Nimonic 105 disks, T42 was six to seven times better than T1. The performance in broaching was disappointing due to lack of toughness.

In broaching, T1 was not too bad at 850 HV but at 800 HV was very poor indeed. M15 and T15 had such poor grindability that both manufacture and maintenance were an expensive business and the problem became so acute that some intricate forms were impossible to produce.

This difficulty was not solved until the arrival of M42 which was offered by the supplier as having the performance of T15, M15, and T42 with the grindability of T1. This was found to be satisfactory in some cases but, until the heat-treatment problems had been sorted out, was susceptible to chipping. A number of heat treatments have now been defined for various applications.

The next section of the paper deals with metallurgical quality requirements. It is stated that Rolls-Royce have standards for composition ranges, central segregation, carbide distribution, cleanliness, decarburization, and heat-treatment response. Not unnaturally, the application of these standards reduce poor tool performance due to low hardness and poor quality tool material.

Faced by a proliferation of inaccurate information from suppliers on heat treatment, Rolls conducted some work themselves to optimize hardening and tempering treatments. Usually, hardness falls off very slowly for the first 10–15°C above peak hardness and because toughness and stability are at a premium, minimum tempering temperatures were fixed at slightly above those required to produce peak hardness. It is said that the number of tempers was fixed by metallographic examination, presumably by the amount of retained austenite. It would be interesting to know what metallographic features in particular Rolls associated with good performance and whether these were correlated with performance in controlled machining studies.

Rolls' experience has enabled them to group the high-speed steels available to ease the problem of selection. These groups are given in Table 1 of the paper.

The first group has as the basic material T1, but because of materials cost associated with the high tungsten content, work has been conducted to replace it with M1 or M2. M2 is tougher for the same hardness and appears to perform as well as T1. The biggest problem is the increase in grinding difficulties associated with the higher vanadium-bearing M2. In the gas turbine business, broaches are usually required in relatively small numbers and the poor grindability can be tolerated. As far as M1 is concerned there is no conclusive evidence as yet that it performs as well as T1. This group of alloys can be nitrided and a case is sometimes useful on drills and taps for nickel alloys.

The basic alloy in group 2 is the 14W–4Cr–4V alloy which appears to have been used for high hardness levels before the introduction of T15 and M15 without having been critically evaluated.

Group 3 containing M15 and T15, has like group 2 a high wear resistance with the attendant difficulties in grinding and tool production. The increase in cost per tool compared with T1 can be anywhere between 5 and 100% depending on the amount of grinding required.

Group 4 materials contain a higher than usual carbon content in relation to the vanadium level. This, together with the cobalt level, increases the hardening response. They are, however, sensitive to overheating on hardening and in this respect M42 gives most problems.

Group 5 materials are the most highly alloyed of the high-speed steels as regards Co content and their toughness is very much reduced at comparative hardness levels when compared with other high-speed steels. The traditional alloy is T6 which has given good performance where the tool form is solid, used over the range 850–900 HV. M44 and T42 are slightly tougher and can be heat treated to 1000 HV. M44 has better grindability than T42. These would only be used for high-speed applications, preferably on small tools as the material cost is high.

Considering all groups, M2 is used as the basic engineering alloy with M42 next in line if the performance of M2 is insufficient. Further work is needed to establish the economic boundary between M42 and T42 or M44.

The next paper, by R. Dickson of the Caterpillar Tractor Company, begins with generalities concerning the four basic requirements of high-speed steel: abrasion resistance, toughness, red hardness, and grindability. The choice depends on which of these factors is considered most important for any particular application.

Little is said, unfortunately, on exactly how this choice is made, and I guess that Caterpillar rely heavily on general experience, supplier's advice, and adopting a policy of ringing the changes until it works. A few general remarks are also made about carbides. The term 'strength' is used quite loosely here to mean resistance to chipping, which is not the same thing as strength in my understanding of the word.

The main part of the paper, however, concerns quality control of high-speed steels and carbides at Caterpillar.

Supplies of high-speed steel and carbides to Caterpillar must be approved metallurgically. Inferior high-speed steels are detected by hardness and microstructural examination. The structure is examined for sulphide and carbide size and distribution as well as retained austenite and grain size. No standards of acceptance are given or information on the effects of deviation from such standards. Cemented carbides are also examined but with more difficulty due to their high hardness. With carbides, hardness, density, and microstructure are examined, the latter including porosity and identification of phases. While a typical approval sheet is shown in Fig.2, nothing is said about how the various levels of acceptance are arrived at.

High-speed steels are rejected for two basic causes, either faulty raw material or faulty heat treatment. Faulty steels are due to (a) voids due to piping in the ingot stage or bursting during rolling or forging; (b) tears caused by working at too low a temperature; (c) poor distribution of carbide, particularly networks leading to sudden brittle failure; (d) poor sulphide distribution; (e) the use of bar stock instead of forgings, leading to insufficient hot working; and (f) wrong material due to poor control of stock by the toolmaker. This last has, incredibly enough, a higher incidence than any of the others.

Faulty heat treatment occurs due to (a) overheating leading to coalescence of carbides and large grain sizes; (b) under-heating; (c) single tempering leaving too much retained austenite; (d) carbon enrichment; (e) decarburization; (f) quench cracking, associated very often with poor tool design.

Cemented carbide rejection is due to segregation or improper sintering.

Other causes of carbide failure are associated with large grain sizes, the epsilon phase, incomplete diffusion, the presence of large rounded crystals, uneven distribution of cobalt binder, the presence of the eta phase, and porosity. The relative importance of these various defects is not discussed and probably not known.

In conclusion, it is said that many metal-cutting problems may be best solved by eliminating deficiencies in the cutting-tool material rather than seeking solutions in tool geometries

have two functions. First, it could be used to check the variability of tool and workpiece materials. Second, it could be used to assemble basic turning data and as such be the basis for future more realistic tests. It is this latter use with which I am most concerned. I am at present undertaking work on the machinability of materials used in the aerospace industry. Several hundred thousand pounds have been spent converting material into chips in order to obtain extensive tool life results. A large percentage of these results are useless, owing to the different wear criteria, etc. that have been used. A standard machining test would be advantageous provided realistic wear criteria are used.

I would like to comment on the wear criteria chosen for carbide cutting tools, since I consider that these do not conform with current industrial practice. The ISO test recommends certain flank wear values (both mean and maximum) and a crater wear value. I would have liked to see 'oxidation pit' included, particularly as recent work by Professor König has indicated that this type of wear can be a fairly reliable criteria for tool life, at high metal-removal rates. In the first instance we have to combine all these criteria, tool life being defined as when one of the criteria reaches its limiting value. This is important since when machining even one workpiece material with several carbide tools, different wear mechanisms are evident. Under certain conditions, experience will allow the rejection of some criteria.

With regard to the actual value of flank wear, I consider these to be too low to give a realistic comparison between different tool and workpiece materials. In my experience the values of 0·3 mm mean flank wear and 0·5 mm maximum flank wear should be of the order of 0·4 and 0·8 mm respectively. The value of 0·3 mm for mean flank wear is a very unfortunate choice, since when machining a wide range of materials (cast iron, stainless steels, Ni and Co base alloys) with ISO K grade carbides, one often gets very rapid wear up to a flank wear level of between 0·2 and 0·3 mm before the wear curve levels off for a considerable time. It is, therefore, essential that the mean flank wear level chosen should be greater than the value of 'levelling off'.

In conclusion, I should like to repeat that I am in favour of a standard machining test provided it conforms to current industrial practice. I should like to ask the authors on what grounds the wear criteria were chosen and what sections of industry were consulted.

Dr R. Edwards (Metro-Cutanit Ltd)

Speaking on behalf of the British Hardmetal Association, I think that British industry should make some defence against the strong contingent of Swedes we have at this conference.

I think it ought to be said that a meeting is to take place between Professor Chisholm and members of the British Hardmetal Association to discuss ways and means in which we might cooperate.

I must say that we have a lot of misgivings about the possibilities of this testing but, at the same time, we are quite prepared to examine every possible way in which we may cooperate, even to the extent, perhaps, of producing standard compositions.

Dr Greenfield mentioned the problems that the users have. The British Hardmetal Association have one very big problem indeed, and that is contact with the user. We have a club, the Hardmetal Association Club, which cooperates technically, but there is no such users' club. I think it would be a good thing if this were formed. By contact between a users' club and the Hardmetal Association, some of these problems could be exchanged in the concentrated if sporadic forums we have from time to time.

The Chairman

I was quite happy in some respects about the ISO specification. As a steelmaker one of the perennial problems is to make steels which are more free machining and to have a bank of standard tools. At least a test which is acceptable to a number of people is much better than using an arbitrary test and an arbitrary tool shape with materials from some fairly arbitrary source. Therefore I do think it is a good step, if only from the steelmaker's point of view, where we want to look at the effect of deoxidation practice and the variations which can occur in the workpiece, so that we can at least provide you with better steel to use.

Dr Pattinson

I would like to thank Mr A. I. W. Moore for his very able presentation of our paper but I should like to take this opportunity to comment on a few of the points he raised.

Mr Moore questioned the choice of the recommended tool wear test and suggested that a sensitive wear technique of short duration was desirable. I should like to mention that the working group do recognize the attributes of accelerated wear tests and considerable discussion about them has taken place within the working group. Unfortunately accelerated wear tests have a singularly bad record for reliability. As Mr Moore himself said, it is important to choose cutting conditions which represent the correct thermal conditions of the machining operation and there is no real evidence that an accelerated wear test can be chosen to achieve equivalent thermal conditions to those which occur in a conventional machining operation.

Mr Moore also suggested that alternative criteria such as cutting forces or power consumption may be preferable to a wear test. Unfortunately no member of the working group has, to the best of my knowledge, thought of cutting-force measurement or power consumption as alternative criteria to a tool wear test.

With regard to Dr G. Barrow's remarks on the choice of tool life criteria, this was the subject of a special study group under the chairmanship of Professor Pekelharing of Delft, and because of his first hand knowledge of the oxidation of carbide tools, the choice of oxidation as a possible tool life criterion was discussed at length. Nobody is more conscious than members of the working group that a tool life criterion that is a functional criterion is preferable to the choice of an arbitary measure of, say a crater depth or a flank wear scar. It is essential, however, to have a definable dimensional change in the tool that can be measured. It was felt by the group that perhaps eventually such a dimensional change associated with the oxidation of tools may be available, but no one seemed able to suggest one to date.

Finally, I should like to thank Dr Edwards for his comments, and note with interest that the Hardmetal Association are hoping to cooperate with Professor Chisholm in the possible production of some international or national standard of carbide tool material.

The Chairman

I am surprised at the apparent lack of interest when it comes to spending money to send people to ISO meetings. I could recommend to Mr Fenner and his colleagues that this might be a good way of spending some of their money.

Mr A. J. Fenner (Ministry of Technology)

I feel that it is necessary to come back and put the record straight. First of all, with regard to the ISO recommendation, my own impression is that the Ministry of Technology does give a lot of indirect support to the ISO, without paying for the visits of industrial representatives to ISO meetings.

If I may refer to the discussion on tool materials, a remark was made which showed that the speaker interpreted me as saying that the Ministry was proposing to spend £110000 on ausforming. In fact, the programme to which I referred is a broad programme which includes ausforming as a fairly small part, and should be of interest to both makers and users of steel cutting tools.

The Chairman

If high-temperature thermomechanical treatment is included as well as ausforming, I would like to say that my comments still apply.

Mr R. Whittaker (BSC, Special Steels Division,)

Before nationalization, I was at the Park Gate Iron and Steel Company and Mr C. Moore was one of my staff. As a steelmaker, I came to this conference mainly to listen to what the toolmakers had to say, but there have been a number of comments about the producers of materials which are cut, so I feel I should comment.

Regarding the ISO Committee dealing with this recommended test, I think we should note that it is a recommended test, not yet a standard test. When we first got to know about this, we found that there were no representatives on the committee from the steel industry, and we requested that Mr Moore should join, since he was leading a research team into the machining characteristics of free-cutting steels. I think he has made one or two very useful contributions, as we find from time to time that many of our disciplines can become rather parochial and people do not always think of the wider implications.

Dr Pattinson said that there was only one person going from this country to the meeting in Brussels, but that is wrong. Mr Moore attended three earlier meetings and, now that we have been nationalized, I might mention that a considerable amount of activity has been going on regarding machinability research. It has been decided that research into machinability in respect of BSC activities, will be concentrated at the Swinden Laboratories at Rotherham, which is the Central Research Department for the Special Steels Division.

Swinden Laboratories are currently forming a machinability research centre, and we are hoping that they will be able to coordinate a lot of the work which has been done in the past by different companies and that we can eventually resolve many of our problems.

With Mr Moore leaving BSC, it is necessary that we maintain continuity with the ISO committee, and Dr Keane has been nominated to attend the Brussels conference.

We are anxious ourselves to have a test which is meaningful. We have looked at many tests in the past, and so have other people, and as Mr Moore said earlier, machinability research has been going on for more than 70 years, and we have not yet got a standard test. This is something that we need; we know that we are not going to come up with the perfect test, but if we can get some international agreement, we can then compare and contrast our figures and, I sincerely hope, make progress.

Mr A. I. W. Moore (PERA)

I feel I should make it clear, with regard to Dr Pattinson's comments about cutting-force measurement, that I am not in favour of using cutting-force measurements as a basis for comparing cutting-tool performance. My point was that when a standard procedure for a test has been established, it is quite easy to measure cutting forces if one wishes to do so.

Mr W. S. Caisley (Firth Brown Tools Ltd)

During the last session and indeed during the whole conference, constant reference has been made to the relative merits of tungsten-base v. molybdenum-base high-speed steel.

There are arguments for and against each class of steel, but it has been suggested that molybdenum-base high-speed steel will replace tungsten-base because of price advantages. This price advantage may be the case at the present time, but whether or not it will remain so in the future is a matter for debate and I think it is important to point out that, on many applications, Mo base HSS performs equally as well as W base HSS, and on some applications is superior.

My company has recently carried out a series of tests where currently used W base HSS have been compared against equivalent Mo base HSS in toolbit form on continuous turning tests.

Three workpiece materials were chosen, En9, En24, and cast iron. On En9, M2 proved 30% superior to T1, M35 proved equal to T4, and M36 proved marginally superior to T5. However, neither M15 nor M42 achieved parity in performance with T15. Of the two Mo base steels, the M42 proved somewhat superior to M15, but even so gave some 20% poorer performance compared with T15.

On En24, again M2, M35, and M36 proved equal or superior to T1, T4, and T5 respectively over a range of feeds and speeds, M15 and M42 matched the performance of T15 at speeds up to 90 sft/min, but above this T15 showed its superiority.

On cast iron all molybdenum grades performed marginally better than tungsten grades with the exception of M15, the performance of which was slightly lower than T15. However, M42 performed extremely well in comparison to T15.

Mr Horne

Mr Caisley, I feel, gave a very fair summing up of the relative merits of the various tool steels under discussion.

As was made clear in my paper, my company have for some years given consideration to the possibility of a changeover to molybdenum steels. In more recent times it has appeared to us that in the long term it may prove to be as much a matter of necessity as of choice. It is perhaps significant that delivery dates for high-vanadium tungsten steels have become somewhat extended. From a technical standpoint the case for M2 as against T1 seems to be proven.

In this connexion a previous speaker mentioned M7 as a possible alternative to T1, implying, I feel, that it had not been sufficiently investigated. My company have tested this material over several years, and found it had great merit in certain applications but not sufficient to be preferred to M2.

On testing the higher vanadium grades of tool steels it was evident that T15 was superior to M15. M42 appeared to be most nearly comparable to T15, but is somewhat prone to chipping. Chipping creates maintenance problems. Wear is preferable to chipping since the amount of regrinding necessary is less with worn tools than with chipped tools.

Mr Moore asked for a comment on the relative grindability of the higher vanadium tool steels. While not wishing to disagree with the figures in the papers referred to, I can say that over the years my company has developed techniques of grinding these steels without undue difficulty.

Mr Moore also asked whether M2 might replace T1 for inserted-blade cutters. This, as I indicated in my paper, would depend on two factors. First, that we were satisfied, both from a technical and economic viewpoint, that it was a correct decision, and second, that the availability of tungsten and molybdenum steels could well precipitate the changeover.

The Chairman

Everyone has been talking about requiring consistency of tool steels. Mr Horne in his paper made a passing reference to electroslag but said that its superiority in cleanness was not yet proven. However, I would have thought one should have commented both on the structural aspects as well as the cleanness.

Dr Craik said earlier that you get roughly what you deserve. May I add that you get roughly what you pay for, and the sales of electroslag-refined tool steel have not exactly soared. I wonder if some of the tool users and makers can tell us why? Do they see in the future a growing requirement for quality? Do they see electroslag playing a part ? If you are shy of talking about that, you are quite welcome to bring in the possibilities of using powder metallurgy to find Utopia.

Dr R. L. Craik (BISRA – The Corporate Laboratories, BSC)

In the discussion on tool failure, the rapporteur reported on certain developments in tool steels. There is the electroslag route and the powder-metallurgy route. We are now facing a situation where the market is demanding value for money, and value for money extends right through from the raw materials to the finished product, and I think the representative from Firth Brown Tools made an excellent point when he demanded consistency. This is exactly what value for money means.

It seems to me that the conclusions to be drawn from the session on tool steels can be roughly summarized as follows. There is powder metallurgy, which is perfectly capable of making small bar stock of excellent consistency. The processing is a means whereby all the vicissitudes of segregation can be eliminated; it enables the achievement of a fine grain size. I think there may be some question whether fine grain size is an absolute necessity. The point is that in small size tools, the final product demands consistency to be very high.

There is another area of tools which are very much larger, where we are not so interested perhaps in having fine grain size, but nevertheless we are much more concerned with having adequate ductility and consistency in expensive large size components. It seems to me that the production scene in process metallurgy is faced with two alternatives. There is large-scale electroslag refining, which is developing very rapidly in this country and will give clean steels; there is no shadow of doubt about this, so I think that the quality in that direction is there. There are means of operating electroslag refining which give further refinements of structure. It seems to me that the process metallurgist, particularly in this industry, is giving the steel user the opportunity of saying that he can be met on both scores, for large size tool components and small scale if he so wishes.

It is on these two points that I would like some comment to be made, because, in investing in the future, one has to consider where the market lies. Dr Ridal rightly introduced this whole question of what is the market and what is the market growth. This is always a very difficult question to answer, and I think that it is in the hands of tool users and toolmakers to give some indication of this.

I think, from the point of view of the steel industry, it is now coming into the position in all its facets, both public and private, of being able to do this if investment can be made available in the light of the future market. Mr Tomalin at the beginning gave a clear indication of the value of the market, some £35 m. in 1969, and this represents a considerable increase over previous years, particularly as far as export is concerned. I think that, with devaluation in this country and revaluation elsewhere, a very interesting market situation can develop.

Mr B. Hawley (Firth Brown Tools Ltd)

Referring to Table I on page 204 in the paper by McBroom and Smith, we have carried out a similar analysis which reveals that 11·6% of bars inspected over the last 12 months were rejected for quality defects. A breakdown of the causes for rejection is as follows:

	Firth Brown Tools Ltd %	GKN %
Cracked bars	2	5
Inclusions	4·5	9
Macrodefects		6·5
Chemical defects (analysis)	1	0·5
Network patterns of carbides	48 }Segn	10
Carbide banding	22·5 }	
Excessive carbide particle size	1	—
Mixed grades	—	8
Underannealing	6	28
Decarburization	15	16
Bars bent	—	9
Bars oversize	—*	8
*not included	100	100

While the figures cannot be directly compared they serve to show the high incidence of rejections by my company for defects associated with carbide segregation. This is in spite of the fact that our tolerances for segregation do not appear as stringent as those of GKN shown in Fig.3. It should be pointed out however that 11·6% rejection rate refers only to percentage of bars inspected and not to total tonnage.

The Chairman

Mr Chesters of David Brown talks about 700 lb blanks, for very large 13in diameter hobs. Did he not take the opportunity of looking at electroslag material, which seems to be ideal for this purpose?

In the paper by McBroom and Smith of GKN, Figs.4–7 show excellent micrographs of electroslag-refined material. They tell us what the bar size and quality was but they make no reference whatsoever to the size of ingot from which this bar was obtained. I would have thought that this was significant. I wonder whether they would like to comment, and to tell us what their experience was in more detail?

Mr Chesters

When Mr Moore enquired about the quantitative effect of structure on cutting performance he raised an important issue since there can be little point in including stringent structural clauses in a specification unless the effect of such deviations on manufacture or performance are known and that there is proof that the advantages justify the increased cost of quality control. In this modern world there is little virtue in marketing expensive tools in high-quality steel if they do not show more than a marginal increase in cutting performance over a competitor's product.

However, turning to Table I where the performance indices of a number of steel grades are compared, it will be noted that when the results of test no.5 (hot-worked T1) and test no.9 (as-cast T1) are compared there is a difference in cutting index of about 25%. This difference is not great bearing in mind that the hot-worked material being made from $\frac{7}{16}$in square bar would have a 'near ideal' structure and that the as-cast cutter being machined from the periphery of a large cast blank would have all the 'undesirable' constituents present, but this finding has been confirmed by other results carried out on tools where structure has been the only variable introduced into the tests.

For every hob supplied for cutting industrial gears, there are about fifteen manufactured for the automobile industry and here, the quality-control philosophy is that all hobs which give good performance must be of satisfactory metallurgical quality and that any which do not give this level of performance must have some metallurgical deficiency; this leads to the system of examining all hobs which have a poor life to discover whether the structural deficiencies are such that a claim for a free replacement could be justified. In most cases, such claims have been justified and statistical analysis has suggested that in general, performance is related to structural condition. However, there have been many instances where hobs giving outstanding performance have been examined and their structural condition has been found to be substandard.

At many of the metallurgical conferences such as this, metallurgists tend to attribute all shortcomings in performance to metallurgical faults even though they are not prepared to quote quantitative values. Since the cutting tests have shown that a change from ideal to non-ideal structure can only account for a loss of 25% in cutting performance, it is obvious that when tools have very short lives there must be other factors present, and metallurgists do not always take into account factors such as the rigidity of the machining concerned or the possibility that the operator does not always change his hob when resharpening is required. From cutting-test graphs, where the amount of metal removed is plotted against the wear, it is often shown that up to a certain stage the wear is progressive and then suddenly the wear violently accelerates. Should the machine operator allow his hob to operate past the accelerated wear point, damage can be done to the underlying material. Another point worth recording is that, while the greatest care is taken during hob making, re-sharpening processes are often very suspect and, wherever there is abuse during regrinding, the subsequent performance is often adversely affected. In fact, there have been many instances where heavy grinding has induced such a high level of residual stress that fatigue cracks have initiated during subsequent machine cutting operations.

Mr Moore has asked whether I have any quantitative data on grindability of the various grades of steel, and the simple answer is that I have not. On the other hand, it is known that in terms of metal removal, certain grades are more difficult to grind than others, but, provided that special grinding techniques are developed, the problem becomes merely one of economics. The more difficult problem is to find why certain components in a batch are more difficult to grind than the remainder, and this is receiving close attention. This kind of problem is made more difficult to solve from the metal-lurgical point of view since it is well known that with mass-production methods, one is faced with the eternal struggle between tight times and high bonus rates, and often it is possible to show that grindability becomes a serious problem just before annual holidays or when a wise and knowledge-able foreman is absent from work.

Mr McBroom

On the question about ingot size, the reason we do not give this is that when we buy a bar, we do not get a ticket with it telling us what the ingot size was.

Dr Craik and the Chairman have been asking about the potential for electroslag refining and what market may exist. I think that the premium paid for electroslag refining will depend on what premium is being paid for selecting mater-ial anyway, and what relationship may exist between sup-plier and customer. Certainly there is a demand among users for improved quality, or as high a quality as one can get. I think the reasons for this stem first from the sort of rejection figures mentioned. These provide real evidence that we need to push for higher quality. Other reasons are tied up with the questions of whether there is any relationship between qual-ity defects and performance, and what basis the consumer has for applying tight specifications against the manufacturer.

Taking the first point, one can certainly relate the dis-appearance of specific failures in practice with the absence of certain quality defects. I think that Mr Milwain has refer-red to some of these, and if I can use a trim die as a cutting tool for this purpose, I can say that the absence of inadequate heat-treatment response in trim dies does certainly improve their performance. On the second point, I can also say that failures attributable to defects are not encountered as often in works where tool application procedures are well developed as they are when such procedures are not well developed. This provides the basis for quality-control procedures, pro-cedures which are often only part of a much wider inte-grated system, covering such aspects as correct selection, correct application, and one thing which I do not think has been emphasized sufficiently in this conference, correct usage. One definition of the ideal tool material is one which can be subjected to as much abuse as you can give it and still stand up.

I would like to close by adding a comment on the M2–T1 controversy, and in some respects to defend the BSI com-mittee, who, after a great deal of deliberation, came up with the not very exciting recommendation that T1 could be compared against M2. In this context, I am very pleased to see that data comparing the tungsten and molybdenum grades is coming out because in any attempt to rationalize the number of high-speed steels in use, this kind of data is extremely valuable. We had the market prognosis from both the molybdenum and tungsten camps in the discussion on tool materials, and it is perhaps difficult to assess just what way we ought to be going in the future, whether exclusively to molybdenum, or to include tungsten steels, but in any event, I would put it to you that there are probably too many high-speed steels on the market, and that savings might accrue from some rationalization of the number of grades. The Ministry of Technology, I think, would be better spend-ing their money in this area than £110000 on ausforming.

The Chairman

Referring to Figs.4 and 8 in Mr Chesters' paper, I would remind you that this banding and Chinese script do not occur in powder metallurgy. I would be grateful if someone could say categorically that they are not interested in powder metallurgy if that is what their silence signifies.

Mr G. M. Sturgeon (BISRA – The Corporate Laboratories, BSC)

I am disappointed by the response to the paper by Dulis *et al.* on the use of powder-metallurgy techniques for the produc-tion of high-speed tool steels. The development of powder technology in this field could well provide a breakthrough which would benefit both the user and the manufacturer. In the case of the user this could result in a more consistent product with improved tool life and a reduction in the number of sudden tool failures in service. The manufacturers on the other hand could look forward to reducing processing costs owing to a reduction in redundant work, shorter pro-cessing cycles, and a reduction of stock in hand.

It may be that this development is being restricted because we cannot make powder to the high standards mentioned in Dulis's paper.

Mr G. A. Wood (Wickman Wimet Ltd)

I would like to refer to Mr Dickson's paper and to the explan-ation he gives for various defects observed in hardmetal. He states that cobalt lakes resulting from uneven distribution of the cobalt binder phase produce brittle areas devoid of binder material. If a large proportion of the cobalt is present in the form of lakes, then obviously some areas contain less than the nominal amount, but it is virtually impossible to produce areas in sintered hardmetal without any cobalt in them at all. Cobalt lakes can result from undermilling of the original powder and from undersintering but a more usual cause is a hole, created possibly by accidental contamination which is just a little too large for the surface tension forces to completely close during sintering with the result that cobalt flows into the void.

The cobalt lake shown in Fig.10 of the paper is typical of one produced by this latter mechanism.

It is also difficult to see the connexion between A and B porosity and carbon absorption of the cobalt phase. Sintered hardmetal with η-phase and therefore very deficient in carbon, and with considerable amounts of free graphite and therefore with excess carbon, normally have similar low levels of A and B porosity.

The comments on C-type porosity give the impression that hardmetal powders start life with a liberal sprinkling of soot distributed through their indifferent grey mass. Although the tungsten carbide used in the powders does have a slight excess of carbon, this rarely exceeds 0.06% and is there deliberately to compensate for the small carbon losses which occur in tip manufacture. The slight excess of carbon en-sures that the sintered tips are free from η-phase.

It is commendable that users of hardmetal should relate shortcomings in performance to hardmetal quality, but there is the danger that unless the user is as expert with the mater-ials as the manufacturer, the wrong conclusions may be drawn. This particular point has already been made by Dr Greenfield.

Mr Chesters also made the point, although not related to hardmetal, that tool manufacturers and tool users should not work in isolation. This particular point cannot be emphasized too strongly.

Table 1 Comparison of some typical ISO-P10 carbide grades in current production

Grade	Composition, % WC	TiC	TaC	Mo₂C	Co	Ni	BHMA code
A	76	15	—	—	9	—	846
B	70	18	4	—	8	—	537
C	52	17	21·5	—	9·5	—	Not given
D	77	12	4	—	7	—	845
E	—	80	—	10	—	10	919

Table 2 Typical carbide property data supplied for a single US grade*

Primary usage: metal removal
This grade has excellent wear resistance as well as toughness achieved by unique grain structure. Withstands temperatures of high-speed machining operations.

Chemical composition, wt-%	71·0WC–12·5TiC–12·0TaC–4·5Co
Hardness	92·1–92·8 HRA
Density	12·0 gm/cm³
Transverse rupture strength	200000 lb/in²
Ultimate compressive strength	640000 lb/in²
Ultimate tensile strength	—
Modulus of elasticity	82 m. lb/in²
Proportional limit	170000 lb/in²
Ductility (% elongation)	—
Impact resistance (Charpy)	7·0 in lb
Abrasion resistance $\left(\dfrac{1}{\text{vol. loss}}\right)$	7
Electrical conductivity (% copper at 25°C)	4·3%
Electrical resistivity	41·0 μΩ-cm

Thermal conductivity

Temp., °C	200	300	400	500	600	650
cal/cm/s/°C	0·084	0·086	0·088	0·091	0·093	0·093

Coefficient of thermal expansion

From room temp. to °F	400	750	1100	1500	1800
Expansion/°F × 10⁻⁶	2·91	3·14	3·28	3·44	3·59

*Grade 350 cemented carbide data from Carboloy grades/properties handbook, 1966, US General Electric Co., Detroit

Mr K. J. A. Brookes (Hardmetals Consultant)

I should like to follow up the important point made by our distinguished rapporteur, Dr Greenfield, that users can only apply pressure to manufacturers if their knowledge is on the same level. Specifically, I want to discuss the unnecessarily great discrepancy between the levels of information available to carbide manufacturers and carbide users.

The papers presented at this conference show only too clearly the current situation. In every case where a carbide composition has been mentioned in a technical or research paper, the author has failed to relate it to any commercial product. Whenever a commercially available grade has been mentioned, it has been given a virtually meaningless ISO notation, without reference to composition or nominal properties.

There are, of course, numerous methods of specifying carbide grades:

(i) by raw material and manufacturing method
(ii) by composition, microstructure, and basic properties
(iii) by performance
(iv) by a limited range of secondary properties.

The manufacturer needs to lay down his own carbide specifications. To do this, he must first determine the optimum characteristics for a particular application, then establish the vital parameters and evaluate the effects of possible variations. He must then control each of these parameters within commercially acceptable limits.

Each manufacturer has all the necessary information at his disposal with respect to his own carbides. From conferences such as this, from research publications, and from investigations of competitive products, he obtains a great deal of additional knowledge.

The intelligent, metallurgically trained user also has need of carbide specifications:

(i) he, too, wants to determine the optimum carbide characteristics for a particular use
(ii) he wants to know the effects of variations in these characteristics on the efficiency of his cutting process
(iii) he wants to measure these variations
(iv) he needs to know or establish the commercial limits of these variations for any particular grade
(v) he needs to establish the direction and amount of change desirable in these properties for any required alteration in working conditions
(vi) he needs to compare carbide grades from a variety of sources, both as to the properties that should be obtained from a given composition and structure, and as to the properties actually attained in commercial production. In other words, he needs to distinguish between defects in quality and errors in grade selection.

During the conference, we have read and heard a great deal of the parameters controlling carbide functions, and the users present may well have been disheartened to notice how few of these are illumined by the fragments of information doled out by most British manufacturers. To quote a few of the conference papers, these vital factors include: 'some not clearly defined strength characteristic'; TiC and TaC contents taken separately ('For equivalent crater resistance, TaC–TiC hardmetal has a higher strength than that with only TiC'); 'composition of the carbide phases'; 'amount and composition of the binder phase, both chemical and structural'; 'mean cobalt path length'; 'carbide grain size distribution'; 'contiguity and substructural properties of the carbides'.

Without exception, the technical contributions bring out the vital but substantially indeterminate functions of composition and microstructure. It is clear that whatever parameters we evaluate should as closely as possible be a measure of these two basic, but complex, characteristics.

This, then, is what the user needs. What is he offered?

The so-called ISO system does not, of course, define carbides but, instead, defines applications. To employ the ISO system is to rely implicitly on the manufacturer, since an ISO code merely gives that manufacturer's choice for the particular machining operation. Different manufacturers may select closely similar carbides for the same application, or grades that differ violently in every important respect. Table I shows some typical variations in carbide grades which meet a typical ISO designation. The luckless average user must find it difficult, if not impossible, to differentiate between quality and compatibility.

The BHMA (British Hardmetal Association) code provides three parameters: hardness, as a measure of wear resistance (which it is not); transverse rupture stress, to indicate shock resistance (which it does not); and a combined TiC/TaC figure to represent cratering resistance (which it does in a woolly and approximate fashion). One of the most interesting aspects of the papers at this conference is the consistent way in which they demonstrate the inadequacy of the BHMA code; how limited is the information provided compared with the absolute minimum required for technical specification and control. Mr Horne of Richard Lloyd unfortunately stated that there is direct correlation between BHMA code and ISO grading; there is, of course, none whatsoever.

Perhaps the most important lesson of the papers to the average user is that, irrespective of his metallurgical training he can usually make little practical use of the information provided. I say usually because some manufacturers have indeed provided both structural and compositional information for many years: Table 2 is a typical American example of such data. Their numbers are increasing far too slowly, however, but on a world-wide basis they include a number of the most important suppliers. Certainly the education of their

Table 3 Cooperation with customers; manufacturers producing cutting-tool carbides for the British market and providing full details of compositions and other properties for unrestricted publications

	British producers	Foreign producers
No. in UK market	25	11
No. fully cooperative	6	7
% fully cooperative	24	64

customers has not been to their own disadvantage, neither has it been of vast importance to their competitors, who are well equipped to determine carbide properties for themselves.

Regrettably, most major suppliers in the British market do not give this vital information. Those who say they will make it available to selected users miss the point that, unless it is in the public domain and generally discussed, their privileged recipients of information will never really understand the subject. Table 3 compares the number and percentage of British manufacturers cooperating fully with their customers in this way, with the equivalent foreign manufacturers.

It is sad to note that, to a large extent, the user papers on carbides at this conference are based on an inadequacy of information. The quoted Ford system of carbide classification is unfortunately far less advanced than that devised by Oscar Strand of General Motors more than 20 years ago, and in essence differs little from the simplified Buick standards that developed into the American JIC code. In addition, however, the Buick standard sheets gave full details of composition, physical properties, and microstructure, illustrated with typical photomicrographs. Even so long ago, this was considered to be the minimum requirement for adequate consumer control.

It is well known that many suppliers publish no technical data because they merely stamp their own name on someone else's product. Many, with varying degrees of success, seek to hide their own technical inadequacy, sometimes little advanced from the standards of German wartime production. Some are contemptuous of their customers, claiming that they will misunderstand or misuse any information they are given. However, this is not so in such diverse environments as the USSR and the USA, in both of which countries full disclosure has for many years been the rule.

May I, therefore, make this appeal to the British and European market leaders, the firms that take the lion's share of the current market? Having achieved success, your aim should surely be to enlarge the market as a whole. Only education of the ultimate user can do this with maximum efficiency. Only intelligent understanding of his tool materials will create real confidence in the user, will remove from sintered carbides the false aura of magic and mystery that has lingered far too long. Don't give away your manufacturing techniques, these are the secrets that really matter, but publish full compositions, typical microstructures, typical and guaranteed properties. In the long run the user will learn to employ your materials more effectively, to detect faults more intelligently, to optimize both operations and tool materials. With understanding and confidence will come increased carbide consumption, to everyone's advantage.

Mr E. Lundgren (Sandvikens Jernverks Aktiebolag)
I work in the Tool Development Department at the Sandvik Company in Sweden, and am concerned with the design of the tool and the cutting edge. The latest developments involving inserts with as-sintered chipbreakers have meant close contact with our Metallurgical Research Department. At the same time the closest possible contact must be maintained with the tool user. The graphs and microstructures that have been shown are, of course, of great interest, but the primary concern of the tool user is the achievement of the lowest possible production cost. This concern is readily

understood when it is realized that the estimated total cost of machining in the UK is about £2000 m. For this reason many of the diagrams used in this conference would have been of still greater value if the parameter of the production cost had been used where appropriate.

It is, of course, very important to strive for the most wear resistant cutting materials, but reduced wear can also result from optimizing cutting geometry.

Catastrophic tool failure often occurs before the tool is fully worn. When this happens, the tool cost is insignificant when compared with the cost of lost production. Therefore, it is of equal importance to have sufficient cutting-edge strength as overall wear resistance. This can be achieved by suitable cutting-edge geometry including edge reinforcement using either a controlled edge radius or a primary land. The new form-sintering technique makes it possible to design indexable inserts in which more functional cutting-edge and chipbreaker geometry give more reliable continuity of machining with consequent reduction in machining costs.

The tool cost is usually insignificant when compared with the machine cost. For this reason the cutting capacity of the tool is more important than the price. This is demonstrated by our T-Max rhomboidal copying insert which offers only two cutting edges and consequently a relatively high cutting-edge cost. Nevertheless this insert gives very low production costs because of its reliability. The same applies for the single-sided P-insert with triple chipbreaker compared with the double-sided one with single chipbreaker, the former giving more reliable cutting performance and much improved chip control, thus minimizing interruptions to machining.

When considering tool wear test standardization, it is important to define which tool wear factors are to be covered in the test, e.g. flank wear, crater wear, plastic deformation, etc. The influence of tool shape on tool wear must be considered, and if the flat plain insert is to be used in the standard test, the results will not be valid for the inserts with sintered-in chipbreakers which will become more and more widely used in the future. Therefore, the test under production conditions, which allows real production cost to be finally calculated, will still be of the greatest value.

Dr D. H. Houseman (University of Technology, Loughborough)
I would like to discuss the possibility of using the radioactive tests Mr Moore mentioned at the beginning as standardized machinability tests. Mr Moore rightly mentioned that the normal method of using radioactivity in tool wear measurement involves radioactivity in the works and has therefore not caught on, in that it creates a health problem and is rather time consuming.

There is another method, which is to use the radioactivation analysis technique where *no* radioactivity is involved in the works. Very small amounts of certain rare earths are incorporated during melting; these become intensely radioactive, so that, in short-term irradiation, that radioactivity 'shines out' and swamps all the other radioactivity and is thus easily detected. As an illustrative example, consider motor car tyres, where several different isotopes may be put in at different concentration levels. No radioactivity is involved at the manufacturing stage but if the tyre is retreaded a sample can be taken from the wall and the tread by irradiating them together and you can show it is a retread.

This approach can be applied to tool steels, incorporating a small amount of rare earth, a few parts per million, costing only a few shillings per ton. Machining tests may be carried out at leisure, and all the chip samples can be collected together and sent to Harwell for simultaneous in-pile irradiation at the rate of hundreds of samples in an hour. Activity counting gives a figure for the wear of the tool per unit weight of metal removed. Knowing the cost of the tool, one gets the cost per unit weight of metal removed. At the end of the day, surely that is what matters. Another advantage is that the radioactivity in the material quickly decays and the sample can be used again and again, so that you can store the samples and compare them with tools produced in the future, over weeks, months, or years. Thus, when a new material is developed, or a different kind of tool, comparison measurments are easily made. Because of the small sample size, storage is no problem.

This provides a way of comparing results over a period of decades. It also has the advantage that it is extremely economical in scientific manpower. We have passed through the period where our stock of scientific manpower is growing; people are moving out of science and technology more quickly than they are moving in and, in ten years' time, managers will have to give attention to the economical use of scientific manpower. It is then that I think that this kind of test will be considered seriously, but I am afraid it will not be at the moment.

Dr E. M. Trent

I am disappointed that this is the last word, because I have been hoping throughout to stimulate discussion, among this unique body of people collected here at great expense, on precisely the subject matter of the conference, which was, I thought, to look at tool materials as a whole.

We have collected here users and manufacturers of high-speed steels, of carbides, of ceramics, and of diamond tools, and we have not spent any time looking at the overall problem of tool materials and how they are developing.

If you look at the development of tool materials, they have evolved under the incentive of increased productivity which has expressed itself largely as a demand for higher cutting speeds. Up to the moment, there have been three great developments: from carbon tool steel to high-speed steel, with quadrupled rates of production; from high-speed steel, by altering the metallurgy, to cemented carbides which again quadrupled rates of production; and now, having discovered, with carbide tools, that it is possible to cut with negative rake, we are all poised for the great leap forward into the stone age with the use of ceramic tools, synthetic stones it is true. Do we want this leap forward? This transition has been a long time in incubation. I think that the answer to this can come only from the users of tool materials.

It is quite easy to look at the details of high-speed steels, at the differences between molybdenum and tungsten high-speed steels, or at cemented carbide and the differences between titanium carbide and tungsten carbide, but if we try to look at tool materials as a whole, we come into an area where much longer development times and greater expenditure are required. We do not have these meetings very often, and if we are to look forward to the next 15 years, we want the users to answer such questions as whether they are still interested in much higher cutting speeds, or are they, under the influence of numerical control, interested only in more consistency, which means primarily toughness? What in fact do users as a whole want? There obviously is not now time at this conference for them to attempt to answer those questions. Perhaps they could contribute in written discussion, or perhaps some organization could be set up to inform the tool manufacturers of the users requirements in these general terms.

Mr G. J. Hill (GEC Power Engineering Ltd)

This conference has clearly demonstrated the need for a formal liaison on technical matters between the suppliers and users of cutting tools. The comments that Dr Edwards made, as regards the desire of the British Hardmetal Association to consult a group of user companies, are therefore welcomed. This liaison could possibly overcome the opposition which some tool suppliers have expressed towards the general adoption of the proposed ISO machining test.

From discussions and visits to suppliers' works it is evident that considerable attention is being given to the quality con-

trol of tool materials. However, there is still evidence from our own examinations and those of other users that the presence of defects of both a physical and metallurgical character can be frequently observed and it is therefore important to achieve further improvements in quality.

WRITTEN CONTRIBUTIONS
Mr A. E. Longley (Rolls-Royce Ltd)

The paper on high-speed steel bar quality, by McBroom and Smith, is the most intriguing of the conference, pointing the way to the final assessment of the influence of segregation of carbides in high-speed steel. It seems to me that the authors are now in a position to quantify segregation by microprobe analytical techniques. Having done this it would then be possible to relate segregation quantitatively to cutting-tool performance and establish once and for all the economic limits of segregation elimination.

Dr G. Barrow (University of Manchester Institute of Science and Technology)

With regard to the paper by Chisholm and Pattinson, I personally welcome some form of standardization in machining tests if only to make more use of the available data. However, I am a little concerned with some of the ISO recommendations, particularly as they state that the test is intended not only in research but also for use in industry. I fully agree with the suggestion that short-term machining tests should be implemented, but I feel that at this stage there is as yet no acceptable test which covers a wide range of workpiece materials. In view of this a reliable long-term test designed to cater for the majority of tool and workpiece combinations is essential. My major criticism is the recommendation of tool life criteria since it is this aspect above all others which has reduced the value of the considerable amount of machinability data which is available. The points I wish to raise are as follows:

1 *Flank wear criteria* The level of uniform flank wear chosen (0·3 mm) is, in my opinion, somewhat low and is, in fact, a rather poor choice in that many cutting tools, particularly straight tungsten carbide, show a rapid increase in wear to this level and then are essentially stationary until rapid failure of the cutting edge occurs. In view of this I would think the value of 0·4 mm is a much more realistic figure, particularly when considering applications in industry. I would not suggest an increase above 0·4 mm since there is considerable evidence to suggest that above this figure a frequency of premature failure increases. Regarding the value of 0·5 mm given when flank wear is not uniform, I would again suggest that this is somewhat low and that a figure of 0·8 mm is much more realistic.

2 *Crater wear* From an industrial point of view I do not consider that this wear is particularly important since in most practical applications tools do not fail by a crater-wear mechanism. However, in any recommendation one must attempt to consider all possible situations and some value must be placed on it. I would again suggest that the value is somewhat low and that a value of at least 0·15 mm is necessary.

3 *Oxidation wear* I note with interest that the working group has considered oxidation wear but have rejected it from the proposed recommendations. Recent work has indicated that this is a very important type of wear, particularly at high metal-removal rates, and although I agree with the working group that it is not easy at this stage to define when a tool fails by this mechanism, I feel that some ultimate figure could be stated.

When considering these wear criteria I do not think that any one should be chosen but that all should be considered together and that the end point of tool life be defined as when one of the criteria has reached a limiting value.

In view of the above comment I should be obliged if the

authors could indicate on what basis the wear criteria were chosen since, in my opinion, they do not coincide with the values at present under consideration in industry (at least in the UK).

Dr E. J. Pattinson
Dr Barrow comments that the test is intended not only in research but also for use in industry and since no acceptable short-term test exists, he suggests that a long-term test is required. The authors have stated in their paper that the working group do appreciate the difficulties in getting adequate reliability from simple, and therefore cheap, machining tests and have as a consequence attempted to outline a test which, in their opinion, specifies only a minimum set of conditions for research purposes. It has also been stated that the specification for the test conditions are mainly applicable to tool life testing in which the tool wears at a conventional rate and in a conventional manner.

We note with interest Dr Barrow's criticism of the choice of values for the flank wear band and the crater depth and his query as to how these values were arrived at. In particular, the recommended value for crater depth of $KT = 0.1 + 0.3s$ is

considered by Dr Barrow to be too low and he suggests an alternative single value of 0·15 mm. However, if the preferred cutting conditions are chosen which were suggested in Table 4 of our paper the recommended values of KT would range from 0·13 to 0·289 mm depending upon the chosen feed.

With regard to the general criticism of both flank and crater values we would like to point out that the values were only chosen after lengthy discussion with representatives from tool manufacturers and members from a number of European Universities actively engaged in metal cutting and directly associated with the CIRP group engaged on metal-cutting research. In fact the values chosen are identical to those proposed by CIRP in their recent standardization work.

Finally we note the comments made on oxidation wear. We can only repeat that if and when some easily definable dimensional change in the tool can be solely attributed to the phenomenon of oxidation, then this change will be considered seriously as an additional tool life criterion.

Dr Barrow's valued comments will, however, be studied carefully and also be brought to the notice of the working group, possibly by the UK representatives at their next meeting.

Attendance list

G. M. Addison	Rolls-Royce Ltd
H. M. C. Adriaenssens	Metaalinstituut TNO, Holland
P. Ambler	Tungsten Carbide Developments
Dr E. Anderson	Battelle Institute, Switzerland
S. E. Andersson	ASEA, Sweden
D. W. Atkinson	BSC, General Steels Division, Scunthorpe
T. D. Atterbury	GKN Screws and Fasteners Ltd
Dr T. Bagshaw	BSC, Special Steels Division
D. G. K. Bailes	Murex Ltd
W. H. Bailey	BSC, Special Steels Division, River Don Works
H. Banks	'Metalworking Production'
A. Barrie	R. B. Tennent Ltd
Dr G. Barrow	University of Manchester
G. Basile	Italsider, Italy
M. J. Batchelor	BSC, General Steels Division, Scunthorpe
A. D. Baynes	London Scandanivian Metallurgical Co.
D. Beard	BISRA–The Corporate Laboratories, BSC
H. G. Becker	Edelstahlwerk Witten AG, Germany
G. Berwick	Jonas and Colver (Tools) Ltd
Mrs P. Westcott Betenson	Stenographer
Dr W. Betteridge	International Nickel Ltd
M. E. Bidwell	Murex Ltd
H. Bodem	Metallwerk Plansee AG, Austria
Miss S. Boustead	The Iron and Steel Institute
Professor P. M. Braiden	Carnegie-Mellon University, USA
E. A. Brightmore	BSC, Special Steels Division
R. Brocklehurst	Tempered Tools Ltd, Hathersage
G. B. Brook	Fulmer Research Institute Ltd
K. J. A. Brookes	Consultant on carbide technology
R. L. Brooks	Spear and Jackson Ltd
P. J. Brotherton	Lucas Research Centre
Dr C. G. Brown	Borax Consolidated Ltd
Dr R. Brownsword	Lanchester Polytechnic
T. S. Bulmer	Hoy Carbides Ltd
H. Burden	Firth Brown Tools Ltd
N. Burton	Osborn Steels Ltd
Dr M. Bush	International Research and Development Co. Ltd
M. D. Button	Kayser, Ellison, and Co. Ltd
H. R. Caborn	Balfour Darwins Ltd
D. Cadorin	G. Donegani Research Establishment, Italy
W. S. Caisley	Firth Brown Tools Ltd
Dr C. A. Calow	AWRE, Aldermaston
J. A. Catherall	Associated Engineering Developments Ltd
Professor J. Cherry	Cranfield Institute of Technology
D. W. Chester	G. and J. Hall Ltd
W. T. Chesters, OBE	David Brown Gear Industries Ltd
H. C. Child	Jessop-Saville Ltd
Professor A. W. J. Chisholm	University of Salford
W. T. Clark	PERA
Dr M. G. Cockcroft	National Engineering Laboratory
W. Coles	Firth Brown Tools Ltd
J. Cowne	Noble and Samson Ltd
Dr R. L. Craik	BISRA–The Corporate Laboratories, BSC
F. L. Darbyshire	BSC, Special Steels Division
K. E. Davies	Bush Beach Hard Metal Tools Ltd
H. Dearden	The L. S. Starrett Co. Ltd
E. A. Dickenson	Edgar Allen Engineering Ltd
R. Dickson	Caterpillar Tractor Co. Ltd
L. D. Dixon	Sandvik UK Ltd
L. H. Driscoll	The Iron and Steel Institute
W. A. Draper	University of Manchester
Professor D. S. Dugdale	University of Sheffield
E. J. Dulis	Colt Industries, USA
A. M. Edwards	BSC, Special Steels Division, Swinden Laboratories
Dr R. Edwards	Metro-Cutanit Ltd
S. Ekemar	Sandvikens Jernverks AB, Sweden
Dr W. Eliasz	University of Surrey
R. B. Eriksson	Stora Steel Ltd
W. Fairhurst	Climax Molybdenum Co. Ltd
M. Feeney	Marsh Bros
A. J. Fenner	Ministry of Technology
T. F. Foley	BISRA–The Corporate Laboratories, BSC
J. H. E. Fox	BISRA–The Corporate Laboratories, BSC
C. French	Kayser, Ellison, and Co. Ltd
A. Fussey	Danite Hard Metals Ltd
A. Füssl	Schoeller-Bleckmann Stahlwerke AG, Austria
Dr A. Gane	University of Cambridge
H. Gardner	British Timken
L. Gill	Union Carbide (UK) Ltd
S. J. Gilmour	R. B. Tennent Ltd
R. Goodall	GKN Group Technological Centre
S. A. B. Gray	Glacier Metal Co. Ltd
G. Griggs	Bohler Bros and Co. (London) Ltd
A. F. Grogan	Associated Engineering Developments Ltd
Dr P. Greenfield	GEC Power Engineering Ltd
Professor J. Gurland	Brown University, USA
J. W. Haddock	Firth Brown Tools Ltd
A. G. Hague	Nottingham Regional College of Technology
D. Hall	Hoy Carbides Ltd
F. W. Hall	London and Scandinavian Metallurgical Co.
N. Hanlon	Sanderson Kayser Ltd
Dr D. Hardwick	Firth Brown (Tools) Ltd
D. Harris	Raleigh Industries Ltd
G. T. Harris, CBE	Jessop-Saville Ltd
K. A. Hartley	Hillcliff Hard Metals Ltd
Dr D. P. H. Hasseiman	Allied Chemical Corporation, USA
A. Hattersley	F. M. Parkin (Sheffield) Ltd
M. J. S. Hawkins	Stora Steel Ltd
B. Hawley	Firth Brown Tools Ltd
L. H. Haygreen	Esso Research Centre
Dr P. Heliman	Stora Kopparberg, Sweden
C. Hibbert	Osborn-Mushet Tools Ltd
G. J. Hill	GEC Power Engineering Ltd
J. Hill	Hillcliff Hard Metals Ltd
G. Hobson	Ingersoll-Rand Co. Ltd
Dr G. Hoch	Deutsche Edelstahlwerke AG, Germany
P. M. Holmes	Shell-Mex and BP Ltd
D. A. M. Holt	Bush Beach Hard Metal Tools Ltd
A. L. Horne	Richard Lloyd Ltd
N. Hornmark	ASEA, Sweden
Dr D. H. Houseman	University of Technology, Loughborough
G. Hoyle	BISRA–The Corporate Laboratories, BSC (ESRT)
T. L. Hughes	The Iron and Steel Institute
P. L. Hurricks	Swansea Tribology Centre
D. Hurst	Schoeller-Bleckmann Steels (GB) Ltd
J. Illingworth	Lucas Group Research Centre
W. M. Jack	Cleveland Twist Drill Ltd
J. S. Jackson	Production Tool Alloy Co. Ltd
Dr O. Jacura	Acieries de Champagnole, France
J. M. James	BISRA–The Corporate Laboratories, BSC
N. Johnson	Marsh Bros
Dr H. Jonsson	Fagersta Steels Ltd, Sweden
P. Jubb	Brown Firth Research Laboratories
A. Justice	International Twist Drill Co. Ltd
F. M. Karlsen	London and Scandinavian Metallurgical Co. Ltd
A. H. Kaye	Institute for Industrial Research and Standards
Dr D. M. Keane	BSC, Special Steels Division, Swinden Laboratories
D. Keen	Associated Engineering Developments Ltd
J. E. Kelsall	Harrogate Mobile Film Exhibitors
R. G. Kennedy	The Cleveland Twist Drill Company
Professor R. Kieffer	Technische Hochschule Wien, Austria
Dr A. G. King	Zirconium Corporation of America, USA
D. L. King	Osborn Steels Ltd
F. A. Kirk	Osborn Steels Ltd
P. J. A. Kirk	Osborn Steels Ltd
J. Klim	The Metal Box Co. Ltd
Professor O. Knotek	Castolin SA, Switzerland
Professor W. König	Technische Hochschule Aachen, Germany
Dr B. M. Korevaar	Laboratorium voor Metaalkunde, Holland
Dr Krumpholz	Röchlingsche Eisen- und Stahlwerke, Germany
J. Kurzeja	Edelstahlwerke Buderus AG, Germany
E. Lardner	Wickman-Wimet Ltd
Dr D. J. Latham	BISRA–The Corporate Laboratories, BSC
P. Lescop	SNECMA, France
B. J. Linden	Böhler Bros and Co. (London) Ltd
K. B. Lomax	BSC, Special Steels Division, Swinden Laboratories
A. E. Longley	Rolls-Royce (SED) Ltd
R. Lorenz	University of Birmingham
E. M. Lowe	Firth Brown Tools Ltd
Dr E. Lugscheider	Institut für Chemische Technologie Anorganischer Stoffe, Wien, Austria
E. Lungren	Sandvikens Jernverks AB, Sweden
N. H. McBroom	GKN Group Technological Centre
Dr A. B. McIntosh	Four Winds, Hough Lane, Cogshall, Nr. Northwich
Dr H. Mariacher	Deutsche Edelstahlwerke AG, Germany
J. B. Marriott	G. and J. Hall Ltd
R. Marshall	Gleason Works Ltd
P. Martin	SNECMA, France
R. N. Mather	J. Lucas Ltd
Dr M. J. May	BISRA–The Corporate Laboratories, BSC

Dr A. Mazur — Academy of Mining and Metallurgy, Cracow, Poland
L. S. Milan — Danite Hard Metals Ltd
C. P. Mills — ASEA (GB) Ltd
Dr D. R. Milner — University of Birmingham
D. A. Milwain — Rolls-Royce Ltd
Dr A. Moore — AWRE, Aldermaston
A. Moore — Hawker Siddeley Aviation Ltd
A. I. W. Moore — PERA
C. Moore — British Oxygen Co. Ltd
J. Moore — Wild Barfield Ltd
W. R. Morgan — BSC Group Research Centre
R. J. Morris — Gleason Works Ltd
B. Murray — GKN (South Wales) Ltd
I. D. Murray — Wickman Wimet Ltd
Dr T. Mukherjee — BSC, Special Steels Division
R. W. Newey — The Proforma Tool Co. Ltd
G. K. Norton — Edgar Allen Steels Ltd
M. L. Nott — BISRA–The Corporate Laboratories, BSC

A. Ogborne — Clarkson (Engineers) Ltd
R. Oldale — Balfour Darwins Ltd
Dr G. Ostermann — Fried Krupp GmbH, Germany
R. F. Panton — John Harris Tools
G. Parish — David Brown Gear Industries Ltd
Dr E. Pärnama — Fagersta Steels Ltd, Sweden
E. C. Patrick — GEC Power Engineering Ltd
Dr E. J. Pattinson — University of Salford
B. Pegg — Rolls-Royce Ltd
W. C. Perks — Proforma Tool Co. Ltd
G. Persson — Sandvikens Jernverks AB, Sweden
Professor H. J. Pick — University of Aston in Birmingham
C. Pillinger — Sheffield Twist Drill and Steel Co. Ltd
R. Pitt — 'Iron and Steel' Iliffe Science and Technology Publications
S. Postma — Metaalinstituut TNO, Holland
J. R. Powell — BISRA–The Corporate Laboratories, BSC
V. L. Prabhu — Cranfield Institute of Technology
Dr T. Raine — GEC Power Engineering Ltd
P. Redwood — Herbert Small Tools and Equipment Ltd
J. M. Revans — BSC, General Steels Division, Workington

A. Richardson — British Aircraft Corporation Ltd
Dr K. A. Ridal — BSC, Special Steels Division
M. Riddihough — Deloro Stellite Ltd
E. P. Riley — Midland Rollmakers Ltd
I. K. Robson — Hoy Carbides Ltd
B. Roe — International Nickel Ltd
E. Romairone — Italsider, Italy
W. Roper — Coventry Gauge and Tool Co. Ltd
Dr G. W. Rowe — University of Birmingham
G. Ryan — GKN (South Wales) Ltd
Dr A. M. Sage — Highveld Steel and Vanadium Corp. Ltd
A. M. Sanderson — Murex Ltd
R. S. Scarrott — Midland Rollmakers Ltd
Dr Schintimeister — Metallwerk Plansee AG, Austria
Dr G. Schumacher — Fried Krupp GmbH, Germany
D. R. Shail — Sandvik UK Ltd
S. W. K. Shaw — International Nickel Ltd
Dr J. B. Sheehy — BSC, Special Steels Division
G. Skinner — Osborn Steels Ltd

M. S. Skipp — Osborn Steels Ltd
P. Slack — Barworth Flockton Ltd
C. C. Smith — Higher Speed Metals Ltd
Dr T. B. Smith — GKN Group Technological Centre
N. L. Smyth — James Neill and Co. (Sheffield) Ltd
T. J. Soloman — Rolls-Royce (Aero Engine Div.) Ltd
Dr W. S'Pyra — Deutsche Edelstahlwerke AG, Germany
F. Stadler — Gesellschaft für Elektrometallurgie mbH, Germany
J. S. Stafford — Herbert Small Tools and Equipment Ltd
J. H. Stake — Stora Steel Ltd
R. Stanley — Brooke Tool Manufacturing Co. Ltd
J. Stephens — Ingersoll Rand Ltd
D. Stibbs — Borax Consolidated Ltd
Mrs L. W. Strohalm — BISRA–The Corporate Laboratories, BSC
G. M. Sturgeon — BISRA–The Corporate Laboratories, BSC
P. Tamm — Stora Steel Ltd
D. Taylor — James Neill and Co. (Sheffield) Ltd
G. Taylor — Edgar Allen Engineering Ltd
C. J. Thatcher — Richard Lloyd Ltd
A. Tomlinson — BISRA–The Corporate Laboratories, BSC
J. M. Thompson — Rolls-Royce Ltd
J. Tidlund — Fagersta Bruks AB, Sweden
P. F. Tomalin — GEC Power Engineering Ltd
M. H. Trahern — Wickman Wimet Ltd
Dr E. M. Trent — University of Birmingham
L. Triquet — Batelle Institute, Switzerland
P. Turrell — BSC, Special Steels Division
P. Tyden — Uddeholms Aktiebolag, Sweden
N. E. Tyler — Hoy Carbides Ltd
C. Vaughan — Highveld Steel and Vanadium Corp. Ltd
W. Verderber — Stahlwerke Südwestfalen AG, Germany
J. B. Vickers — Firth Brown Tools Ltd
C. Vlantis — Wickman Wimet Ltd
H. Voge — Stahlwerke Südwestfalen AG, Germany
E. A. Walker — Renold Ltd
G. Wastenson — Höganäs AB, Sweden
G. Watt — Spear and Jackson Ltd
D. B. Webb — Jonas and Colver (NOVO) Ltd
C. R. Webster — Technical College, Coatbridge, Lanarks.
Dr H. H. Weigand — Deutsche Edelstahlwerke AG, Germany
A. White — Raleigh Industries Ltd
G. J. A. White — BSC, General Steels Division, Workington

G. Whitehouse — Wickman Wimet Ltd
R. Whittaker — BSC, Special Steels Division
P. F. Whybrow — Edgar Allen Steels Ltd
T. J. Wilkins — BSC, Special Steels Division
E. Williams — Cobalt Information Centre
J. Williams — Osborn Steels Ltd
S. Wilmes — Gebrüder Böhler AG, Germany
G. A. Wood — Wickman Wimet Ltd
P. W. Wood — Jessop-Saville Ltd
T. Worrall — Sanderson Bros and Newbould Ltd
P. K. Wright — University of Birmingham
Mrs V. Wroe — BISRA–The Corporate Laboratories, BSC
D. S. Young — Glacier Metal Co. Ltd

Author index

Discussion pages

Subject index

*Discussion pages

_____

*Discussion pages

*Discussion pages